AF606272

CAMBRIDGE TRACTS IN MATHEMATICS

General Editors

---

185 Rigidity in Higher Rank Abelian Group Actions I

CAMBRIDGE TRACTS IN MATHEMATICS

A complete list of books in the series can be found at www.cambridge.org/mathematics.

Recent titles include the following:

150. Harmonic Maps, Conservation Laws and Moving Frames (2nd Edition). By F. HÉLEIN
151. Frobenius Manifolds and Moduli Spaces for Singularities. By C. HERTLING
152. Permutation Group Algorithms. By A. SERESS
153. Abelian Varieties, Theta Functions and the Fourier Transform. By A. POLISHCHUK
154. Finite Packing and Covering. By K. BÖRÖCZKY, JR
155. The Direct Method in Soliton Theory. By R. HIROTA. Edited and translated by A. NAGAI, J. NIMMO, and C. GILSON
156. Harmonic Mappings in the Plane. By P. DUREN
157. Affine Hecke Algebras and Orthogonal Polynomials. By I. G. MACDONALD
158. Quasi-Frobenius Rings. By W. K. NICHOLSON and M. F. YOUSIF
159. The Geometry of Total Curvature on Complete Open Surfaces. By K. SHIOHAMA, T. SHIOYA, and M. TANAKA
160. Approximation by Algebraic Numbers. By Y. BUGEAUD
161. Equivalence and Duality for Module Categories. By R. R. COLBY and K. R. FULLER
162. Lévy Processes in Lie Groups. By M. LIAO
163. Linear and Projective Representations of Symmetric Groups. By A. KLESHCHEV
164. The Covering Property Axiom, CPA. By K. CIESIELSKI and J. PAWLIKOWSKI
165. Projective Differential Geometry Old and New. By V. OVSIENKO and S. TABACHNIKOV
166. The Lévy Laplacian. By M. N. FELLER
167. Poincaré Duality Algebras, Macaulay's Dual Systems, and Steenrod Operations. By D. MEYER and L. SMITH
168. The Cube-A Window to Convex and Discrete Geometry. By C. ZONG
169. Quantum Stochastic Processes and Noncommutative Geometry. By K. B. SINHA and D. GOSWAMI
170. Polynomials and Vanishing Cycles. By M. TIBĂR
171. Orbifolds and Stringy Topology. By A. ADEM, J. LEIDA, and Y. RUAN
172. Rigid Cohomology. By B. LE STUM
173. Enumeration of Finite Groups. By S. R. BLACKBURN, P. M. NEUMANN, and G. VENKATARAMAN
174. Forcing Idealized. By J. ZAPLETAL
175. The Large Sieve and its Applications. By E. KOWALSKI
176. The Monster Group and Majorana Involutions. By A. A. IVANOV
177. A Higher-Dimensional Sieve Method. By H. G. DIAMOND, H. HALBERSTAM, and W. F. GALWAY
178. Analysis in Positive Characteristic. By A. N. KOCHUBEI
179. Dynamics of Linear Operators. By F. BAYART and É. MATHERON
180. Synthetic Geometry of Manifolds. By A. KOCK
181. Totally Positive Matrices. By A. PINKUS
182. Nonlinear Markov Processes and Kinetic Equations. By V. N. KOLOKOLTSOV
183. Period Domains over Finite and *p*-adic Fields. By J.-F. DAT, S. ORLIK, and M. RAPOPORT
184. Algebraic Theories. By J. ADÁMEK, J. ROSICKÝ, and E. M. VITALE
185. Rigidity in Higher Rank Abelian Group Actions I: Introduction and Cocycle Problem. By A. KATOK and V. NIŢICĂ
186. Dimensions, Embeddings, and Attractors. By J. C. ROBINSON
187. Convexity: An Analytic Viewpoint. By B. SIMON

Rigidity in Higher Rank Abelian Group Actions

Volume I. Introduction and Cocycle Problem

ANATOLE KATOK
Pennsylvania State University

VIOREL NIŢICĂ
West Chester University, Pennsylvania

CAMBRIDGE UNIVERSITY PRESS
Cambridge, New York, Melbourne, Madrid, Cape Town,
Singapore, São Paulo, Delhi, Tokyo, Mexico City

Cambridge University Press
The Edinburgh Building, Cambridge CB2 8RU, UK

Published in the United States of America by Cambridge University Press, New York

www.cambridge.org
Information on this title: www.cambridge.org/9780521879095

First published 2011

Printed in the United Kingdom at the University Press, Cambridge

A catalogue record for this publication is available from the British Library

Library of Congress Cataloguing in Publication data
Katok, A. B.
Rigidity in higher rank Abelian group actions / Anatole Katok, Viorel Nitica.
v. cm. – (Cambridge tracts in mathematics ; 185–)
Contents: v. 1. Introduction and cocycle problem
ISBN 978-0-521-87909-5 (hardback)
1. Rigidity (Geometry) 2. Abelian groups. I. Nitica, Viorel. II. Title.
QA640.77.K38 2011
512′.25–dc22
2011006030

ISBN 978-0-521-87909-5 Hardback

Contents

Introduction: an overview

1. Rigidity in dynamics

In a very general sense, modern theory of smooth dynamical systems deals with smooth actions of "sufficiently large but not too large" groups or semigroups (usually locally compact but not compact) on a "sufficiently small" phase space (usually compact, or, sometimes, finite volume manifolds). Important branches of dynamics specifically consider actions preserving a geometric structure with an infinite-dimensional group of automorphisms, two principal examples being a volume and a symplectic structure. The natural equivalence relation for actions is differentiable (corr. volume preserving or symplectic) conjugacy.

One version of the general notion of rigidity in this context would refer to a certain class $\mathcal{A}$ of actions being described by a finite set of parameters, usually smooth *moduli*. Examples of such classes are all actions in the neighborhood of a given one, or all actions of a continuous group with the same orbits, or all G-extensions of a given action α to a given principal G-bundle. In some situations this is too strong and, rather than classifying all actions from $\mathcal{A}$, one may require that actions equivalent to a given one have a finite codimension in a properly defined sense, e.g., appear in typical or generic finite-parametric families of actions.

A perfect extreme case appears when all actions from $\mathcal{A}$ belong to a single equivalence class. This may be referred to as rigidity in the narrow sense of the word. However, one should point out that for abelian group actions even locally this can only happen in the case of discrete groups, since otherwise one can always compose the original action with a group automorphism close to the identity.

2. Limited extent of rigidity in traditional dynamics

The material presented in this book relies to a considerable extent on the classical theory of (uniform) hyperbolic and partially hyperbolic systems, i.e., the study of *rank-one* cases of $\mathbb{Z}_+$, $\mathbb{Z}$, and $\mathbb{R}$-actions with hyperbolic or partially hyperbolic behavior. Anosov diffeomorphisms and Anosov flows (see Section 1.8.1) are prime examples of such actions.

Anosov diffeomorphisms and Anosov flows display certain elements of rigidity of the *topological* orbit structure, both locally (structural stability, see e.g., [67, Corollary 18.2.2]) and globally (topological restrictions on the ambient manifolds and homotopy invariants [67, Theorem 18.6.1]). Nevertheless, their *differentiable* properties are far from rigid; at best, the classification with respect to differentiable conjugacy is given by infinitely many moduli as in the case of Anosov diffeomorphisms of $\mathbb{T}^2$ ([67, Section 20.4]), or transitive Anosov flows on three-dimensional manifolds ([105]). Similarly, cohomology classes of sufficiently regular (Hölder or smooth) rank-one cocycles over an Anosov system are determined by infinitely many parameters, e.g., the periodic data (see Theorem 4.2.2).

It is also worth pointing out that local differentiable rigidity in the more loose parametric sense does take place for some non-hyperbolic systems, such as circle rotations with Diophantine rotation number or, more generally, for Diophantine translations and linear flows on a torus.

3. Rigidity for actions of higher rank abelian groups

The goal of this monograph and its projected sequel is to give an up-to-date and, as much as possible, self-contained presentation of certain rigidity phenomena that appear for actions of higher rank abelian groups by smooth maps on compact differentiable manifolds. We will consider hyperbolic and partially hyperbolic $\mathbb{Z}^k$- and $\mathbb{R}^k$-actions, $k \geq 2$, or, more generally, $\mathbb{Z}^k \times \mathbb{R}^l$, $k+l \geq 2$.

Certain results for higher rank abelian semigroups $\mathbb{Z}^k_+$, $k \geq 2$, are also discussed and, by a slight abuse of terminology, we will use the phrase *higher rank abelian groups actions* for them as well.

The list of known examples of Anosov (normally hyperbolic) and partially hyperbolic actions of higher rank abelian groups which do not arise from products and other standard constructions is restricted. All such basic examples are differentiably conjugate to algebraic actions. Basic definitions appear in Section 1.6; principal classes of examples are surveyed in Chapter 2. These actions exhibit a remarkable array of measurable and differentiable rigidity properties, markedly different from the rank-one situation. In this and the subsequent volume we are concerned with differentiable properties, namely:

(i) *cocycle rigidity;*
(ii) *local differentiable rigidity*, including *foliation rigidity;*
(iii) *global differentiable rigidity.*

In this volume we describe the scene in sufficient detail, and develop principal methods which are at present used in various aspects of the rigidity theory. Part I serves as an exposition and preparation. Cocycle rigidity, which occupies Part II of this volume, serves both as a model for other rigidity phenomena and as a tool for studying them.

The area of differentiable rigidity is experiencing a rapid development. While local differentiable rigidity for Anosov algebraic actions was proved in the 1990s [81], now we are close to a comprehensive understanding of local differentiable rigidity for all appropriate classes of partially hyperbolic algebraic actions; see [19, 22, 23, 171, 172] for a partial realization of the program. However, a number of key results in the area have not yet appeared in the journals so it is not currently possible to provide a comprehensive treatment in book form. Such a treatment of local differentiable rigidity will appear in the sequel to this book, which will be largely based on the material of the present volume.

We should add that some of the methods developed in the local rigidity theory for partially hyperbolic actions turned out to be applicable to some classes of non-hyperbolic, specifically parabolic actions, thus leading to a totally new phenomena [24].

Global differentiable rigidity has been shown for a certain class of Anosov actions on the torus [151] and for actions satisfying certain dynamical assumptions stronger than just Anosov [64]. A promising opening in the direction of global rigidity is provided by the the non-uniform measure rigidity discussed in the Section 5 below.

4. Mechanisms of rigidity

The following property appears in all known cases and may even turn out to be necessary and sufficient for a (properly adjusted) cocycle rigidity and local differentiable rigidity for an algebraic action of $\mathbb{Z}^k \times \mathbb{R}^l$:

($\mathfrak{R}$) The group $\mathbb{Z}^k \times \mathbb{R}^l$ contains a subgroup L isomorphic to $\mathbb{Z}^2$ such that for the suspension of the restriction of the action to L every element other than the identity acts ergodically with respect to the standard invariant measure obtained from the Haar measure.

The basic source of various kinds of rigidity in actions of higher rank abelian groups with hyperbolic behavior is the interplay between the linear algebra, describing the infinitesimal speeds of growth in various directions in time and space, and the existence of recurrence. The key notion here is that of the *Weyl chamber*, which is a generalization of the classical notion from Lie group theory. The reason for the appearance of rigidity in higher rank (encapsulated in the condition ($\mathfrak{R}$)) is that while the dynamics along the walls of Weyl chambers may be (and often is) highly non-trivial, it acts as isometry on a certain invariant foliation (such as a *Lyapunov* or *coarse Lyapunov* foliation, see Section 1.6.4), and hence ties invariant geometric structures on different leaves of that foliation.

In contrast, in the rank-one case, there are only two Weyl chambers, the positive and negative half-lines, and their common boundary is zero. Hence nothing happens along the "wall."

There are certain differences in the treatment of the continuous, discrete invertible, and discrete non-invertible actions. There are some advantages in looking into discrete time situations, primarily better visualization in low dimensions, and we will take this approach while treating some important examples or model problems. However, there are two decisive reasons for taking $\mathbb{R}^k$ as the main case: (i) the geometry of the Weyl chambers, whose walls are often "irrational" and hence "invisible" in the group itself in the discrete time cases, and (ii) the possibility to reduce the other cases to this one via constructions of *natural extension* and *suspension*, which are described in Section 1.2.

5. Measure rigidity

A related class of rigidity phenomena is the rigidity of invariant measures, sometimes called simply *measure rigidity*. There are two directions here: one dealing with algebraic actions, and the other with actions defined by some global topological conditions or with invariant measures satisfying certain dynamical properties.

a. Algebraic actions

One considers here the same class of algebraic Anosov actions as well as their non-Archimedean counterparts. The goal is to classify all Borel invariant measures for such actions and to show specifically that those measures are all of algebraic nature except for very special exceptional cases. This program, which is has been partially realized in [28, 29, 30, 31, 80, 154], can be considered as a counterpart of cocycle rigidity and local differentiable rigidity.

This work uses the same fundamental structures outlined in the previous section which are responsible for cocycle and local differentiable rigidity. There are, however, important technical differences.

For example, harmonic analysis methods are very fruitful in our setting but so far have been much less productive in the study of invariant measures, since it is very difficult to distinguish invariant measures among usually much more abundant invariant distributions.

Another difference appears in the use of invariant structures on various invariant foliations for the action. In the case of a measure such structures are corresponding conditional measures, which may be trivial if the measure in question has zero entropy for all elements of the action. This makes all existing results in measure rigidity for hyperbolic or partially hyperbolic actions subject to an assumption of positivity of entropy and thus fundamentally incomplete.[1] This is reflected in applications. For example, while full measure rigidity for the Weyl chamber flow on $SL(3,\mathbb{R})/SL(3,\mathbb{Z})$ would imply the Littlewood conjecture in multiplicative Diophantine approximation, the existing results for the rigidity of positive entropy measure only imply that the hypothetical set of counter-examples has a Hausdorff dimension of zero [30].

b. Non-uniform measure rigidity

This is a new direction based on combining geometric ideas of measure rigidity with those of non-uniform hyperbolicity that are mentioned in Section 1.7 [61, 62, 74, 75]. It is still in the process of rapid development and its potential is far from having been realized. However, even the results obtained so far are fairly striking: purely topological conditions on an action lead to the existence of an absolutely continuous invariant measure and a flat affine structure defined on an invariant set of positive volume. Furthermore, there is a smooth conjugacy in the sense of Whitney correspondence on an invariant set of positive volume with a standard algebraic model. This opens a new approach to global differentiable rigidity problems: for globally hyperbolic actions, invariant geometric structures smooth in the sense of Whitney probably can be extended to genuine smooth structures defined everywhere.

[1] This of course stands in contrast with Ratner classification of invariant measures for parabolic unipotent homogeneous actions [84, 148, 178]. It seems that the higher rank hyperbolic case is fundamentally more difficult in this respect.

6. Contrast and similarities with actions of "large" groups

Some of the rigidity phenomena exhibited by smooth actions of higher rank abelian groups with hyperbolic or partially hyperbolic behavior look quite similar to those found in actions of "large" and "rigid" non-abelian groups, such as semisimple Lie groups of $\mathbb{R}$-rank greater than one or irreducible lattices in semisimple Lie groups of $\mathbb{R}$-rank greater than one. Those properties are the main subject of the survey [32], see especially Section 6 there.

For classes of non-abelian groups mentioned above there are fundamental rigidity phenomena already at the measure-theoretic level. The prototype result, fundamental for dynamical applications, is Zimmer's cocycle super-rigidity extension of the Margulis super-rigidity theorem, see [32, Section 6.2], [179]. Based on these fundamental properties, extra geometric, analytical, and dynamical tools allow the study of rigidity properties specific for smooth actions, see [36, 41, 111], for characteristic results in that direction. A recent example of the successful application of the approach based on non-uniform measure rigidity for actions of higher rank abelian groups to rigidity of actions of "large" groups appeared in [76].

Since higher rank abelian groups are amenable, for actions of such groups there are no general rigidity properties at the basic measurable level, such as the classification of measurable cocycles or orbit equivalence [47, Theorem 3.5.4]. Rigidity only appears in the presence of an extra structure, most typically in the smooth case in the presence of certain hyperbolicity. The toolkit has many similarities with that used for going from measurable to differentiable rigidity results for actions of simple Lie groups of $\mathbb{R}$-rank greater than one or irreducible lattices in semisimple Lie groups of $\mathbb{R}$-rank greater than one, see [32, Theorem 6.5.3] as a characteristic example.

7. Background, references, and other sources

We extensively use background material from several areas of mathematics. Let us mention more important ones:

Hyperbolic dynamics occupies the first place here. It is an important field of modern mathematics and expositions of it can be found in numerous places. A standard reference is [67], and a detailed survey of the principal results in the field appears in [46]. Section 1.8 contains formulations of essential results both in the classical case (diffeomorphisms and flows) and for actions of higher rank abelian groups, which are used later.

The theory of Lie groups and Lie algebras extending to symmetric spaces and lattices is essential for understanding classes of algebraic actions which play a central part in our considerations. A comprehensive source for the semisimple case and symmetric spaces is [48]. For the general and nilpotent case one can consult [132]. We choose to present necessary material from this area piecemeal as needed rather than put it in a single place upfront. Necessary material from the general and nilpotent theory is introduced in two places in Section 2.1, while more general results and the semisimple case are reviewed in Section 2.3.3. Lattices in Lie groups, both semisimple and nilpotent, appear throughout our discussion. Fortunately, we do not need much general theory and in most cases co-compact lattices appear as "black boxes." In a number of places they come alive through ingenuous specific constructions.

Classical analysis is essential in the treatment of the regularity of conjugacies and cocycles for the actions considered in this book. A variety of results allow us to conclude the regularity of functions from various seemingly weaker properties as well as to obtain appropriate norm estimates. Some of these results were proved specifically for dynamical applications and were never collected in a single place. This is a crucial part of technical apparatus in rigidity theory and we dedicate the whole of Chapter 4 to a detailed presentation of these results with complete proofs.

Notions and results from other areas appear in a more limited way. Let us mention several of them together with some relevant bibliography: *algebraic number theory* (Section 2.2.4) [16], *algebraic K-theory* (Section 4.4.5) [115], *commutative algebra* (Section 2.4.2) [90], and *theory of unitary representation of Lie groups* (Section 4.4.2) [85].

The area which is the subject of the present book is in the process of active development and expository literature is still quite small. For cocycle rigidity there is a survey by Niţică and Török [128]. Local rigidity is covered in detail in a survey by Fisher [34]. A great overview of measure rigidity by Lindenstrauss [95] contains a few proofs. A more limited and less up-to-date but quite detailed exposition of measure rigidity can be found in the article of Kalinin and Katok [60].

Acknowledgements

A. Katok's research during the period of writing of this book was partially supported by NSF grants DMS 0071339 and DMS 0505539. V. Niţică's research was partially supported by NSF grant DMS 0500832. V. N. would also like to thank the Center of Dynamical Systems and Geometry at the Pennsylvania State University for support on a number of occasions. In particular, this support allowed him to visit Penn State during the Fall of 2006 when large parts of this book were written.

Part I

Preliminaries from dynamics and analysis

1

Definitions and general properties of abelian group actions

1.1 Group actions, conjugacy, and related notions

Let X be a space provided with a certain structure $\mathcal{S}$; the cases of interest for us are measure (such that X is a Lebesgue space) or an equivalence class of such measures, metrizable topology (usually compact), the structure of a (usually compact) differentiable manifold, or the structure of a homogeneous or double coset space of a Lie group. In each of these cases there exists a natural topology in the space of automorphisms of $\mathcal{S}$. By the action of a topological group G in this context we will always mean a continuous homomorphism α into the space of automorphisms of the structure $\mathcal{S}$. In our setting the group G will always be locally compact; in fact we may assume that it is a Lie group which includes both discrete and connected cases.

An *isomorphism* or *conjugacy* between two actions of a group G, say $\alpha\colon G \times X \to X$ and $\alpha'\colon G \times Y \to Y$, is a bijection $h\colon X \to Y$ that preserves or respects the particular structure (diffeomorphism, homeomorphism, measure-preserving, non-singular map, etc.), such that

$$h(\alpha(g, x)) = \alpha'(g, (h(x)) \text{ for all } g \in G, x \in X. \tag{1.1.1}$$

The notion of isomorphism is natural from the categorical point of view and provides the natural starting point for looking into the classification of actions. One should note that, in some particular settings, a weaker structure should be preserved in order to have a meaningful working notion. For example, as we mentioned in Section 2 of the introduction, for smooth dynamical systems in the classical setting (i.e., actions of $\mathbb{Z}$ or $\mathbb{R}$), topological classification is in many situations more tractable than a smooth one. Nevertheless, the main purpose of this book is to investigate certain special situations when the classification of actions up to a differentiable conjugacy becomes feasible.

The action α' of the space Y is a *factor* of the action α on X if there is a surjective map with specified properties $h: X \to Y$ such that (1.1.1) holds. In the measurable setting, the map h can be assumed either measure-preserving or non-singular; in the topological setting, h is continuous; in the differentiable setting, it is usually assumed to be a differentiable covering map. Sometimes such a map h is called a *semi-conjugacy*, but one has to remember that α and α' play non-symmetric roles and that semi-conjugacy is not an equivalence relation between actions.

Let $\mathcal{O}_\alpha$ be the partition of the space X into orbits of the action α. An *orbit equivalence* (measurable, topological, differentiable, etc.) between actions α and β of not necessarily the same group is a bijection $h: X \to Y$ (non-singular, measurable, homeomorphism, diffeomorphism, etc.), such that

$$h(\mathcal{O}_\alpha) = \mathcal{O}_{\alpha'}.$$

Two actions α and β of the same group are obtained by a *time change* from each other if $\mathcal{O}_\alpha = \mathcal{O}_{\alpha'}$. Time changes are closely related to *cocycles* over group actions, see Section 4.1, in particular equation (4.1.1) and calculation (4.1.2).

1.2 Functorial constructions

In this section we describe several standard constructions that are used in dynamics to produce new actions from given ones. These constructions allow for restrictions, extensions, and products of dynamical systems. For a more general overview see [47, Sections 1.3, 2.2, 3.4].

(i) The *restriction* of an action to a subgroup. In the abelian setting this appears, for example, as a restriction of an $\mathbb{R}^k$-action to a connected subgroup isomorphic to $\mathbb{R}^l$ for $1 \le l < k$, or to a lattice $\mathbb{Z}^l$.

(ii) The *Cartesian* or *direct product* of two actions α and β of the groups G and H on the spaces X and Y, correspondingly, is the action $\alpha \times \beta$ of the group $G \times H$ on the space $X \times Y$ given by

$$(\alpha \times \beta)(g, h)(x, y) = (\alpha(g)(x), \beta(h)(y)).$$

The restrictions of the Cartesian product to various subgroups of $G \times H$ are also considered. In particular one can look at the *diagonal action*, i.e., the restriction of the Cartesian product $\alpha \times \alpha$ to the diagonal subgroup of $G \times G$.

(iii) The *quotient actions* of various kinds, including projections to orbit spaces, of finite and other group actions commuting with a given action.

(iv) The *suspension* of a $\mathbb{Z}^k$-action. Suppose $\mathbb{Z}^k$ acts on a space N. Embed $\mathbb{Z}^k$ as a lattice in $\mathbb{R}^k$. Let $\mathbb{Z}^k$ act on $\mathbb{R}^k \times N$ by $z\,(x, m) = (x - z, z\,m)$ and form the quotient space

$$M = \mathbb{R}^k \times N/\mathbb{Z}^k.$$

Note that the action of $\mathbb{R}^k$ on $\mathbb{R}^k \times N$ by $x\,(y, n) = (x + y, n)$ commutes with the $\mathbb{Z}^k$-action and therefore descends to M. This $\mathbb{R}^k$-action is called the suspension of the $\mathbb{Z}^k$-action.

There is a fairly obvious generalization of the suspension construction to actions of $\mathbb{Z}^k \times \mathbb{R}^l$ where only the discrete part of the action is "suspended," producing a $\mathbb{R}^{k+l}$ action as the result.

(v) The *natural extension* of a $\mathbb{Z}^k_+$-action α on X is a $\mathbb{Z}^k$-action α_e on the space X_P of "pasts," i.e., all maps $p : -\mathbb{Z}^k_+ \to X$ such that if $m \in -\mathbb{Z}^k_+$, $n \in \mathbb{Z}^k_+$ and $m + n \in -\mathbb{Z}^k_+$ then

$$p(m + n) = \alpha(n)p(m).$$

The natural extension $\alpha_e : X_P \to X_P$ is defined for any $m \in \mathbb{Z}^k$ by

$$\alpha_e(m)(p) = \alpha(m)p.$$

Notice that when X is a manifold, the space X_P usually is not. An important case is an action $\mathbb{Z}^k_+$ acting by covering maps: in this case X_P has locally the structure of the product of a Euclidean space and Cantor set; *solenoids* (see Section 2.4) provide typical examples of this situation.

1.3 Principal bundles

Now we introduce several notions from differential topology, such as fiber bundles and principal bundles, which are needed for our future discussion of cocycles.

Definition 1.3.1 A *smooth fiber bundle* (E, π, M, F) consists of E, M, F smooth manifolds and a smooth map $\pi : E \to M$ for which the following holds: each $x \in M$ has an open neighborhood U such that $E|_U := \pi^{-1}(U)$ is diffeomorphic to $U \times F$ via a diffeomorphism $\psi_U : E|_U \to U \times F$ which preserves the fibers, that is, $\pi_1 \circ \psi_U = \pi$, where $\pi_1 : U \times F$ is the projection of the first factor.

M is called the *base*, F the *fiber* and π the *canonical projection.* A pair (ψ_U, U) is called *local trivialization.*

Given a collection of local trivializations (ψ_α, U_α) for which $\{U_\alpha\}$ is an open cover of M, one has:

$$(\psi_\alpha \circ \psi_\beta^{-1})(x, y) = (x, \psi_{\alpha\beta}(x, y)),$$

where $\psi_{\alpha\beta} : (U_\alpha \cap U_\beta) \times F \to F$ is smooth and $y \to \psi_{\alpha\beta}(x, y)$ is a diffeomorphism of F. The functions $\psi_{\alpha\beta}(x) := \psi_{\alpha\beta}(x, \cdot)$ are called *transition functions*. They satisfy the following *cocycle equation*:

$$\psi_{\alpha\gamma}(x) \circ \psi_{\gamma\beta}(x) = \psi_{\alpha\beta}(x), \qquad x \in U_\alpha \cap U_\beta \cap U_\gamma,$$

and $\psi_{\alpha\alpha}(x)$ is the identity of F for $x \in U_\alpha$.

A *section* of a fiber bundle (E, π, M, F) is a map $s : M \to E$ such that $\pi \circ s$ is the identity map of M. By changing the regularity of the trivializing functions, one can introduce C^r or *continuous fiber bundles*. One calls (E, π, M, F) a *measurable fiber bundle* if F is a smooth manifold with Borel measurable structure, E and M are measurable spaces, and there exists a measurable isomorphism $\Phi : E \to M \times F$, called a *measurable trivialization*, which preserves the fibers.

A measurable trivialization can be constructed for any topological bundle over a topological manifold. Let μ be a σ-finite measure on M. Then in any family of disjoint sets, only countable many can have non-zero measure. This implies that if $x \in M$ and $B(x, r)$ is a ball of radius r for some metric on M, then arbitrarily close to r there exists r' such that the ball $B(x, r')$ has a boundary of measure zero. One starts with a countable cover of M by balls included in local charts and modifies it to obtain a cover $\{V_n\}_n$ consisting of balls included in local charts that have boundary of measure zero. A refinement of this cover constructed inductively,

$$W_1 = V_1, W_2 = V_2 - \overline{V_1}, \ldots, W_n = V_n - \overline{V_1 \cup \cdots V_{n-1}}, \ldots,$$

gives an open cover of an open subset $U \subset M$ that has full measure. Since each V_i gives a local trivialization of the fiber bundle, one can define now a continuous global trivialization on U which extends to a measurable trivialization on E.

Definition 1.3.2 Let H be a Lie group and (E, π, M, F) a smooth fiber bundle. The fiber bundle is called *H-bundle* if there exists a smooth action $\Psi : H \times F \to F$ and a family $\{\psi_\alpha\}$ of smooth trivializations of E for which the transition maps take values in $\Psi(H)$.

A *vector bundle* over a manifold M is a $GL(V)$-bundle, where V is a vector space that coincides with the fiber.

Example 1.3.3 The tangent bundle of an n-dimensional manifold M is an example of $GL(n, \mathbb{R})$-bundle. In this case the fiber is $\mathbb{R}^n$ with the linear action of $GL(n, \mathbb{R})$ and the transition functions are the Jacobian matrices of coordinate changes.

Definition 1.3.4 A *principal bundle* is an H-bundle with fiber H such that the action of H on itself is given by left translations.

The *trivial H-bundle* is $M \times H$, with H acting on itself.

Example 1.3.5 If H is a closed subgroup of a Lie group G, then the natural projection $p : G \to G/H$ makes G a principal H-bundle over the manifold G/H.

A principal H-bundle P admits a right H-action $P \times H \to H$ on itself given in a trivialization $U \times H$ by right multiplication on the H-component. This is well defined since the transition functions are left translations by elements in H and thus commute with the right translation.

It can be shown that an equivalent definition of a principal H-bundle is the following:

Definition 1.3.6 Let M be a manifold and H a Lie group. A *principal H-bundle $P \to M$* consists of a manifold P and an action of H on M satisfying the following conditions:

(i) H acts freely on P on the right: $(\xi, h) \in P \times H \to \xi h$;
(ii) M is the quotient space of P by the equivalence relation induced by H; let $\pi : P \to M$ be the natural projection;
(iii) P is locally trivial, that is, for every point in M there exists a neighborhood U and a diffeomorphism $\phi : \pi^{-1}(U) \to U \times H$ such that $\phi(\xi) = (\pi(\xi), \psi(\xi))$, where $\psi : \pi^{-1}(U) \to H$ satisfies $\psi(\xi h) = \psi(\xi)h$ for all $\xi \in \pi^{-1}(U), h \in H$.

Example 1.3.7 Let G be a Lie group, Γ a discrete subgroup of G, and $\rho : G \to H$ a representation. Let $P = (G \times H)/\Gamma$, where Γ acts on the product by $(g, h)\gamma = (g\gamma, \rho(\gamma)^{-1}h)$. Then P is a principal H-bundle over M.

A principal H-bundle over a manifold M will be denoted by $P(M, H)$ or simply P.

Definition 1.3.8 An *H-map* $F : P_1(M_1, H) \to P_2(M_2, H)$ between principal H-bundles is a continuous mapping that satisfies $F(\xi h) = F(\xi)h$, for all $\xi \in P_1, h \in H$.

Since an H-map takes fibers of the bundle P_1 into fibers of P_2, it also induces a map $f : M_1 \to M_2$. If H is a compact Lie group, one can choose Riemannian metrics on P_1, P_2 such that the restriction of F to any fiber becomes an isometry. Note also that if $f : M_1 \to M_2$ is a homeomorphism, then any H-map $F : P_1 \to P_2$ covering f is a bundle isomorphism. Moreover, any principal H-bundle over a smooth manifold is isomorphic to a smooth principal bundle. See [50, §4.3]. We will only consider smooth principal H-bundles.

1.4 Cocycles

The notion of cocycle over a group action is a fundamental tool for understanding the action. In this section we present the basic definitions related to this notion. An in-depth discussion of the history, motivations, and methods will appear throughout Chapters 4, 5, and 6.

In this book we are mostly concerned with Hölder and differentiable cocycles over differentiable hyperbolic and partially hyperbolic actions. The basic definitions in the measurable, topological, and differentiable settings are similar; see [47] for a comparative treatment of cocycles in various settings. See also [179] for an extensive treatment of cocycles in the measurable setting, primarily over actions of "sufficiently large" groups.

Definition 1.4.1 Let X be a topological space. Let $\alpha : G \times X \to X$ be a continuous (or discrete) action of a continuous (or discrete) group G on X. If H is a topological group then a *cocycle* (or a 1-cocycle) over the action α with values in H is a continuous (or measurable) function $\beta : G \times X \to H$ satisfying

$$\beta(g_1 g_2, x) = \beta(g_1, \alpha(g_2, x))\beta(g_2, x), \tag{1.4.1}$$

for any $g_1, g_2 \in G$.

Most of the time the topological space X will be a smooth manifold. In this situation, an important example of a cocycle is the derivative cocycle. Let M be an n-dimensional compact Riemannian manifold on which the group G acts by diffeomorphisms. Denote by TM the tangent bundle of M and define

$$\beta(x, g) := (Dg)|_x : TM_x \to TM_{gx}.$$

One observes that the chain rule for differentiation is exactly the cocycle equation (1.4.1). If M has a trivial tangent bundle (an example is the n-dimensional torus), then we can choose a smooth section in the principal bundle of n-frames on M and β becomes a smooth cocycle with values in $GL(n, \mathbb{R})$. For an arbitrary manifold M it is shown in Section 1.3 that TM has a measurable trivialization. Thus β can be identified with a measurable cocycle $G \times M \to GL(n, \mathbb{R})$.

More generally, one can consider extensions given by principal bundles. Let G be a group that acts on P via *bundle automorphisms*, that is, the G action on P factors to a G action on M. The relationship between bundle automorphisms and cocycles is as follows. Given a cocycle $\beta : G \times M \to H$, one can construct a G action on the $P = M \times H$ by

$$g(x, h) = (g(x), \beta(x, g)h), \quad x \in M, h \in H, g \in G. \tag{1.4.2}$$

Conversely, if G acts on P via bundle automorphisms, then with respect to any trivialization of P (as explained in Section 1.3, a measurable one always exists), one can construct a cocycle $\beta : G \times M \to H$ that describes this action. If $\sigma : M \to P$ is the section, the cocycle β satisfies

$$g\sigma(x) = \sigma(g)\beta(g, x), \quad x \in M, g \in G.$$

If one works in a category in which a trivialization is not available, then the G-action is given by formula (1.4.2) only in coordinate charts. See [42] or [130] for more details.

Other cocycles naturally associated to the dynamics are the Radon–Nikodym cocycle for transformations with quasi-invariant measures, and the Jacobian cocycle, which appears in the case of differentiable dynamics.

Given an action $\alpha : G \times X \to X$ and a cocycle $\beta : G \times X \to H$, one can construct an extension α_β of α to the trivial H-bundle $X \times H$ over X defined by

$$\alpha_\beta(g, (x, h)) = (\alpha(g, x), \beta(g, x)h).$$

Note that the cocycle equation (1.4.1) implies that the extension α_β is actually an action of G on the trivial H-bundle, i.e.:

$$\alpha_\beta(g_1 g_2) = \alpha_\beta(g_1)\alpha_\beta(g_2).$$

One can construct more general extensions if the group H acts itself as a group of natural transformations on a space N. For example, H can be the group of homeomorphisms of a topological manifold, or the group of diffeomorphisms of a smooth manifold.

The natural equivalence relation on a class of cocycles is the cohomology. Two cocycles β_1 and β_2 are called continuously (or smoothly, measurably, etc.)

cohomologous if there exists a continuous (or smooth, measurable) map $P : X \to H$ such that

$$\beta_2(g, x) = P(gx)\beta_1(g, x)P(x)^{-1}, \tag{1.4.3}$$

for all $g \in G, x \in X$. The map P is called a *transfer map.*

For two continuous cohomologous cocycles β_1 and β_2 the induced actions α_{β_1} and α_{β_2} are topologically equivalent in the sense that there exists a continuous conjugacy map h on $X \times H$ such that

$$\alpha_{\beta_2}(g) = h \circ \alpha_{\beta_1}(g) \circ h^{-1}.$$

The conjugacy h is given by $h(x, h) = (x, P(x)h)$. Similar notions can be defined in measurable or differentiable settings.

Note that the cocycle equation (1.4.1) implies that a cocycle independent of the variable x is given by a homomorphism $\pi : G \to H$. In this case the extension coincides to a diagonal action.

Definition 1.4.2 A cocycle is cohomologous to a constant cocycle (cocycle not depending on x) if there exists a homomorphism $\pi : G \to H$ and a transfer map P such that

$$\beta(g, x) = P(gx)\pi(g)P(x)^{-1} \tag{1.4.4}$$

In particular, a cocycle is a *coboundary* if it is cohomologous to a trivial cocycle $\pi(g) = id_H$, i.e., if the following equation, which we will call *Livshitz's cohomology equation*, holds:

$$\beta(g, x) = P(gx)P(x)^{-1}. \tag{1.4.5}$$

Obvious obstacles to trivialization of a continuous cocycle are *closing conditions*, namely for any $x \in M$ which is fixed under some $g \in G$ the condition

$$\beta(g, x) = id_H \tag{1.4.6}$$

must hold. As we will see in Section 4.2, for Hölder cocycles over hyperbolic dynamical systems the closing conditions are also sufficient for the trivialization of a cocycle. Moreover, as we will see in Section 4.4 there are certain group actions for which the closing conditions are automatically satisfied for every cocycle of zero average.

The degree of regularity of a cocycle depends on the regularity of the action, and can be measurable, continuous, Hölder, or C^K, $1 \leq K \leq \omega$. Sometimes it is necessary or convenient to consider equivalence of cocycles in a weaker sense then the natural ambient structure, that is, the transfer map may have lower regularity then the cocycles themselves.

Definition 1.4.3 An action α is $C_H^{a,b}$*-cocycle rigid* if *any* C^a cocycle over α with values in H is cohomologous to a constant cocycle via a C^b transfer map.

In classical rank-one systems, cocycle rigidity is a rare phenomenon and appears only in totally non-hyperbolic situations such as Diophantine translation on the torus [66, Section 11.2][1] while in the higher rank setting it seems to be prevalent. We will discuss this later in Chapters 4, 5, and 6.

The following definitions were introduced in [66] in order to summarize properties of various classes of cocycles. These comprise two notions defined previously as C^a-stability (cohomology classes being closed in C^a topology) and C^b-effectiveness (C^b regularity of the transfer map).

Definition 1.4.4 An action α is $C_H^{a,b}$*-cocycle stable* if any class of cohomologous C^a cocycles over α with values in the group H is closed and the transfer map for two cohomologous cocycles is of class C^b, where $a, b > 0$ or $a, b \in \{\infty, \omega\}$. We use the notation C_H^a when $a = b$.

1.5 Roots and Weyl chambers for linear actions

Let ρ be an action of an abelian group A, isomorphic to $\mathbb{Z}^k \times \mathbb{R}^l$, by linear transformations of $\mathbb{R}^m$, or, equivalently, an embedding $\rho : A \to GL(m, \mathbb{R})$. Such an action will be called a *linear action*. Let $\lambda : A \to \mathbb{C}$ be a character, or an eigenvalue of the action, i.e., for every $a \in A$, and for some vector $v \in \mathbb{R}^m \setminus \{0\}$ then

$$\rho(a)v = \lambda(a)v.$$

The space $\mathrm{Ker}(\rho - \lambda \mathrm{Id})^m \stackrel{\text{def}}{=} R_\lambda$ is called the *root space* corresponding to the eigenvalue λ. It follows from a version of the Jordan normal form theorem that the space $\mathbb{R}^m$ splits into the direct sum of the root spaces corresponding to different real eigenvalues and the real parts of the sums of the roots spaces corresponding to the pairs of complex conjugate eigenvalues.

Definition 1.5.1 For an eigenvalue λ let $\chi(\lambda) = \log |\lambda|$. Any such χ is called a *Lyapunov exponent* of the action ρ. Let E_χ be the sum of all root spaces R_λ such that $\chi(\lambda) = \chi$. The space E_χ is called the *Lyapunov space* for the

[1] And conjecturally only in those [66, Conjecture 11.6].

exponent χ. The dimension of the Lyapunov space E_χ is called the *multiplicity* of the Lyapunov exponent χ.

For a given element of the action, the sum of all Lyapunov spaces for the exponents which have positive (corr. negative) values at this element is called the *expanding* (or *unstable*) (corr. *contracting* (or *stable*)) space for that element.

Lyapunov exponents can be extended by linearity to $\mathbb{R}^{k+l}$ so we will always assume that Lyapunov exponents are defined on $\mathbb{R}^{k+l}$.

Definition 1.5.2 The kernel of a non-zero Lyapunov exponent is called a *Lyapunov hyperplane*.

Connected components of the complement to the union of Lyapunov hyperplanes are called the *Weyl chambers* for the linear action. An element of an action is called *regular* if it does not lie in any of the Lyapunov hyperplanes.

Thus Weyl chambers are the connected components of the set of regular elements.

Definition 1.5.3 A linear action is called *hyperbolic* if none of the Lyapunov exponents is identically equal to zero. A linear action is called *partially hyperbolic* if there is at least one non-zero Lyapunov exponent.

Remark 1.5.4 In this book we will exclusively deal with volume preserving actions and their perturbations. If such an action is partially hyperbolic it has at least two distinct non-zero Lyapunov exponents.

1.6 Algebraic actions

1.6.1 The linear part

An algebraic action is an action α by diffeomorphisms, usually on a compact manifold, whose infinitesimal behavior can be described by a single linear action called the *linear part* of α. Of particular interest in dynamics and rigidity theory are those algebraic actions that have the linear part hyperbolic (called *algebraic Anosov actions*), or partially hyperbolic. All known algebraic actions of $\mathbb{Z}^k \times \mathbb{R}^l$ are constructed using projections of translation automorphisms or affine transformations of Lie groups to various coset spaces.

Representative of algebraic actions are actions of $\mathbb{Z}^k$ by automorphisms of a torus $\mathbb{T}^n = \mathbb{R}^n/\mathbb{Z}^n$. Those are often convenient as models, in particular because the linear algebra associated with them is fairly flexible. Direct generalizations of these include affine abelian actions and abelian actions on infra-nilmanifolds.

Another important class of algebraic actions consists of $\mathbb{R}^k$-actions by left translations on homogenous spaces of semisimple Lie groups, such as $SL(n, \mathbb{R})$, see Section 2.3.3. Those actions appear in applications to geometry and number theory. Notice that in general actions of continuous groups with hyperbolic properties are more abundant than such actions of discrete groups. A crude explanation is the presence of the orbit direction along which naturally no hyperbolicity appears, so that continuous actions are "less hyperbolic" than discrete ones.

We will present only general definitions now and postpone the survey of examples until the next chapter.

1.6.2 Affine actions of $\mathbb{Z}^k$ and $\mathbb{Z}^k_+$

Let H be a connected, simply connected Lie group. Let $\mathrm{Aut}(H)$ be the group of continuous automorphisms of H. An *affine automorphism* of H is an element (g, ϕ) of the semi-direct product $\mathrm{Aff}(H) = H \rtimes \mathrm{Aut}(H)$ with $g \in H$ called the *translational part* and $\phi \in \mathrm{Aut}(H)$ called the *linear part*. The product in $\mathrm{Aff}(H)$ is given by $(g, \phi)(h, \psi) = (g\phi(h), \phi\psi)$. An element $(g, \psi) \in \mathrm{Aff}(H)$ acts on $x \in H$ by $(g, \phi)(x) = g\phi(x)$.

We note that Aff (H) can be identified with the group G of diffeomorphisms of H which map right invariant vector fields on H to right invariant vector fields on H. Denote by $\mathrm{Aff}_R(H)$ the subgroup of G consisting of diffeomorphisms f that preserve the right invariant vector fields. $\mathrm{Aff}_R(H)$ can be identified with the group H acting on itself by right translations. Indeed, if $g \in H$ such that $R_g^{-1} \circ f(e) = e$, where e is the identity in H, then $R_g^{-1} \circ f$ induces the identity map on $\mathfrak{h}$, the Lie algebra of H, and has to be the identity of H. Consider now σ a smooth section of the frame bundle $F(H)$ such that for each $x \in \mathbb{R}^d$, $d = \dim(H)$, $m \to \sigma(m)x$ is a right invariant vector field. Consider the homomorphism

$$A : G \to \mathrm{Aut}(\mathrm{Lie}(H)) \subset \mathrm{GL}_d(\mathbb{R}),$$

such that $Tf_m\sigma(m) = \sigma(f(m))(Af)$. Note that the Lie algebra $\mathfrak{h}$ is identified with $\mathbb{R}^d$ equipped with the bracket $[e, e'] = \sigma^{-1}[\sigma e, \sigma e']$. Since the kernel of A is the group G, and since H is simply connected, one can identify any map $\Phi \in G$ with a composition of an automorphism ϕ of H with left multiplication L_g by an element of $g \in H$.

Let $\Lambda \subset H$ be a co-compact lattice in H and let Aff (H/Λ) be the set of diffeomorphisms of H/Λ which lift to elements of Aff (H).

Let Γ be a discrete group, and define an action ρ of G on H/Λ to be *affine algebraic* if $\rho(g)$ is given by some homomorphism $\Gamma \to$ Aff (H/Λ). In the subsequent discussion Γ will be $\mathbb{Z}^k$. Let $\mathfrak{h}$ be the Lie algebra of H. Identifying $\mathfrak{h}$ with the right invariant vector fields on H, any affine algebraic action determines a homomorphism $\sigma : \Gamma \to \operatorname{Aut} \mathfrak{h}$. Denote by σ the *linear part* of this action. We will also consider quotient actions induced by the above on finite quotients of H/Λ. An affine algebraic action of $\mathbb{Z}^k$ is called *Anosov* or *hyperbolic* (corr. *partially hyperbolic*) if its linear part is hyperbolic (corr. partially hyperbolic). As we will see, in order to admit an Anosov affine map (and hence an affine Anosov action of any discrete group) the group H must be nilpotent, see Theorem 2.1.1.

One may also consider in a similar fashion non-invertible affine actions of $\mathbb{Z}^k_+$. If Λ is a normal subgroup this construction gives an action by *endomorphisms* of the factor-group H/Λ. The notions of an Anosov and partially hyperbolic action are defined similarly to the invertible case since the linear part acts by invertible linear transformations. A particular case which appears only in the non-invertible situation is worth mentioning. A linear map is called expanding if all of its eigenvalues are of absolute value greater than one, or, equivalently, all Lyapunov exponents are positive. A $\mathbb{Z}^k_+$ action is *expanding* if all of its elements have expanding linear parts.

1.6.3 Homogeneous and double coset actions

Let A be a subgroup, isomorphic to $\mathbb{R}^k$, of a connected Lie group H. Let Λ be a lattice in H, not necessarily co-compact. The group A acts naturally on the quotient H/Λ by left translations. Suppose that C is a compact subgroup of H which commutes with A. Then the $\mathbb{R}^k$-action on H/Λ descends to an action on $C \setminus H/\Lambda$. The general *algebraic* $\mathbb{R}^k$*-action* ρ is a finite factor of such an action.

Let $\mathfrak{c}$ be the Lie algebra of C. The *linear part* of ρ is the representation of $\mathbb{R}^k$ on $\mathfrak{c} \setminus \mathfrak{h}$ induced by the adjoint representation of $\mathbb{R}^k$ on the Lie algebra $\mathfrak{h}$ of H. Let $\mathfrak{a}$ be the Lie algebra of A. The linear part of ρ leaves every element of $\mathfrak{a}$ fixed. Thus we can consider the factor of the linear part of ρ on $\mathfrak{c} \setminus \mathfrak{h} \oplus \mathfrak{a}$. We will call this action the *reduced linear part* of ρ.

Let us note that the suspension of an algebraic $\mathbb{Z}^k$-action is an algebraic $\mathbb{R}^k$-action (cf. [78, Section 2.2]).

An algebraic $\mathbb{R}^k$-action is called *partially hyperbolic* if its linear part is partially hyperbolic. Such an action is called *Anosov* if its reduced linear part is hyperbolic.

One can also consider the restriction of the action of A to a subgroup of A isomorphic to $\mathbb{Z}^l \times \mathbb{R}^{k-l}$. Such an action is partially hyperbolic if the A-action is partially hyperbolic, but it has l extra zero Lyapunov exponents.

The Lyapunov exponents, Lyapunov hyperplanes, Weyl chambers, and regular elements for affine and algebraic actions defined above are defined as those for their linear parts. A more detailed discussion of these notions in a more general setting appears in Section 1.7.2.

Since for an $\mathbb{R}^k$-action Lyapunov exponents in $\mathfrak{a}$ are zeroes, the multiplicity of the zero exponent for such an action is at least k; it is equal to k if and only if the action is Anosov.

1.6.4 Invariant distributions and their integrability

Given an algebraic action, the root spaces, Lyapunov spaces, and other invariant subspaces of the Lie algebra or its factors extend in the right invariant way to invariant fields of subspaces (or distributions) for the action. We will extend the terminology for the invariant spaces (such as Lyapunov, stable, etc.) to those distributions.

In the algebraic setting, these distributions are smooth, thus the integrability of those distributions as well as their direct sums are determined by the usual Frobenius bracket criterion. In the particular case of a homogeneous action on H/Λ, a right invariant distribution is uniquely integrable if and only if its tangent space at the identity is a Lie subalgebra of $\mathfrak{h}$. If a distribution is uniquely integrable, then its integral manifolds form an *invariant homogeneous foliation*. We will call those foliations by the same names as their tangent distributions, i.e., stable, unstable, Lyapunov, etc.

In general, Lyapunov distributions may not be integrable. Examples will be discussed in Section 2.3. Let us list the principal cases of integrability for Lyapunov distributions and their sums:

(i) Naturally a Lyapunov distribution is integrable if the Lyapunov exponent is *simple*, i.e., has multiplicity one. In this case the distribution is a smooth line field and integrability follows from the existence and uniqueness of solutions for autonomous ODE.

(ii) Stable and unstable distributions for a hyperbolic or partially hyperbolic element are always integrable. For algebraic systems this statement reduces to a computation in the Lie algebra that shows that the bracket of two stable (or unstable) vectors is a stable (unstable) vector. This extends to the non-algebraic situation, see Theorem 1.8.14.

(iii) Intersections of stable or unstable distributions for different hyperbolic elements of an abelian action are also integrable; this again extends to the case of general partially hyperbolic actions, see Definition 1.8.16.

(iv) The smallest non-trivial subspaces which can be obtained as such intersections correspond to sums of Lyapunov distributions obtained from all exponents proportional to a given one with positive coefficients of proportionality. These integrable *coarse Lyapunov distributions* play a central role in rigidity theory.

(v) Furthermore, within each coarse Lyapunov distribution there is a filtration of integrable "fast" distributions obtained as follows. Within each collection of positively proportional Lyapunov exponents there is the "fastest" exponent χ such that all other exponents from the collection can be written as $\rho_i \chi$, $i = 1, \ldots, m$, where $1 = \rho_1 > \rho_2 > \cdots > \rho_m$. Then for every $j = 1, \ldots, m$ the sum of Lyapunov distributions corresponding to the "faster" exponents $\rho_i \chi$, $i = 1, \ldots, j$ is integrable. See Section 1.7.2 for a more detailed discussion in greater generality.

(vi) Finally, one case of integrability which does not directly extend to more general situations is that of the *neutral* distribution, i.e., the Lyapunov distribution for the zero Lyapunov exponent. The neutral distribution is jointly integrable with any stable distribution and with any intersection of those.

One should note that, for general smooth non-algebraic actions discussed below in Section 1.8, stable and unstable distributions exhibit only Hölder regularity transversally to the leaves, even though they are as regular as the action along the leaves and each one of them is integrable. This fact greatly reduces the applicability of Frobenius criterion in dealing with sums of distributions that are parts of stable/unstable distributions.

1.6.5 Resonances

The general reason for the non-integrability of Lyapunov distributions and their sums is the presence of *resonances*, i.e., linear relations between Lyapunov exponents with integer coefficients of a particular kind. We will only consider three simple, representative examples here, and will leave a more general discussion to Section 2.3.4.

(i) *Exponents of the opposite sign.* If both χ and $-\chi$ are Lyapunov exponents, the sum of their Lyapunov distributions is often not integrable. Typical examples appear in geodesic flows on symmetric spaces of negative curvature ($\mathbb{R}$-actions), and for Weyl chamber flows for split simple Lie groups of rank $k \geq 2$ ($\mathbb{R}^k$-actions, see Section 2.3.4). This symmetry follows immediately from reversibility of flows and actions in question: the flip $v \to -v$ in the acting group ($\mathbb{R}$ or $\mathbb{R}^k$) produces an isomorphic action.

(ii) *Exponents with positive integer proportionality coefficient.* If χ is an exponent with multiplicity greater than one, than the bracket of two vector fields from its Lyapunov distribution has to have the Lyapunov exponent equal to 2χ. If this one is among the Lyapunov exponents, then the Lyapunov distribution for χ may be (and often is) non-integrable. See Section 2.3.4 for a specific example and a detailed discussion of the resonances.

(iii) *A relation of a different kind.* In Example 2.1.6 one has $\chi_i = \lambda_i$, $i = 1, 2, 3$ and hence $\chi_1 + \chi_2 = \chi_3$, and the non-trivial commutation relation $[X_1, X_2] = X_3$ leads to the non-integrability of the distribution generated by two positive exponents χ_1 and χ_2.

1.7 Measurable and non-uniform differentiable setting

We proceed in this section to a general treatment of Lyapunov characteristic exponents which can be defined for the linear extensions of actions by measure-preserving transformations of a finite measure space. The natural situation where this appears is the linear extension of a smooth action to the tangent bundle provided by the derivative of the action. One then considers an invariant measure for the action. While from the topological point of view the tangent bundle is often non-trivial there is always a piece-wise smooth section on a set of full measure (see Section 1.4). This allows us to consider the bundle as the direct product from the measure theory point of view.

For the algebraic actions considered in Section 1.6, Lyapunov exponents and derivative notions are independent of the measure and are determined by the linear part of the action. Notice that for homogeneous spaces of Lie groups of the form H/Λ as in Section 1.6.3 the tangent bundle is always trivial since its framing is given by a basis of right invariant vector fields. This is in general not true any more for double coset spaces.

In subsequent parts of this book we will not use the above general setting as such. However, it provides the most general way in which certain features of the algebraic situation, namely, precise rates of exponential growth/decay, persist. The issues of resonances, integrability of invariant "distributions" (invariant measurable fields of subspaces), and suchlike appear in a way surprisingly similar to the algebraic situation. This approach has already been used in a crucial way in the new development in global measure rigidity [61, 62, 74] and is very likely to play a central role in the future development of global differentiable rigidity.

1.7.1 Multiplicative ergodic theorem

Let (X, μ) be a probability Lebesgue space. An action $\alpha : \mathbb{R}^k \times X \to X$ is said to be ergodic if the group $\mathbb{R}^k$ acts by measure-preserving transformations of the space (X, μ) and, moreover, any α-invariant function $f \in L^2(X, \mu)$ is constant.

Let $A : \mathbb{R}^k \times X \times \mathbb{R}^m \to X \times \mathbb{R}^m$ be a *linear extension* of the action α. Such an extension is determined by a *matrix-valued cocycle* $\mathcal{A} : \mathbb{R}^k \times X \to GL(m, \mathbb{R})$ as follows:

$$A(s, x, t) = (\alpha(s, x), \mathcal{A}(s, x)(t)),$$

where, by the group property, one has

$$\mathcal{A}(s_1 + s_2, x) = \mathcal{A}(s_2, \alpha(s_1, x))\mathcal{A}(n_1, x) \tag{1.7.1}$$

(cf. the discussion in Section 1.4).

Theorem 1.7.1 (Multiplicative ergodic theorem for $\mathbb{R}^k$-actions) *Suppose for each $s \in \mathbb{R}^k$,*

$$\log \|\mathcal{A}(s, x)\| \in L^1(X, \mu). \tag{1.7.2}$$

Then there exist linear functionals $\chi_1, \dots, \chi_l$ on $\mathbb{R}^k$ and for μ-a.e. point $x \in X$ a decomposition of the fiber ${\mathbb{R}^m}_x$ over x:

$$\mathbb{R}^m_x = E_{\chi_1}(x) \oplus \cdots \oplus E_{\chi_l}(x), \tag{1.7.3}$$

such that for $i = 1, \dots, l$ and for any $v \in E_{\chi_i}(x)$ one has

$$\lim_{s\to\infty} \frac{\log \|\mathcal{A}(s, x)(t)\| - \chi_i(s)}{\|s\|} = 0, \tag{1.7.4}$$

and

$$\lim_{s\to\infty} \frac{\log \det \mathcal{A}(s, x) - \sum_{i=1}^{l} m_i \chi_i(s)}{\|s\|} = 0, \tag{1.7.5}$$

where $m_i = \dim E_{\chi_i}(x)$. Moreover, the set of points where this decomposition is defined is α-invariant and the decomposition itself is A-invariant.

Definition 1.7.2 The functionals $\chi_1, \ldots, \chi_l$ are called the *Lyapunov characteristic exponents* of A. The dimension m_i of the space $E_{\chi_i}(x)$ is called the *multiplicity* of the exponent χ_i.

The decomposition (1.7.3) is called the *(fine) Lyapunov decomposition* at the point x.

Proof [Sketch of proof] Theorem 1.7.1 can be easily deduced from the standard Oseledets multiplicative ergodic theorem for linear extensions of a single measure-preserving transformation [6, Theorem 3.4.3] by a simple induction process. Namely, one first applies the Oseledets theorem to the first generator of the $\mathbb{R}^k$-action. The Lyapunov decomposition for it is invariant under the whole action so one can apply the Oseledets theorem to the restriction of the extension of the second generator to each element of the Lyapunov decomposition, and so on. Once one has obtained a decomposition for which limits exist for multiples of all generators, the existence of the limits (1.7.4) and (1.7.5) and the linearity of the exponents follow easily from the cocycle relation (1.7.1). See [6, Theorem 3.6.6] for somewhat more detailed arguments. □

Points where the assertions of the theorem are satisfied will be called *regular*. The set of all regular points will be usually denoted by the letter Λ, sometimes with extra indices.

1.7.2 Lyapunov hyperplanes and Weyl chambers

Now we can generalize and discuss in greater detail certain notions previously introduced in Sections 1.5 and 1.6 for special cases.

Definition 1.7.3 The hyperplane $\ker \chi \subset \mathbb{R}^k$, where χ is a non-zero Lyapunov exponent, is called a *Lyapunov hyperplane*.

The subspace $\chi^{-1}(-\infty, 0)$ (corr. $\chi^{-1}(0, \infty)$) is called a negative (corr. positive) Lyapunov half-space.

For $k = 2$ we will call Lyapunov hyperplanes *Lyapunov lines*.

An element $s \in \mathbb{Z}^k$ is called *regular* if s does not belong to any of the Lyapunov hyperplanes. A regular element for a hyperbolic linear extension of a $\mathbb{Z}^k$-action is called *hyperbolic*.

Definition 1.7.4 A Weyl chamber is a connected component of the complement to the union of all Lyapunov hyperplanes or, equivalently, connected components of the set of regular elements.

Each Weyl chamber is an open convex polyhedral cone in $\mathbb{R}^k$. Inside a Weyl chamber every non-zero Lyapunov exponent has a constant sign. Conversely, the locus of points in $\mathbb{R}^k$ for which each non-zero Lyapunov exponent has a particular sign is either empty or is a Weyl chamber. Thus any Weyl chamber can be characterized as a minimal non-empty intersection of positive and negative Lyapunov half-spaces.

Definition 1.7.5 A linear extension of an ergodic $\mathbb{R}^k$-action is called partially hyperbolic if there is at least one non-zero Lyapunov exponent and hyperbolic if all Lyapunov exponents are different from zero.

A hyperbolic linear extension is called totally non-symplectic, or *TNS*, if non of the Lyapunov exponents is proportional to another with the negative coefficient of proportionality.

For a partially hyperbolic element $s \in \mathbb{R}^k$ and a regular point x we set

$$E_s^+(x) = \bigoplus_{i:\chi_i(s)>0} E_{\chi_i}(x) \quad \text{and} \quad E_s^-(x) = \bigoplus_{i:\chi_i(s)<0} E_{\chi_i}(x). \tag{1.7.6}$$

These subspaces are called correspondingly the *expanding* (or *unstable*) and the *contracting* (or *stable*) subspaces for s at the point x. These spaces are the same for any s within a Weyl chamber and change from one Weyl chamber to another. Thus the fiber at x decomposes

$$E_s^+(x) \oplus E_s^-(x) \oplus E_0(x). \tag{1.7.7}$$

This is sometimes called the *hyperbolic decomposition* for s. The most robust common refinement of hyperbolic decompositions for different regular elements s, or, equivalently, for different Weyl chambers, will play an important role in the subsequent discussions. It is called the *coarse Lyapunov decomposition* and its elements are the minimal non-empty intersections of expanding subspaces for various Weyl chambers.

Equivalently, let us associate to every half-space H all Lyapunov exponents for which H is the positive Lyapunov subspace. Suppose there are m_H such exponents. Naturally, all these exponents are proportional with positive coefficients of proportionality and thus one can find a unique exponent χ_H such that all exponents have the form $\rho_i \chi_H$, $i = 1, \ldots, m_H$, where $1 = \rho_1 > \rho_2 > \cdots > \rho_{m_H}$. Then put

$$E_H(x) = \bigoplus_{i=1}^{m_H} E_{\rho_i \chi_i}(x),$$

and let $H_1, \ldots, H_r$ be different positive Lyapunov half-spaces. The coarse Lyapunov decomposition at the point x is

$$E_{H_1}(x) \oplus \cdots \oplus E_{H_r}(x) \oplus E_0(x). \tag{1.7.8}$$

Notice that a hyperbolic linear extension is totally non-symplectic if and only if there are no complementary half-spaces among $H_1, \ldots, H_r$.

Inside each space $E_H(x)$ there is a flag of fast subspaces

$$E_{H,j}(x) \stackrel{\text{def}}{=} \bigoplus_{i=1}^{j} E_{\rho_i \chi_i}(x). \tag{1.7.9}$$

1.7.3 Invariant manifolds

Now we will consider the case of the extension of a smooth action to the tangent bundle given by the differentials. We will denote the differential of a map f by Df. We will assume that phase space is a manifold M provided with a Riemannian metric. If M is compact the condition (1.7.2) is satisfied for any Borel measure μ and for any Riemannian metric, and Lyapunov exponents and other notions described above are independent of the metric. For a non-compact M the same is true if the support of μ is compact. Otherwise we will assume that condition (1.7.2) holds. We will also assume that all maps are uniformly of class $C^{1+\gamma}$ for some $\gamma > 0$, i.e., the derivative of each map satisfies the Hölder condition with exponent γ and fixed constant (which may depend on the map)

Consider an action α of $\mathbb{Z}^k \times \mathbb{R}^l$ by such diffeomorphisms. The following result is a simple corollary of the Hadamard–Perron theorem for non-uniformly hyperbolic dynamical systems [6, Chapter 7]. We refer to the same source for a detailed discussion of the precise meaning of the notions involved.

Theorem 1.7.6 *For each $i = 1, \ldots, r$, $j = 1, \ldots, M_i$ the family $E_{H_i,j}$ uniquely integrates μ almost everywhere to a measurable family of smooth manifolds $W_{H_i,j}$.*

For the rank one case $k = 1$ there are only two Weyl chambers, the positive and negative half-lines, and we will denote the corresponding families of subspaces and invariant manifolds by $E_{+,j}$, $E_{-,j}$, $W_{+,j}$, and $W_{-,j}$.

Remark 1.7.7 Notice that individual Lyapunov subspaces E_χ may not be integrable even if they depend smoothly on x. Such examples appear already in the homogeneous situations when the corresponding Lyapunov subspace in the Lie algebra is not a subalgebra, as was discussed in Section 1.6.4.

1.8 Uniform differentiable setting

Now we consider a situation intermediate between those discussed in two previous sections, namely Anosov (normally hyperbolic) and partially hyperbolic actions. The basic structures here, namely stable, unstable, and neutral distributions, are assumed to be continuous. Hence for any invariant measure the setting of the previous section is applicable and one can define Lyapunov exponents, check their non-vanishing and consider derivative notions as above. However, continuity of the basic structures does not guarantee any uniformity in this case[2] and hence the structures associated with Lyapunov exponents are usually just measurable, and in fact exponents are different for different invariant measures. Thus, in this setting there is much less analogy at the topological or differentiable level with the algebraic one than in the general measurable setting of the previous section.

On the other hand, hyperbolic and partially hyperbolic dynamics are principal tools in rigidity theory and results outlined in this section will be extensively used later in this book.

1.8.1 Anosov diffeomorphisms and flows

A good introduction to the theory of Anosov systems and, more generally, to hyperbolic dynamics, is given in [67]. The proofs of all the basic results for diffeomorphisms stated below without proofs can be found there. The proofs for flows are mostly quite similar.

We consider a compact differentiable manifold M and a C^1 diffeomorphism $f : M \to M$. We will denote by TM the tangent bundle of M, and by $Df : TM \to TM$ the derivative of f.

Definition 1.8.1 The diffeomorphism f is said to be an *Anosov diffeomorphism* if there exists a smooth Riemannian metric $\|\cdot\|$ on M, a number $\lambda \in (0, 1)$, and a continuous splitting $TM = E^s \oplus E^u$ of the tangent bundle into Df-invariant sub-bundles E^s and E^u, such that

$$\begin{aligned} \|Dfv\| &\leq \lambda\|v\|, \ v \in E^s, \\ \|Df^{-1}v\| &\leq \lambda\|v\|, \ v \in E^u. \end{aligned} \tag{1.8.1}$$

Remark 1.8.2 The Riemannian metric in Definition 1.8.1 is said to be *adapted.*

[2] It does in some higher rank situations but only as a result of highly non-trivial arguments, see [81].

Let us denote by d_M the distance on M induced by the adapted metric. For $\delta > 0$ and $x \in M$ we denote by $B(x, \delta)$ the ball of radius δ centered in M.

Theorem 1.8.3 (Stable and unstable manifolds theorem) *Let M, f, E^s, E^u, and λ be as in Definition 1.8.1. Then for each $x \in M$ there is a pair of embeded C^1-discs $W^s_{loc}(x)$, $W^u_{loc}(x)$, called the local stable manifold and the local unstable manifold at x, respectively, such that:*

(i) $T_x W^s_{loc}(x) = E^s(x)$, $T_x W^u_{loc}(x) = E^u(x)$;
(ii) $f(W^s_{loc}(x)) \subset W^s_{loc}(fx)$, $f^{-1}(W^u_{loc}(x)) \subset W^u_{loc}(f^{-1}x)$;
(iii) for any $\mu \in (\lambda, 1)$, there exists a constant $C > 0$ such that for all $n \in \mathbb{N}$,

$$d_M(f^n x, f^n y) \leq C\mu^n d_M(x, y), \text{ for } y \in W^s_{loc}(x),$$
$$d_M(f^{-n} x, f^{-n} y) \leq C\mu^n d_M(x, y), \text{ for } y \in W^u_{loc}(x);$$

(iv) there exists a constant $\beta > 0$ such that for all $x \in M$,

$$W^s_{loc}(x) = \{y \in M \,|\, f^n(y) \in B(f^n(x), \delta) \text{ for all } n \geq 0\},$$
$$W^u_{loc}(x) = \{y \in M \,|\, f^{-n}(y) \in B(f^{-n}(x), \delta) \text{ for all } n \geq 0\}.$$

The local stable (unstable) manifolds can be extended to global stable (unstable) manifolds $W^s(x)$ and $W^u(x)$:

$$W^s(x) = \cup_{n=0}^{\infty} f^{-n}(W^s_{loc}(f^n(x))),$$
$$W^u(x) = \cup_{n=0}^{\infty} f^{n}(W^s_{loc}(f^{-n}(x))),$$

which are well defined, smooth injectively immersed and given also by

$$W^s(x) = \{y \in M | d_M(f^n(x), f^n(y)) \to 0, \text{ as } n \to \infty\},$$
$$W^u(x) = \{y \in M | d_M(f^{-n}(x), f^{-n}(y)) \to 0, \text{ as } n \to \infty\}.$$

These global manifolds are the leaves of global foliations W^s and W^u of M.

Theorem 1.8.4 (Local product structure) *Given an Anosov diffeomorphism $f : M \to M$, there are constants $\Delta > 0$, $K > 0$ such that for any $x, y \in M$ with $d_M(x, y) < \Delta$, the intersection $W^s_{loc}(x) \cap W^u_{loc}(y)$ contains exactly one element, whose distance to both x and y is at most $K d_M(x, y)$.*

Definition 1.8.5 Let $f : M \to M$ diffeomorphism. Let δ be a positive number. A sequence $(x_0, x_1, \ldots, x_{n-1})$ of points in M is called a *periodic δ-pseudo-orbit for f* if

$$d_M(f x_j, x_{j+1}) \leq \delta, \text{ for } j = 0, 1, \ldots, n-1,$$

where $x_n = x_0$. A *periodic orbit* is a δ-pseudo-orbit for $\delta = 0$.

Theorem 1.8.6 (Anosov closing lemma) *Given an Anosov diffeomorphism $f : M \to M$, there are constants $K > 0$, $\delta_0 > 0$ such that given a periodic δ-pseudo-orbit $(x_0, x_2, \dots, x_{n-1})$ with $\delta < \delta_0$, there is a periodic orbit of the same length $(x, fx, \dots, f^{n-1}x)$, $f^n x = x$, such that*

$$d_M(x_k, f^k x) \le K\delta, \quad 0 \le k \le n.$$

One observes that the constants K appearing in Theorems 1.8.4 and 1.8.6 can be taken to be identical if desired.

We need also the following consequence of the previous results.

Lemma 1.8.7 *Let M, f, λ, Δ, and K be as in the previous results, fix $\mu \in (\lambda, 1)$ and let C be provided by Theorem 1.8.3(iii). There exists a constant $c > 0$ with the following property: if $x \in M$ and $N \in \mathbb{N}$ are such that $d_M(f^N x, x) < \Delta/K$, there exist y and $z \in M$ with the following properties:*

(i) $f^N y = y$;
(ii) $d_M(f^k x, f^k z) \le c\mu^k d_M(f^N x, x)$ *for* $k = 0, 1, \dots, N-1$; *and*
(iii) $d_M(f^k z, f^k y) \le c\mu^{N-k} d_M(f^N x, x)$ *for* $k = 0, 1, \dots, N-1$.

Proof The Anosov closing lemma applied to the pseudo-orbit of x $(x, fx, \dots, f^{N-1}x)$ yields a point $y \in M$ such that $f^N y = y$ and

$$d_M(f^k x, f^k y) \le K d_M(f^N x, x) < \Delta, \ k = 0, 1, \dots, N.$$

The local product structure theorem then provides a unique point z in the intersection of $W^s(x)$ and $W^u(y)$ such that

$$d_M(x, z) \le K d_M(x, y) \le K^2 d_M(f^N x, x),$$

and, likewise,

$$d_M(y, z) \le K^2 d_M(f^N x, x).$$

Inequalities (iii) in Theorem 1.8.3 imply that

$$d_M(f^k x, f^k y) \le C\mu^k d_M(x, z) \le CK^2\mu^k d_M(f^N x, x),$$

for all $k \ge 0$. We still have to verify the inequalities (3). To do this, observe that $f^N z \in W^u(f^N y) = W^u(y)$ and, therefore,

$$d_M(f^{-k} f^N z, f^{-k} f^N y) \le C\mu^k d_M(f^N z, f^N y), \ k \ge 0.$$

Now,

$$\begin{aligned} d_M(f^N z, f^N y) &\le d_M(f^N z, f^N x) + d_M(f^N x, f^N y) \\ &\le CK^2 d_M(z, x) + d_M(f^N x, f^N y) \qquad (1.8.2) \\ &\le CK^3 d_M(f^N x, x) + K d_M(f^N x, x), \end{aligned}$$

and the lemma follows with $c = C(CK^3 + K) + 2$. □

Almost all previous definitions and results have their analogs for flows.

Definition 1.8.8 Let M be a compact manifold and $\phi : \mathbb{R} \times M \to M$ a C^1-flow. The flow ϕ is said to be an *Anosov flow* if there exists a Riemannian metric $\|\cdot\|$ on M, a constant $0 < \lambda < 1$, and a continuous splitting $TM = E^s \oplus E^c \oplus E^u$ of the tangent bundle into Df-invariant sub-bundles E^s, E^c and E^u, such that for all $x \in M$,

(i) $\frac{d}{dt}|_{t=0}\phi^t \in E^c_x \setminus \{0\}$, $dim\ E^c_x = 1$;
(ii) $\|D\phi^t v\| \le \lambda^t \|v\|$, $v \in E^s$; and
(iii) $\|D\phi^{-t} v\| \le \lambda^t \|v\|$, $v \in E^u$.

Remark 1.8.9 The Riemannian metric in Definition 1.8.8 is said to be *adapted.*

The analog of Theorem 1.8.3 is the following result.

Theorem 1.8.10 *Let M, ϕ, E^s, E^c, E^u, and λ be as in Definition 1.8.8. Then for each $x \in M$ there is a pair of embedded C^1-discs $W^s_{loc}(x)$, $W^u_{loc}(x)$, called the local (strong) stable manifold and the local (strong) unstable manifold at x, respectively, such that:*

(i) $T_x W^s_{loc}(x) = E^s(x)$, $T_x W^u_{loc}(x) = E^u(x)$;
(ii) $\phi^t(W^s_{loc}(x)) \subset W^s_{loc}(\phi^t x)$, $\phi^{-t}(W^u_{loc}(x)) \subset W^u_{loc}(\phi^{-t} x)$ *for* $t > 0$;
(iii) For any $\mu \in (\lambda, 1)$, there exists a constant $C > 0$ such that for all $n \in \mathbb{N}$,

$$d_M(\phi^t x, \phi^t y) \le C\mu^t d_M(x, y), \text{ for } y \in W^s_{loc}(x), t > 0,$$
$$d_M(\phi^{-t} x, \phi^{-t} y) \le C\mu^t d_M(x, y), \text{ for } y \in W^u_{loc}(x), t > 0;$$

(iv) There exists a constant $\beta > 0$ such that for all $x \in M$,

$$W^s_{loc}(x) = \{y \in M | \phi^t(y) \in B(\phi^t x, \beta), t > 0, \text{ and } \lim_{t\to\infty} d_M(\phi^t(x), \phi^t(y)) = 0\},$$
$$W^u_{loc}(x) = \{y \in M | \phi^{-t}(y) \in B(\phi^{-t} x, \beta), t > 0, \text{ and } \lim_{t\to\infty} d_M(\phi^{-t}(x), \phi^{-t}(y)) = 0\}.$$

The local stable (unstable) manifolds can be extended to global stable (unstable) manifolds $W^s(x)$ and $W^u(x)$:

$$W^s(x) = \cup_{t>0}\phi^{-t}(W^s_{loc}(\phi^t(x))),$$
$$W^u(x) = \cup_{t>0}\phi^{t}(W^s_{loc}(\phi^{-t}(x))),$$

which are well defined, smooth injectively immersed and also given by

$$W^s(x) = \{y \in M | d_M(\phi^t(x), \phi^t(y)) \to 0, \text{ as } t \to \infty\},$$
$$W^u(x) = \{y \in M | d_M(\phi^{-t}(x), \phi^{-t}(y)) \to 0, \text{ as } t \to \infty\}.$$

These global manifolds are the leaves of global foliations W^s and W^u of M.

One can also define weak stable and weak unstable foliations with leaves given by $W^{cs}(x) = \cup_{t\in\mathbb{R}}(W^s(x))$ and $W^{cu}(x) = \cup_{t\in\mathbb{R}}(W^u(x))$, which have as tangent distributions $E^{cs} = E^c \oplus E^s$ and $E^{cu} = E^c \oplus E^s$.

The analog of Theorem 1.8.6 is the following.

Theorem 1.8.11 *Let M be a compact manifold and $\phi : \mathbb{R} \times M \to M$ a C^1 Anosov flow with contraction constant λ. There exist positive constants δ_0, C such that if $x \in M$ and $t \in \mathbb{R}$ with*

$$d_M(\phi^t x, x) < \delta < \delta_0,$$

then there exists a closed ϕ-orbit $\mathcal{O}$, a point $y \in \mathcal{O}$, and a differentiable map $\gamma : [0, t] \to \mathbb{R}^k$ such that for all $s \in [0, t]$ we have

(i) $d_M(\phi^s(x), \phi^{\gamma(s)}y) \le C\delta$;
(ii) $\phi^{\gamma(t)}(y) = \phi^\tau(y)$, *where* $\tau \le C\delta$; *and*
(iii) $\|\gamma' - 1\| < C\delta$.

1.8.2 Partially hyperbolic diffeomorphisms

Let L be a linear transformation between two normed linear spaces. The *norm*, respectively *conorm*, of L are defined as

$$\|L\| := \sup\{\|Lv\|; \|v\| = 1\}, \qquad m(L) := \inf\{\|Lv\|; \|v\| = 1\}.$$

The following notion was introduced in [12].

Definition 1.8.12 Let X be a compact differentiable manifold. A C^1 diffeomorphism $f : X \to X$ is called *partially hyperbolic* if the derivative $Df : TX \to TX$ leaves invariant a continuous splitting $TX = E^s \oplus E^c \oplus E^u$, $E^s \neq 0 \neq E^u$, and there exists a Riemanninan metric such that Df contracts

E^s by a constant $0 < \lambda_- < 1$, Df^{-1} contracts E^u by a constant $0 < \lambda_+ < 1$, and the inequalities

$$\|D_p^s f\| < m(D_p^c f) \text{ and } \|D_p^c f\| < m(D_p^u f)$$

hold for all $p \in X$.

The following notion was introduced in [51].

Definition 1.8.13 Assume that f is a partially hyperbolic C^r diffeomorphism, $1 \leq r < \infty$, that leaves invariant a C^1-foliation $\mathcal{L}$ tangent to the neutral direction E^s. Then f is said to be *r-normally hyperbolic at $\mathcal{L}$* if:

$$m(D_p^u f) > \|D_p^c f\|^k \text{ and } \|D_p^s f\| < m(D_p^c f)^k, \tag{1.8.3}$$

for all $0 \leq k \leq r$ and $p \in X$.

The following results are proved in [51].

Theorem 1.8.14 *Let X be a compact manifold, $f \in \mathit{Diff}(X), r \geq 1$, a diffeomorphism that is r-normally hyperbolic at a C^r-foliation $\mathcal{L}_f$.*

(i) The distributions E^s and E^u are integrable. The corresponding foliations are called stable, *respectively* unstable, *and are denoted by W^s, respectively W^u. The foliations are Hölder, and their leaves $W^s(x)$ and $W^u(x)$ are C^r and depend continuously on $x \in X$ in C^r-topology. The leaves can be characterized as follows:*
Let $0 < \lambda'_- < \lambda_-, 0 < \lambda'_+ < \lambda_+$, where λ_-, λ_+ are as in Definition 1.8.12. Then $y \in W^s(x)$ if and only if:

$$\lim_{n\to\infty} (\lambda'_-)^{-n} d_M(f^n(y), f^n(x)) = 0, \tag{1.8.4}$$

and $y \in W^u(x)$ if and only if

$$\lim_{n\to\infty} (\lambda'_+)^{-n} d_M(f^{-n}(y), f^{-n}(x)) = 0. \tag{1.8.5}$$

(ii) If $g \in \mathit{Diff}(X)$ is C^1 close to f, then g is r-normally hyperbolic at a unique C^r-foliation $\mathcal{L}_g$. Moreover, the stable and unstable leaves of g converge in C^r to those of f as g converges to f in the C^r-topology.

(iii) If $\mathcal{L}_f$ is a C^r-foliation and $g \in \mathit{Diff}(X)$ is C^r close to f, then there exists a leaf-conjugacy $H \in \mathit{Homeo}(X)$ between $(f; \mathcal{L}_f)$ and $(g; \mathcal{L}_g)$. In addition H maps the leaves of $\mathcal{L}_f$ to those of $\mathcal{L}_g$ and

$$\mathcal{L}_g(H \circ f(x)) = \mathcal{L}_g(g \circ H(x)).$$

The map H is a C^r-diffeomorphism of each leaf of $\mathcal{L}_f$ onto its image, varying continuously in C^r with the leaf. For $x \in X$, $\mathcal{L}_g(H(x))$ is *uniquely characterized by the fact that its g-orbit does not stray away from the f-orbit of $\mathcal{L}_f(x)$. Modulo the choice of a normal bundle to $\mathcal{L}_f$, H is uniquely determined. If g converges to f in the C^r-topology then H converges to the identity in the C^r-topology along the leaves of $\mathcal{L}_f$ and to Id_X in C^0-topology.*

Remark 1.8.15 The phrase above "never strays away" means that the iterates $g^n(\mathcal{L}_g(H(x))$ stay within a tubular neighborhood of predetermined small size of $f^n(\mathcal{L}_f(x))$, for each $n \in \mathbb{Z}$.

1.8.3 Anosov actions of higher rank abelian groups

Definition 1.8.16 An action α of a higher rank abelian group A on a compact manifold M is called *partially hyperbolic* if there exist an element $g \in A$ that acts on M as a partially hyperbolic diffeomorphism.

If in addition, the neutral distribution E^c_g is uniquely integrable to a foliation $\mathcal{F}$ then the action of g is *normally hyperbolic* with respect to the foliation $\mathcal{F}$ in the sense of Definition 1.8.12.

An element $g \in A$ is said to be *regular* if the corresponding neutral direction is contained in the neutral direction of any other partially hyperbolic element in A.

If g is such that E^c_g coincides with the subbundle $T\mathcal{O}$ tangent to the orbit foliation of the A-action, than g is normally hyperbolic with respect to the orbit foliation and is regular. An action containing such an element is said to be *Anosov* or *hyperbolic*.

An action is called *genuinely partially hyperbolic* if it is partially hyperbolic, but not hyperbolic.

Remark 1.8.17 It is not known whether all partially hyperbolic higher rank abelian action have regular elements. However, the neutral distribution E^c_g always contains $T\mathcal{O}$. Conjecturally, regular elements are dense in A.

We call E^s_g and E^u_g the stable, respectively unstable, distribution of g. They are integrable and the corresponding foliations will be denoted by $\mathcal{W}^s_g$ and $\mathcal{W}^u_g$ respectively.

Note that since M is compact, these notions do not depend on the ambient Riemannian metric. Note that the splitting and the constants in the definition above depend on the normally hyperbolic element $g \in A$.

Now given a $\mathbb{Z}^k$-action α suppose at least one element $g \in \mathbb{Z}^k$ acts by an Anosov diffeomorphism on M. Then the suspension of α (see page ..) is an Anosov $\mathbb{R}^k$-action. Indeed, g, thought of as an element of $\mathbb{R}^k$, is a regular element.

The following result is a consequence of Theorem 1.8.14.

Theorem 1.8.18 *Let M be a closed manifold, and $\alpha : A \times M \to M$ an action with a normally hyperbolic element g. If $\alpha^* : A \times M \to M$ is a second action of A sufficiently close to α in the C^1-topology then g is also normally hyperbolic for α^*. The stable and unstable manifolds of $\alpha^*(g)$ tend to those of $\alpha(g)$ in the C^k-topology as α^* tends to α in the C^k-topology. Furthermore, there is a Hölder homeomorphism $\phi : M \to M$ close to Id_M such that ϕ takes the leaves of the orbit foliation of α^* to those of α.*

Let us call an orbit $\mathbb{R}^k \cdot x$ of a locally free $\mathbb{R}^k$-action *closed* if the stationary subgroup S of x (and hence of each point of that orbit) is a lattice in $\mathbb{R}^k$. Thus any closed orbit is naturally identified with the k-torus $\mathbb{R}^k/S$.

Proposition 1.8.19 *Any orbit of an Anosov $\mathbb{R}^k$-action whose stationary subgroup contains a regular element a is closed, and furthermore, a fixes any point y in the closure of the orbit.*

Another standard fact about Anosov $\mathbb{R}^k$-actions is an Anosov type closing lemma, which is a straightforward generalization of a similar statement for Anosov flows [51].

Theorem 1.8.20 (Closing lemma) *Let $g \in \mathbb{R}^k$ be a regular element of an Anosov $\mathbb{R}^k$-action α on a closed manifold M. There exist positive constants δ_0, C, and λ depending continuously on α in the C^1-topology and g such that:*

if for some $x \in M$ and $t \in \mathbb{R}$

$$dist(\alpha(tg)x, x) < \delta_0,$$

then there exists a closed α-orbit $\mathcal{O}$, a point $y \in \mathcal{O}$, and a differentiable map $\gamma : [0, t] \to \mathbb{R}^k$ such that for all $s \in [0, t]$ we have

(i) $dist(\alpha(sg)(x), \alpha(\gamma(s))y) \le C\, e^{-\lambda(\min(s,t-s))}\, dist(\alpha(tg)(x), x)$;
(ii) $\alpha(\gamma(t))(y) = \alpha(\delta)(y)$ *where* $\| \delta \| \le C\, dist(\alpha(tg)(x), x)$; *and*
(iii) $\| \gamma' - g \| < C\, dist(\alpha(tg)x, x)$.

Let us point out a fundamental difference between the properties of "closing" (i.e., approximating a certain part of an orbit by a part of a closed orbit) in rank-one and higher rank cases. In the former case (i.e., for $\mathbb{Z}_+$, $\mathbb{Z}$, and $\mathbb{R}$ actions) an almost closed orbit segment is approximated by a *complete* closed orbit, albeit possibly run over several times. In the higher rank case close return in a *particular direction* guarantees presence on an initial condition nearby whose orbit closes in *all directions*. While a closing time (again not necessarily the first) in the original direction is close to that of the original close return the closing times in the remaining directions cannot be controlled. Even if the same initial condition almost closes in several directions which generate the orbit the approximating closed orbits for different directions will in general be different and again the return times in the complementary directions will not be controlled.

2

Principal classes of algebraic actions

2.1 Automorphisms of tori and (infra)nilmanifolds

2.1.1 Nilpotency of the ambient group

Let us start with a fundamental result that restricts the array of examples of algebraic actions of interest to us. We show that for an Anosov algebraic action ρ of a discrete group G, $\rho : G \to \mathrm{Aff}\,(H/\Lambda)$, the Lie group H has to be nilpotent. Thus, the most general case of an affine Anosov $\mathbb{Z}^k$-action takes place on an infranilmanifold, that is, a finite quotient of a nilmanifold N/Λ, where N is a connected simply connected nilpotent Lie group and $\Lambda \subset N$ is a co-compact lattice.

The proof of the following theorem is taken from [41, Proposition 3.13].

Theorem 2.1.1 *Let H be a connected, simply connected Lie group. Assume there exists $\Phi \in Aff\,(H)$ such that the linear part of Φ is hyperbolic. Then H is nilpotent.*

Proof Let $\mathfrak{h}$ be the Lie algebra of H. We recall from Section 1.6.2 that any map $\Phi \in \mathrm{Aff}\,(H)$ is a composition of an automorphism ϕ of H with left multiplication L_g by an element of $g \in H$. Let ψ be the automorphism of $\mathfrak{h}$ induced by $\mathrm{Ad}(g) \circ \phi_*$. Let ψ_s be the semisimple component of the Jordan decomposition of ψ, which is also an automorphism of $\mathfrak{h}$. Since, with respect to a right invariant metric on H, $\mathrm{Ad}(g) \circ \phi_*$ and $(R_{g^{-1}})_*\mathrm{Ad}(g) \circ \phi_* = \phi_*$ have the same norm for any $v \in \mathfrak{h}$, Φ Anosov implies that ψ, and hence ψ_s, cannot have eigenvalues of modulus one.

If $\mathfrak{s}$ is the (solv)radical of $\mathfrak{h}$, that is, the maximal solvable ideal of $\mathfrak{h}$, then ψ_s induces an automorphism of $\mathfrak{h}/\mathfrak{s}$, which is also denoted by ψ. Since $\mathfrak{h}/\mathfrak{s}$ is a semisimple Lie algebra, some finite power ψ_s^q of ψ_s coincides to $\mathrm{Ad}(h+\mathfrak{s})$ for some $h \in \mathfrak{h}$. Moreover, $\mathrm{Ad}(h + \mathfrak{s})$ must contain eigenvalues of modulus one.

Since the eigenvalues of ψ_s^q are powers of those of ψ, this gives a contradiction unless $\mathfrak{h}/\mathfrak{s}$ is trivial. So one can assume that H is solvable.

To show that $\mathfrak{h}$ is nilpotent one shows that it coincides with its nil-radical (the maximal nilpotent ideal in $\mathfrak{h}$). It is enough to do this for the complexification of $\mathfrak{h}$, which it is also denoted by $\mathfrak{h}$. Let $\mathfrak{n}$ be the nil-radical of $\mathfrak{h}$. Then $[\mathfrak{h}, \mathfrak{h}] \subset \mathfrak{n}$. If $\mathfrak{h} \neq \mathfrak{n}$, then there exists $X \in \mathfrak{h}$ such that $X \notin \mathfrak{n}$ and X is an eigenvector for ψ_s with eigenvalue λ, $|\lambda| \neq 1$. To finish the proof it is enough now to show that $\mathbb{R}X + \mathfrak{n}$ is a nilpotent ideal.

The nilpotent algebra $\mathfrak{n}$ has two natural filtrations. The first one is given by the descending central series, which is finite. Let $\mathcal{C}^0\mathfrak{n} = \mathfrak{n}$, $\mathcal{C}^i\mathfrak{n} = [\mathfrak{n}, \mathcal{C}^{i-1}\mathfrak{n}]$, and k be the first integer such that $\mathcal{C}^k\mathfrak{n} = 0$. Then

$$\mathfrak{n} = \mathcal{C}^0\mathfrak{n} \supset \mathcal{C}^1\mathfrak{n} \supset \cdots \supset \mathcal{C}^k\mathfrak{n} = 0,$$

and $[X, \mathcal{C}^i\mathfrak{n}] \subset \mathcal{C}^i\mathfrak{n}$.

To define the second filtration, order the eigenvalues $\{\lambda_1, \ldots, \lambda_r\}$ of ψ_s on $\mathfrak{n}$ in increasing order if $|\lambda| > 1$, and in decreasing order if $|\lambda| < 1$. If V_i is the eigenspace of the eigenvalue λ_i, and $W_I = \oplus_{j=i+1}^r V_i$, then

$$\mathfrak{n} = W_0 \supset W_1 \supset W_1 \supset \cdots \supset W_r = 0.$$

We show now that $\mathbb{R}X + \mathfrak{n}$ has nilpotency degree kr. Let $Y \in \mathcal{C}^l(\mathbb{R}X + \mathfrak{n})$ with $l > kr$, that is,

$$Y = [a_lX + N_l, [\cdots [a_2X + N_2, a_1X + N_1]\cdots],$$

where $a_i \in \mathbb{R}$ and $N_i \in \mathfrak{n}$. An easy computation shows that Y can be written as a linear combination of terms of the form

$$y = \mathrm{ad}(Y_l)\mathrm{ad}(Y_{l-1})\cdots\mathrm{ad}(Y_2)(Y_1),$$

where either $Y_i = X$ or $Y_i \in \mathfrak{n}$. Since $l > kr$, either k of the Y_is lie in $\mathfrak{n}$, or there exists a string of r consecutive Y_is all equal to X. In the first situation use $[X, \mathcal{C}^i\mathfrak{n}] \subset \mathcal{C}^i\mathfrak{n}$ to conclude that $y = 0$. In the second case use that $[X, W_i] \subset W_{i+1}$ to conclude again that $y = 0$. In either case $Y = 0$. □

2.1.2 Ergodic automorphisms of the torus

An automorphism of the torus $\mathbb{T}^m$ is determined by an $m \times m$ matrix A with integer entries and determinant ± 1. Our standard notation for this automorphism is F_A. The group of all such matrices, which is isomorphic to the group of automorphisms of the torus $\mathbb{T}^m$, is denoted by $GL(m, \mathbb{Z})$.

Recall that the transformation F_A is *ergodic* with respect to Lebesgue measure μ on $\mathbb{T}^m$ if and only if any L^2 function that is F_A-invariant is constant.

The dual (character) group of $\mathbb{T}^m$ is $\mathbb{Z}^m$. By looking at the dual action of A on the characters one can easily characterize the ergodic automorphisms of a torus. The dual to F_A is the automorphisms $A^* : \mathbb{Z}^m \to \mathbb{Z}^m$ given by $A^* = (A^t)^{-1}$.

Proposition 2.1.2 *An automorphism F_A is ergodic with respect to the Lebesgue measure if and only if none of the eigenvalues of the matrix A is a root of unity.*

Proof Let $\phi \in L^2(\mu, \mathbb{T}^m)$, that is, F_A-invariant. The ϕ can be written as a Fourier series

$$\phi = \sum a_n e_n, \quad e_n(x) = e^{2\pi i <n,x>}.$$

From the F_A-invariance of ϕ it follows that $a_{A^*n} = a_n$ for all $n \in \mathbb{Z}^m$. But $\sum_{n\in\mathbb{Z}^m} |a_n|^2 < \infty$, so $a_n \neq 0$ is equivalent to the fact that the set $\{(A^t)^k n\}_{k\in\mathbb{Z}}$ is finite. Now, the last set is finite if and only if $1 \in \operatorname{spec}(A^k)$ for any k, or if $n = 0$. So ϕ has to be constant. □

Remark 2.1.3 It is an immediate corollary of the previous proof that the ergodicity of F_A is equivalent to one of the following: the periodic points of F_A are exactly points all of whose coordinates are rational; or every orbit of the dual map A^*, except that of zero, is infinite.

One can show that the ergodicity of F_A is equivalent to F_A being Bernoulli with respect to the Lebesgue measure. This equivalence is proved in [83]. See [3] for a simple proof.

Proposition 2.1.4 *Any ergodic automorphism of a torus is partially hyperbolic.*

Proof Assume first that the matrix A of the linear part is semisimple (no non-trivial Jordan blocks). If all eigenvalues have absolute value 1 then $A^{n_k} \to \mathrm{Id}$ for a certain sequence $n_k \to \infty$. Since all powers of A are integer matrices this implies that for a large enough k $A^{n_k} = \mathrm{Id}$, so F_A cannot be ergodic.

If there are Jordan blocks, then there is an invariant rational subspace L such that A restricted to L is semisimple. Since L is rational its intersection with the integer lattice is a lattice in L. Hence, the restriction of A to L is an integer matrix expressed in that basis. Now the previous argument applies. □

The classification of ergodic automorphisms from a measure theory point of view is given by their entropy, which is equal to the sum of positive Lyapunov characteristic exponents. This follows from the Ornstein isomorphism theorem [133] and the fact that every ergodic automorphism of a torus is Bernoulli with respect to the Lebesgue measure [83].

Since for $A \in \mathrm{GL}(n, \mathbb{R})$ the Lyapunov exponents are equal to the logarithms of the absolute values of eigenvalues of the matrix A, entropy is determined by the conjugacy class of A over $\mathbb{Q}$ (or over $\mathbb{C}$). As a consequence, all ergodic automorphisms of a torus which are conjugate over $\mathbb{Q}$ are measurably conjugate with respect to the Lebesgue measure.

2.1.3 Anosov diffeomorphisms on nilmanifolds

All known examples of Anosov diffeomorphisms are topologically conjugate to affine Anosov diffeomorphisms of infranilmanifolds (including nilmanifolds and tori as special cases). Moreover, it was proved by Franks [39] and Manning [109] that an arbitrary Anosov diffeomorphism of an infranilmanifold is topologically conjugate to one of this type. It is a well-known conjecture that, up to topological conjugacy, affine Anosov diffeomorphisms are the only Anosov diffeomorphisms on infranilmanifolds. It has been verified in dimensions two and three where the only possible ambient manifold is a torus.[1]

Now we present several examples of affine Anosov diffeomorphisms on nilmanifolds of higher dimension.

Preliminaries on nilpotent Lie groups and lattices

Let G be a connected simply connected Lie group. A subgroup $\Gamma \subset G$ is called *lattice* if the orbit space G/Γ has finite volume, and is called *uniform* or *co-compact* if in addition G/Γ is compact. If Γ is torsion free uniform lattice, then the action of Γ by left translations on G is properly discontinuous, i.e., for every compact set $C \subset G$ the set $\{g \in \Gamma | gC \cap C \neq \emptyset\}$ is finite, and the orbit space G/Γ is a well-defined compact manifold.

Let N be a connected simply connected nilpotent Lie group, and $\mathfrak{n}$ its nilpotent Lie algebra. It is known that N and $\mathfrak{n}$ are diffeomorphic via the exponential map, and the groups of automorphisms of N and $\mathfrak{n}$ can be identified.

Recall that a Lie algebra $\mathfrak{g}$ is a (real) vector space endowed in addition with a bilinear operation, called a *bracket*, $\mathfrak{g} \times \mathfrak{g} \ni (x, y) \to [x, y] \in \mathfrak{g}$, which satisfies the Jacobi identity, $[x, [y, z]]+[z, [x, y]]+[y, [z, x]] = 0$, and is antisymmetric, $[x, y] = -[y, x]$. If a basis for $\mathfrak{g}$ is fixed, say $x_1, \dots, x_n$, then the bracket is uniquely determined by the *structure constants*, which are defined by

[1] Nothing like that holds for Anosov flows where already in dimension three there is a variety of essentially non-algebraic possibilities. This is a manifestation of the somewhat less rigid nature of the continuous group actions mentioned before.

$$[x_i, x_j] = \sum_k c_{ij}^k x_k.$$

Let $\Gamma \subset N$ be a lattice. It follows from the results of Malcev [107] that N admits a lattice if and only if there exists a basis in its Lie algebra $\mathfrak{n}$ for which the structure constants are rational. The following two properties are specific for lattices in nilpotent Lie groups: first, any lattice Γ in a connected nilpotent Lie group is uniform; second, Γ is a finitely generated torsion free nilpotent subgroup of N. The manifold N/Γ is called a *nilmanifold.*

Conversely, for any discrete finitely generated torsion free nilpotent group Γ there exists a unique connected simply connected nilpotent Lie group $\Gamma_{\mathbb{R}}$, called the *Malcev completion* of Γ, in which Γ is a uniform lattice. We briefly recall the construction of the Malcev completion.

The lower central series of Γ is defined inductively via $c_1(\Gamma) = \Gamma$, $c_{i+1}(\Gamma) = [c_i(\Gamma), \Gamma]$, where $[\cdot, \cdot]$ is the notation for the centralizer. The group Γ is called *r-step nilpotent* if $c_r(\Gamma) \neq 1$, but $c_{r+1}(\Gamma) = 1$. The *isolator* of a subgroup $S \subset \Gamma$ is

$$\sqrt[\Gamma]{S} := \{\gamma \in \Gamma | \gamma^k \in S \text{ for some } k \in \mathbb{N}\}.$$

If Γ is r-step nilpotent, the sequence

$$\begin{aligned}\Gamma_{r+1} = 1 \subset \Gamma_r &= \sqrt[\Gamma]{c_r(\Gamma)} \subset \Gamma_{r-1} = \sqrt[\Gamma]{c_{r-1}(\Gamma)} \subset \cdots \subset \Gamma_2 \\ &= \sqrt[\Gamma]{c_2(\Gamma)} \subset \Gamma_1 = \sqrt[\Gamma]{c_1(\Gamma)} = \Gamma\end{aligned}$$

forms a central series with $\Gamma_i / \Gamma_{i+1} \cong \mathbb{Z}^{k_i}$ for some $k_i \in \mathbb{N}$.

Fix now

$$a_{r,1}, \ldots, a_{r,k_r}, a_{r-1,1}, \ldots, a_{r-1,k_{r-1}}, \ldots, a_{2,1}, \ldots, a_{2,k_2}, a_{1,1}, \ldots, a_{1,k_1},$$

a set of generators for Γ such that for any integer $1 \leq i \leq r$, the classes $\bar{a}_{i,1}, \ldots, \bar{a}_{i,k_i} \in \Gamma_i / \Gamma_{i+1}$ freely generate the free abelian group Γ_i / Γ_{i+1}.

Any $\gamma \in \Gamma$ can be written as a product:

$$\gamma = a_{r,1}^{v_{r,1}} \cdots a_{r,k_r}^{v_{r,k_r}} a_{r-1,1}^{v_{r-1,1}} \cdots a_{r-1,k_{r-1}}^{v_{r-1,k_{r-1}}} \cdots a_{2,1}^{v_{2,1}} \cdots a_{2,k_2}^{v_{2,k_2}} a_{1,1}^{v_{1,1}} \cdots, a_{1,k_1}^{v_{1,k_1}},$$

with

$$\mathbf{v} = (v_{r,1}, \ldots, v_{r,k_r}, v_{r-1,1}, \ldots, v_{r-1,k_{r-1}}, \ldots, v_{2,1}, \ldots, v_{2,k_2}, v_{1,1}, \ldots, v_{1,k_1}),$$

which is a vector in $\mathbb{Z}^{k_1+\cdots+k_r}$. The notation $\gamma(\mathbf{v})$ shows the dependence of γ on $\mathbf{v}$. By [107] it is known that the product in Γ is given by polynomial functions in $\mathbf{v}$, that is, there exists a polynomial $P : \mathbb{Z}^{2(k_1+\cdots+k_r)} \to \mathbb{Z}^{k_1+\cdots+k_r}$ such that $\gamma(\mathbf{v}_1)\gamma(\mathbf{v}_2) = \gamma(P(\mathbf{v}_1, \mathbf{v}_2))$, for all $\mathbf{v}_1, \mathbf{v}_2 \in Z^{k_1+\cdots+k_r}$. The Malcev completion $\Gamma_{\mathbb{R}}$ is defined to be the set of all formal products

$$n(\mathbf{w}) = a_{r,1}^{w_{r,1}} \dots a_{r,k_r}^{w_{r,k_r}} a_{r-1,1}^{w_{r-1,1}} \dots a_{r-1,k_{r-1}}^{w_{r-1,k_{r-1}}} \dots a_{2,1}^{w_{2,1}} \dots a_{2,k_2}^{w_{2,k_2}} a_{1,1}^{w_{1,1}} \dots, a_{1,k_1}^{w_{1,k_1}},$$

with

$$\mathbf{w} = (w_{r,1}, \dots, w_{r,k_r}, w_{r-1,1}, \dots, w_{r-1,k_{r-1}}, \dots, w_{2,1}, \dots, w_{2,k_2}, \\ w_{1,1}, \dots, w_{1,k_1}),$$

which is a vector in $\mathbb{R}^{k_1+\cdots+k_r}$. The product in $\Gamma_{\mathbb{R}}$ is given by

$$n(\mathbf{w}_1)n(\mathbf{w}_2) = n(P(\mathbf{w}_1, \mathbf{w}_2)),$$

for all $\mathbf{w}_1, \mathbf{w}_2 \in \mathbb{R}^{k_1+\cdots+k_r}$.

A similar completion can be done over $\mathbb{Q}$ and is denoted by $\Gamma_{\mathbb{Q}}$. A group is called *radicable* if any element of it has roots of any order. In particular, $\Gamma_{\mathbb{Q}}$ is the torsion free radicable nilpotent group containing Γ as a subgroup, such that each element in $\Gamma_{\mathbb{Q}}$ has some positive power lying in Γ.

Hyperbolic automorphisms

In order to exhibit concrete examples of Anosov diffeomorphisms on nilmanifolds, we restrict the discussion to the class of nilpotent Lie algebras with integer structure constants. Let $\mathfrak{n}$ be a Lie algebra as above. A $\mathbb{Z}$-*subalgebra* in it is the set of all $\mathbb{Z}$-linear combinations of an integer basis of $\mathfrak{n}$.

To find an Anosov diffeomorphism on a nilmanifold N/Γ one finds a hyperbolic automorphism A of $\mathfrak{n}$ for which there exists a basis of $\mathfrak{n}$ in which the matrix of A is hyperbolic, has integer entries, and a determinant ± 1. Since $\mathfrak{n}$ has integer structure constants, there exists a $\mathbb{Z}$-Lie subalgebra $\mathfrak{n}_1$ (of finite index) in $\mathfrak{n}$ such that $\Gamma = \exp(\mathfrak{n}_1)$ is a lattice in N. If $\{u_1, \dots, u_n\}$ is a basis in $\mathfrak{n}$ with integer structure constants, take $\mathfrak{n}_1 = \mathbb{Z}mu_1 + \cdots + \mathbb{Z}mu_n$, where m is chosen so that the denominators in the Campbell–Hausdorff formula divide the products of the constant structures of $\{mu_1, \dots, mu_n\}$. Now there is an integer k such that $A^k(\mathfrak{n}_1) = \mathfrak{n}_1$. Then the lift of A^k via the exponential map is a hyperbolic automorphism of N that induces an Anosov diffeomorphisms on N/Γ.

Note that the existence of a hyperbolic automorphism A of $\mathfrak{n}$ with integer matrix in a basis of $\mathfrak{n}$ is a necessary condition for the existence of a hyperbolic automorphism of N that preserves a lattice Γ. Indeed, let ϕ be a hyperbolic automorphism of N that induces an automorphism of Γ. Then it follows from [118, Theorem 2] that there is a subgroup $\Gamma_1 \subset \Gamma$ of finite index such that $\log(\Gamma_1)$ is a $\mathbb{Z}$-subalgebra of $\mathfrak{n}$. Then there is an integer k such that $\phi^k(\Gamma_1) = \Gamma_1$ and the pull back of ϕ^k via the exponential map is a hyperbolic automorphism of $\mathfrak{n}$ with integer matrix.

In order to simplify the verification that certain maps on nilmanifolds are hyperbolic, we need the following lemma. Note that for an N-connected simply connected Lie group the lower central series is defined inductively via $c_1(N) = N, c_{i+1}(N) = [c_i(N), N]$. Then $c_i(N)/c_{i+1}(N) \cong \mathbb{R}^{k_i}$ and any $\sigma \in Aut(N)$ induces automorphisms $\sigma_i \in Aut\,(c_i(N)/c_{i+1}(N))$.

Lemma 2.1.5 *Let N be a connected simply connected Lie group. Let $\sigma \in Aut(N)$. Let $d\sigma$ be the derivative of σ. Then the set of eigenvalues of $d\sigma$ is equal to the set of all eigenvalues of the automorphisms σ_i.*

Proof Let the nilpotency degree of N be c. Choose a basis

$$x_{c,1}, x_{c,2}, \ldots, x_{c,k_c}, x_{c-1,1}, \ldots, x_{1,k_1} \tag{2.1.1}$$

in $\mathfrak{n}$ such that the elements $x_{c,1}, x_{c,2}, \ldots, x_{c,k_c}, \ldots, x_{i,k_i}$ form a basis of $c_i(\mathfrak{n})$. Define $X_{i,j} = \exp(x_{i,j})$. The set $X_{i,1}c_{i+1}(N), X_{i,2}c_{i+1}(N), \ldots, X_{i,k_i}c_{i+1}(N)$ form a basis of the vector space $c_i(N)/c_{i+1}(N)$. The matrix of $d\sigma$ with respect to the basis (2.1.1) is

$$\begin{pmatrix} A_c & * & \cdots & * \\ 0 & A_{c-1} & \cdots & * \\ \cdots & \cdots & \cdots & \cdots \\ 0 & 0 & \cdots & * \\ 0 & 0 & \cdots & A_1 \end{pmatrix},$$

with $A_i \in GL(k_i, \mathbb{R})$. Note that

$$d\sigma(x_{i,p}) + c_{i+1}(\mathfrak{n}) = \sum_{q=1}^{k_1} (A_i)_{q,p}(x_{i,p}) + c_{i+1}(\mathfrak{n}). \tag{2.1.2}$$

Using now the Campbell–Baker–Hausdorff formula one has

$$\begin{aligned}\sigma(X_{i,p})c_{i+1}(N) &= \sigma(\exp(x_{i,p}))c_{i+1}(N) = \exp(d\sigma(x_{i,p}))c_{i+1}(N) \\ &= x_{i,1}^{(A_i)_{1,p}} x_{i,2}^{(A_i)_{2,p}} \cdots x_{i,k_1}^{(A_i)_{k_i,p}} c_{i+1}(N),\end{aligned}$$

so σ_i is also represented by the matrix A_i. □

First examples

The first example of an Anosov diffeomorphism that is not an automorphism of a torus, which we present below, was found by Borel, answering a question of Smale [162].

Example 2.1.6 A familiar example of connected simply connected nilpotent Lie group is the Heisenberg group:

$$H = \left\{ \begin{pmatrix} 1 & x & z \\ 0 & 1 & y \\ 0 & 0 & 1 \end{pmatrix} | x, y, z \in \mathbb{R} \right\}.$$

The group H can be identified to its Lie algebra

$$\mathfrak{h} = \left\{ \begin{pmatrix} 0 & x & z \\ 0 & 0 & y \\ 0 & 0 & 0 \end{pmatrix} | x, y, z \in \mathbb{R} \right\},$$

which is generated by

$$X = \begin{pmatrix} 0 & 1 & 0 \\ 0 & 0 & 0 \\ 0 & 0 & 0 \end{pmatrix}, Y = \begin{pmatrix} 0 & 0 & 0 \\ 0 & 0 & 1 \\ 0 & 0 & 0 \end{pmatrix}, Z = \begin{pmatrix} 0 & 0 & 1 \\ 0 & 0 & 0 \\ 0 & 0 & 0 \end{pmatrix}$$

subject to the relation $[X, Y] = Z$. The Campbell–Baker–Hausdorff formula is

$$X \cdot Y = X + Y + \frac{1}{2}[X, Y].$$

Consider now the direct sum Lie algebra $\mathfrak{n} = \mathfrak{h} \oplus \mathfrak{h}$, which is the Lie algebra of $N = H \times H$ and has a basis $\mathfrak{b}_1$ consisting of the following 6×6 matrices

$$X_1 = \begin{pmatrix} X & 0 \\ 0 & 0 \end{pmatrix}, X_2 = \begin{pmatrix} Y & 0 \\ 0 & 0 \end{pmatrix}, X_3 = \begin{pmatrix} Z & 0 \\ 0 & 0 \end{pmatrix},$$

$$Y_1 = \begin{pmatrix} 0 & 0 \\ 0 & X \end{pmatrix}, Y_2 = \begin{pmatrix} 0 & 0 \\ 0 & Y \end{pmatrix}, Y_3 = \begin{pmatrix} 0 & 0 \\ 0 & Y \end{pmatrix}$$

subject to the relations $[X_1, X_2] = X_3$, $[Y_1, Y_2] = Y_3$.

If $\lambda \in \mathbb{R}$, define an automorphism ϕ_λ of $\mathfrak{n}$ by

$$\phi_\lambda(X_i) = \lambda^i X_i$$
$$\phi_\lambda(Y_i) = \lambda^{-i} Y_i$$

for $i = 1, 2, 3$.

If $a \in \mathbb{Z}$, $a \geq 2$, then the roots λ, λ^{-1} of the polynomial $x^2 - 2ax + 1$ are

$$\lambda = a + (a^2 - 1)^{1/2}, \quad \lambda^{-1} = a - (a^2 - 1)^{1/2},$$

$0 < \lambda^{-1} < 1 < \lambda$ and ϕ_λ is a hyperbolic automorphism.

Consider now a new basis $\mathfrak{b}_2$ in $\mathfrak{n}$ given by

$$X_1 + Y_1, (a^2-1)^{1/2}(X_1 - Y_1),$$
$$X_2 + Y_2, (a^2-1)^{1/2}(X_2 - Y_2),$$
$$X_3 + Y_3, (a^2-1)^{1/2}(X_3 - Y_3).$$

The only non-trivial relations for $\mathfrak{b}_2$ are

$$[X_1 + Y_1, X_2 + Y_2] = X_3 + Y_3,$$
$$[X_1 + Y_1, (a^2-1)^{1/2}(X_2 - Y_2)] = (a^2-1)^{1/2}(X_3 - Y_3),$$
$$[(a^2-1)^{1/2}(X_1 - Y_1), X_2 + Y_2] = (a^2-1)^{1/2}(X_3 - Y_3),$$
$$[(a^2-1)^{1/2}(X_1 - Y_1), (a^2-1)^{1/2}(X_2 - Y_2) = (a^2-1)(X_3 - Y_3),$$

which implies that $\mathfrak{b}_2$ is a $\mathbb{Z}$-basis.

Observe now that if $\begin{pmatrix} \lambda & 0 \\ 0 & \lambda^{-1} \end{pmatrix}$ is the matrix of a linear transformation in the basis $\{X_i, Y_i\}$, then $\begin{pmatrix} a & a^2-1 \\ 0 & a \end{pmatrix} \in \mathrm{SL}(2, \mathbb{Z})$ is the matrix associated to the same linear transformation in the basis $\{X_i + Y_i, (a^2-1)^{1/2}(X_i - Y_i)\}$. Hence the matrix associated to ϕ_λ in the basis $\mathfrak{b}_2$ is

$$\begin{pmatrix} a & a^2-1 & 0 & 0 & 0 & 0 \\ 0 & a & 0 & 0 & 0 & 0 \\ 0 & 0 & a & a^2-1 & 0 & 0 \\ 0 & 0 & 0 & a & 0 & 0 \\ 0 & 0 & 0 & 0 & a & a^2-1 \\ 0 & 0 & 0 & 0 & 0 & a \end{pmatrix} \in \mathrm{SL}(6, \mathbb{Z}).$$

Thus the automorphism ϕ_λ induces an Anosov diffeomorphism on N/Γ, $\Gamma = \exp(C)$ where C is the $\mathbb{Z}$-subalgebra generated by $\mathfrak{b}_2$.

More examples

Borel examples were generalized in [92] to a larger class of nilmanifolds. We present this construction here.

Example 2.1.7 Let $\mathfrak{n}$ be a Lie algebra over $\mathbb{R}$ that is graded, that is there exist subspaces $\mathfrak{n}_i$ of $\mathfrak{n}$ such that $\mathfrak{n} = \mathfrak{n}_1 \oplus \cdots \oplus \mathfrak{n}_k$ and $[\mathfrak{n}_i, \mathfrak{n}_j] \subset \mathfrak{n}_{i+j}$. Any graded Lie algebra is nilpotent. Assume that there is a $\mathbb{Z}$-basis $\{X_1, \ldots, X_d\}$ of $\mathfrak{n}$ that is compatible with the graduation, i.e., each $X_i \in \mathfrak{n}_j$ for some j. Then the direct sum Lie algebra $\mathfrak{n} \oplus \mathfrak{n}$ has an Anosov automorphism.

Let $\{Y_i\}_i$ be a copy of the basis and consider

$$\mathfrak{b}_1 = \{X_1, \ldots, X_d, Y_1, \ldots, Y_d\}$$

a basis in $\mathfrak{n} \oplus \mathfrak{n}$. Then the relations in $\mathfrak{n} \oplus \mathfrak{n}$ are

$$[X_i, X_j] = \sum_{k=1}^{d} p_{ij}^k X_k, \quad [Y_i, Y_j] = \sum_{k=1}^{d} p_{ij}^k Y_k, \quad p_{ij}^k \in \mathbb{Z}. \tag{2.1.3}$$

For any $\lambda \in \mathbb{R}$ define an automorphism ϕ_λ of $\mathfrak{n} \oplus \mathfrak{n}$ given by $\phi_\lambda(X_i) = \lambda^j X_i$ and $\phi_\lambda(Y_i) = \lambda^{-j} Y_i$, where $X_i \in \mathfrak{n}_j$. Take $a \in \mathbb{Z}$ and λ as in the previous example and consider a new basis $\mathfrak{b}_2$ for $\mathfrak{n} \oplus \mathfrak{n}$ given by

$$\{X_1 + Y_1, (a^2 - 1)^{1/2}(X_1 - Y_1), \ldots, X_d + Y_d, (a^2 - 1)^{1/2}(X_d - Y_d)\}.$$

Using (2.1.3) it follows that $\mathfrak{b}_2$ is a $\mathbb{Z}$-basis. Moreover, the computations done in the previous example shows that the matrix of ϕ_λ with respect to the basis $\mathfrak{b}_2$ is hyperbolic and block diagonal, with 2×2 blocks on the diagonal from $SL(2, \mathbb{Z})$.

Note that Smale's example is supported by a 2-step nilpotent Lie algebra. An example of k-step nilpotent Lie algebra supporting Anosov automorphisms can be constructed in the following way: let $\mathfrak{n}$ be the Lie algebra generated by $\mathfrak{b} = \{X_1, \ldots, X_{k+1}\}$ subject to the relations:

$$[X_1, X_2] = X_3, \ [X_1, X_3] = X_4, \ldots, \ [X_1, X_k] = X_{k+1}.$$

It is clear that $\mathfrak{n}$ is graded by $\mathfrak{n} = \mathbb{R}X_1 \oplus \cdots \mathbb{R}X_{k+1}$ and the basis $\mathfrak{b}$ is a $\mathbb{Z}$-basis. So the construction above gives an Anosov automorphism of $\mathfrak{n}$.

A different class of examples of Anosov automorphisms on nilmanifolds was found by Auslander and Scheuneman [4], using free k-step nilpotent Lie algebras. This class of examples actually allows for higher rank abelian groups actions on nilmanifolds. We present this class of examples later in Section 2.2.9.

Finding Anosov diffeomorphisms on nilmanifolds is a field of active research. More examples and references can be found in [25], where it is shown that for every $n \geq 17$ there exists an n-dimensional 2-step connected simply connected nilpotent Lie group N which is indecomposable (that is, not a direct product of lower dimensional nilpotent Lie groups), and a lattice Γ in N such that N/Γ admits an Anosov diffeomorphism.

2.1.4 Anosov diffeomorphisms on infratori and infranilmanifolds

Slightly more generally, one can introduce infranilmanifolds, which are finitely covered by nilmanifolds. Let N be a connected simply connected Lie group and let $\text{Aff}(N)$ be the group of affine transformations of N. An *almost crystallographic group* is a subgroup E of $\text{Aff}(N)$ such that its subgroup $E \cap N$

of pure translations is a uniform lattice in N, and moreover $E \cap N$ is of finite index in E. The finite quotient group $E/(E \cap N)$ is isomorphic to the image of E under the projection $\mathrm{Aff}(N) \to \mathrm{Aut}(N)$, and so it can be viewed as a subgroup of $\mathrm{Aut}(N) \subset \mathrm{Aff}(N)$. Any almost crystallographic group acts properly discontinuously on N and the orbit space N/E is compact. When E is torsion free the orbit space is a compact manifold. Such a manifold is called *infranilmanifold*.

Note that if $N = \mathbb{R}^n$ then infranilmanifolds and almost crystallographic groups become tori, flat Riemannian manifolds, and crystallographic groups. A flat Riemannian manifold is sometimes called an *infratorus*.

Example 2.1.8 We describe an example from [63] of an Anosov automorphism of an orientable finite factor of the four-dimensional torus $\mathbb{T}^4$ which is an infratorus, but not a torus.

One considers the group Γ of isometries of $\mathbb{R}^4 = \mathbb{R}^2 \times \mathbb{R}^2$ generated by the integral translations $\mathbb{Z}^4 = \mathbb{Z}^2 \times \mathbb{Z}^2$ and an element γ such that for $(x, y) \in \mathbb{R}^2 \times \mathbb{R}^2$, $\gamma(x, y) = (x + v, -y)$, where $v = \begin{pmatrix} 0 \\ 1/2 \end{pmatrix}$. Note that $\gamma^2 \in \mathbb{Z}^4$, and $\mathbb{Z}^4$ is a normal subgroup of index 2 in Γ.

It is easy to see that the group Γ acts on $\mathbb{R}^4$ without fixed points. Hence $M = \mathbb{R}^4/\Gamma$ is a flat manifold whose double cover is $\mathbb{T}^4$. Note that M is orientable since both $\mathbb{Z}^4$ and γ preserve the orientation of $\mathbb{R}^4$. Note also that M is not a torus since Γ is not abelian. Indeed, if $\beta(x, y) = (x, y+y')$, where $y' \neq (0, 0)$, then $\beta \circ \gamma \neq \gamma \circ \beta$.

Let $\mathbf{A}$ be the direct product of an Anosov automorphism A of $\mathbb{R}^2$ with itself:

$$\mathbf{A} = \begin{pmatrix} A & 0 \\ 0 & A \end{pmatrix} : \mathbb{R}^4 \to \mathbb{R}^4, \qquad \text{where} \qquad A = \begin{pmatrix} 3 & 2 \\ 1 & 1 \end{pmatrix}.$$

To show that the action of $\mathbf{A}$ on $\mathbb{R}^4$ projects to M we verify that for any $(x, y) \in \mathbb{R}^4$, $\mathbf{A}(\Gamma(x, y)) = \Gamma(\mathbf{A}(x, y))$. Since $\det \mathbf{A} = 1$, $\mathbf{A}(\mathbb{Z}^4) = \mathbb{Z}^4$. Thus it suffices to check that $\mathbf{A}(\gamma(x, y)) \in \mathbb{Z}^4(\gamma(\mathbf{A}(x, y))$ and hence $\mathbf{A}(\mathbb{Z}^4(\gamma(x, y))) = \mathbb{Z}^4(\gamma(\mathbf{A}(x, y))$. This can be seen as follows:

$$\mathbf{A}(\gamma(x, y)) - \gamma(\mathbf{A}(x, y)) = \mathbf{A}(x + v,\ -y) - \gamma(Ax, Ay)$$
$$= (Ax + Av,\ -Ay) - (Ax + v, -Ay) = (Av - v,\ 0)$$
$$= \left(\begin{pmatrix} 1 \\ 1/2 \end{pmatrix} - \begin{pmatrix} 0 \\ 1/2 \end{pmatrix},\ 0\right) = \left(\begin{pmatrix} 1 \\ 0 \end{pmatrix},\ 0\right) \in \mathbb{Z}^2 \times \mathbb{Z}^2.$$

If E is a torsion free almost crystallographic group, i.e., a Bieberbach group, then an equivalent description for it is given in [94]: E is a discrete group that has a finitely generated torsion free nilpotent normal subgroup Γ such that Γ

is of finite index in E and Γ is maximal nilpotent amongst all subgroups of E. One can associate to the pair $\Gamma \subset E$ an extension $1 \to \Gamma \to E \to F \to 1$. The finite group F is called the *holonomy* of E, Γ is called the *translational subgroup* of E, and $\Gamma_{\mathbb{R}}/\Gamma$ is the *covering nilmanifold* of the infranilmanifold $\Gamma_{\mathbb{R}}/E$. There always exists a commutative diagram

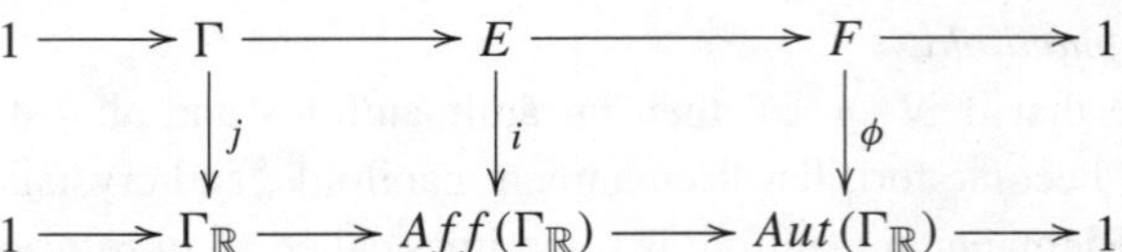

where $j : \Gamma \to \Gamma_{\mathbb{R}}$ is the canonical inclusion of Γ into its Malcev completion, $i : E \to \mathrm{Aff}(\Gamma_{\mathbb{R}})$ realizes E as a genuinely almost crystallographic group, and $\phi : F \to \mathrm{Aut}(\Gamma_{\mathbb{R}})$ is the lift of the homomorphism $\psi : F \to \mathrm{Out}(\Gamma_{\mathbb{R}})$ induced by a normalized section $s : F \to E$.

In [140] Porteous completely characterized flat Riemannian manifolds supporting Anosov diffeomorphisms. Each such manifold M defines an effective action $T : F \to GL(n, \mathbb{Z})$ of the holonomy group on the translational subgroup $\mathbb{Z}^n$ of M. The manifold M supports an Anosov diffeomorphism if and only if each $\mathbb{Q}$-irreducible component of T which is of multiplicity one is irreducible over $\mathbb{R}$.

We are ready to show several explicit examples of Anosov diffeomorphisms on infranilmanifolds. We follow closely [108], where the reader can find more details. The first example is inspired by an example of Shub [160]. It is shown in [108] that the original example of Shub defines only an orbifold rather than an infranilmanifold.

Example 2.1.9 Consider the group E presented by

$$E := < a, b, c, d, e, f, \alpha \| [c, a] = e^2, [d, a] = f^2, [c, b] = f^2, [d, b] = e^6,$$
$$\alpha a = a^{-1}\alpha, \alpha b = b^{-1}\alpha, \alpha c = c^{-1}\alpha, \alpha d = d^{-1}\alpha$$
$$\alpha e = e\alpha, \alpha f = f\alpha, \alpha^2 = ef >$$

Note that the commutators between the generators that do not appear in the presentation are zero.

E is part of the extension $1 \to \Gamma \to E \to F \to 1$, where Γ is a finitely generated torsion free nilpotent normal subgroup, of finite index in E, maximal in E, and $F = \mathbb{Z}_2$. The element of E with nontrivial image in F is α. A presentation for Γ is given by

$$\Gamma :< a, b, c, d, e, f \| [c, a] = e^2, [d, a] = f^2, [c, b] = f^2, [d, b] = e^6 > .$$

Let $\Gamma_{\mathbb{R}}$ be the Malcev completion of Γ. $\Gamma_{\mathbb{R}}$ is a six-dimensional connected simply connected Lie group. An integer $\mathbb{Z}$-basis in its Lie algebra is given by the logarithms of a, b, c, d, e, f. So E is an almost-Bieberbach group that defines an infranilmanifold $M = \Gamma_{\mathbb{R}}/E$.

We check that E is torsion free. Using the commutation relations, any element $g \in E$ can be written in the form $g = \alpha^{\epsilon} e^s f^t a^x b^y c^z d^r$, where s, t, x, y, z, r are integers and $\epsilon \in \{0, 1\}$. If $\epsilon = 0$ then g belongs to the nilpotent subgroup N and cannot be torsional unless it is trivial. If $\alpha = 1$, then due to the multiplication formula

$$g^2 = e^{1+2s+2x+6y-2z-6r} f^{1+2s+2x-2r+2y-z}. \tag{2.1.4}$$

g^2 belongs to N and is nontrivial, so again g is not torsion.

The group E embeds into $\mathrm{Aff}(\Gamma_{\mathbb{R}}) \cong \Gamma_{\mathbb{R}} \rtimes \mathrm{Aut}(\Gamma_{\mathbb{R}})$ by sending $a \to (a, 1)$, $b \to (b, 1)$, $c \to (c, 1)$, $d \to (d, 1)$, $e \to (e, 1)$, $f \to (f, 1)$, $\alpha \to (e^{1/2} f^{1/2}, \phi^{\alpha})$, where $\phi^{\alpha} \in \mathrm{Aut}(\Gamma_{\mathbb{R}})$ is given by

$$\phi^{\alpha}(a) = a^{-1}, \phi^{\alpha}(b) = b^{-1}, \phi^{\alpha}(c) = c^{-1},$$
$$\phi^{\alpha}(d) = d^{-1}, \phi^{\alpha}(e) = e, \phi^{\alpha}(f) = f.$$

Consider now the automorphism of Γ, which lifts uniquely to an automorphism of $\Gamma_{\mathbb{R}}$:

$$\nu : a \to a^2 b^{-1}, b \to a^{-3} b^2, c \to c^7 d^4, d \to c^{12} d^7, e \to e^2 f, f \to e^3 f^2.$$

To check that ν is an automorphism, observe that ν preserves the relations, and that the restriction of ν to each of the abelian subgroups $< a, b >$, $< c, d >$, $< e, f >$ is an automorphism. Note that ν and ϕ^{α} commute on $\Gamma_{\mathbb{R}}$. Moreover, it is easy to check that $(1, \nu)$ normalizes E.

In the block diagonal representation required in Lemma 2.1.5 ν is given by

$$A = \begin{pmatrix} 2 & 3 & 0 & 0 & 0 & 0 \\ 1 & 2 & 0 & 0 & 0 & 0 \\ 0 & 0 & 2 & -3 & 0 & 0 \\ 0 & 0 & -1 & 2 & 0 & 0 \\ 0 & 0 & 0 & 0 & 7 & 12 \\ 0 & 0 & 0 & 0 & 4 & 7 \end{pmatrix}.$$

Since the eigenvalues of A are $2 - \sqrt{3}, 2 + \sqrt{3}, 7 - 4\sqrt{3}, 7 + 4\sqrt{3}$, the first two with multiplicity two, ν defines an Anosov diffeomorphism on M.

Example 2.1.10 Let E be the group presented by

$$E := < a, b, c, d, e, f, \alpha | [b, a] = d^2, [c, a] = e^2, [c, b] = f^2,$$
$$\alpha a = a^{-1}\alpha, \alpha b = b^{-1}\alpha, \alpha c = c^{-1}\alpha, \alpha d = d\alpha$$
$$\alpha e = e\alpha, \alpha f = f\alpha, \alpha^2 = d > .$$

Note that the commutators between the generators that do not appear in the presentation are zero.

E is part of the extension $1 \to \Gamma \to E \to F \to 1$, where Γ is a finitely generated torsion free nilpotent normal subgroup, of finite index in E, maximal in E, and $F = \mathbb{Z}_2$. The element of E with non-trivial image in F is α. A presentation for Γ is given by

$$\Gamma :< a, b, c, d, e, f | [b, a] = d^2, [c, a] = e^2, [c, b] = f^2 > .$$

Let $\Gamma_{\mathbb{R}}$ be the Malcev completion of Γ. $\Gamma_{\mathbb{R}}$ is a six-dimensional connected simply connected Lie group. An integer $\mathbb{Z}$-basis in its Lie algebra is given by the logarithms of a, b, c, d, e, f. So E is an almost-Bieberbach group that defines an infranilmanifold $M = \Gamma_{\mathbb{R}}/E$.

The group E embeds into $\mathrm{Aff}(\Gamma_{\mathbb{R}}) \cong \Gamma_{\mathbb{R}} \rtimes \mathrm{Aut}(\Gamma_{\mathbb{R}})$ by sending $a \to (a, 1)$, $b \to (b, 1)$, $c \to (c, 1)$, $d \to (d, 1)$, $e \to (e, 1)$, $f \to (f, 1)$, $\alpha \to (d^{1/2}, \phi^\alpha)$, where $\phi^\alpha \in \mathrm{Aut}(\Gamma_{\mathbb{R}})$ is given by

$$\phi^\alpha(a) = a^{-1}, \phi^\alpha(b) = b^{-1}, \phi^\alpha(c) = c^{-1}, \phi^\alpha(d) = d, \phi^\alpha(e) = e, \phi^\alpha(f) = f.$$

Consider now the automorphism of Γ, which lifts uniquely to an automorphism of $\Gamma_{\mathbb{R}}$:

$$\nu : a \to abc^2, b \to ab^2c^2, c \to abc^3, d \to df^{-2}, e \to def^{-1}, f \to ef^2.$$

To check that ν is an automorphism, observe that ν preserves the relations, and that the restriction of ν to each of the abelian subgroups $< a, b, c >$, $< d, e, f >$ is an automorphism. Note that ν and ϕ^α commute on $\Gamma_{\mathbb{R}}$. Moreover, it is easy to check that $(1, \nu)$ normalizes E.

In the block diagonal representation required in Lemma 2.1.5 ν is given by

$$A = \begin{pmatrix} 1 & 1 & 0 & 0 & 0 & 0 \\ 0 & 1 & 1 & 0 & 0 & 0 \\ -2 & -1 & 2 & 0 & 0 & 0 \\ 0 & 0 & 0 & 1 & 1 & 1 \\ 0 & 0 & 0 & 1 & 2 & 2 \\ 0 & 0 & 0 & 2 & 2 & 3 \end{pmatrix}.$$

Since the eigenvalues of A are non-zero and not of absolute value 1, ν defines an Anosov diffeomorphism on M.

2.2 Actions of $\mathbb{Z}^k$, $k \geq 2$ on tori and nilmanifolds

2.2.1 On algebraic conjugacy

Definition 2.2.1 Let α and α' be actions of $\mathbb{Z}^k$ by automorphisms of $\mathbb{T}^m$ and $\mathbb{T}^{m'}$ correspondingly.

The actions α and α' are *algebraically isomorphic* if $m = m'$ and there is a group automorphism $h : \mathbb{T}^m \to \mathbb{T}^m$ such that

$$\alpha'(g) \circ h = h \circ \alpha(g), \text{ for all } g \in \mathbb{Z}^k.$$

The action α' is called *an algebraic factor of* α if there exists a surjective homomorphism $h : \mathbb{T}^m \to \mathbb{T}^{m'}$ such that

$$\alpha'(g) \circ h = h \circ \alpha(g), \text{ for all } g \in \mathbb{Z}^k.$$

In particular, if h is everywhere finite to one, then α' is called a *finite factor* or a *factor with finite fibers* of α.

The factor action α' is called a *rank-one factor* of α if $\alpha'(\mathbb{Z}^k)$ has a subgroup of finite index which consists of powers of a single map.

The actions α and α' are *weakly algebraically isomorphic* if each one is an algebraic factor of the other. In this case $m = m'$ and each factor map has finite fibers.

These algebraic notions have natural measure-theoretic counterparts. We will mainly discuss the algebraic setting, but will from time to time give references for the measurable properties.

We show now the relationships between the conjugacy over $\mathbb{C}$, $\mathbb{Q}$, and $\mathbb{Z}$ for algebraic actions by automorphisms of a torus. Any $\mathbb{Z}^k$-action α by automorphisms of $\mathbb{T}^m$ generated by $F_{A_1}, \dots, F_{A_k}$, where $A_1, \dots, A_k$ are integral matrices of determinant ± 1, defines an embedding $\rho_\alpha : \mathbb{Z}^k \to GL(n, \mathbb{Z})$ by

$$\rho_\alpha^{\mathbf{n}} = A_1^{n_1} \dots A_k^{n_k},$$

where $\mathbf{n} = (n_1, \dots, n_k) \in \mathbb{Z}^k$. Conversely, any embedding $\rho : \mathbb{Z}^d \to GL(n, \mathbb{Z})$ defines an action by automorphisms. Similarly, one can consider actions by endomorphisms induced by general embeddings $\rho : (\mathbb{Z}^+)^d$ into the semi-group of invertible matrices with integer entries.

Two actions α and α' are conjugate via an automorphism (algebraically isomorphic) if and only if the corresponding embeddings ρ_α and $\rho_{\alpha'}$ are conjugated over $\mathbb{Z}$. This implies conjugacy over $\mathbb{Q}$, which is equivalent to conjugacy over $\mathbb{C}$. Note that conjugacy over $\mathbb{C}$ is determined by the eigenvalue structure.

In general, the opposite is not true. The conjugacy over $\mathbb{Z}$ is determined not just by the linear algebra, as is the case for the conjugacy over $\mathbb{Q}$, but also by the algebraic number theory data. Classification, up to conjugacy over $\mathbb{Z}$, of matrices in $SL(n, \mathbb{Z})$ which are irreducible and conjugate over $\mathbb{Q}$, has to do with the *class numbers of the algebraic fields*. The simplest case of this situation appears for $n = 2$. In this case the trace determines conjugacy over $\mathbb{Q}$, and in particular the entropy. However, if the class number of the corresponding number field is greater than 1, there are matrices with the given trace that are not conjugate over $\mathbb{Z}$. For example, one can consider

$$\begin{pmatrix} 5 & 2 \\ 2 & 1 \end{pmatrix} \text{ and } \begin{pmatrix} 5 & 4 \\ 1 & 1 \end{pmatrix}.$$

This is in contrast with the case of measurable isomorphism, which is completely determined by the entropy. For the case of automorphisms of the torus, concrete metric isomorphisms in the case of equal entropies are constructed using Markov partitions by Adler and Weiss [1]. These are more specific then those produced by the general Ornstein isomorphism theory, and yet not algebraic.

One should note that every action α by automorphisms of a torus has many algebraic factors with finite fibers. These factors are in one-to-one correspondence with lattices $\Gamma \subset \mathbb{R}^k$ which contain the standard lattice $\Gamma_0 = \mathbb{Z}^k$, and which satisfy $\rho_\alpha^{\mathbf{n}}(\Gamma) \subset \Gamma$ for all $\mathbf{n} \in \mathbb{Z}^{\mathbf{k}}$.

We denote the factor action associated to a particular lattice $\Gamma_0 \subset \Gamma$ by α_Γ. In the case of actions by automorphisms the embeddings ρ_α are invertible. Hence $\rho_\alpha^{\mathbf{n}}(\Gamma) = \Gamma$ for all $\mathbf{n} \in \mathbb{Z}^{\mathbf{k}}$.

Let $\Gamma_0 \subset \Gamma$. Take any basis in Γ and let $S \in GL(n, \mathbb{Q})$ be the matrix which maps the standard basis in Γ_0 to this basis. Then obviously the factor-action α_Γ is equal to the action $\alpha_{S\rho_\alpha S^{-1}}$. In particular, ρ_α and ρ_{α_Γ} are conjugate over $\mathbb{Q}$, although not necessarily over $\mathbb{Z}$.

For any positive integer q, the lattice $\frac{1}{q}\Gamma_0$ is invariant under any automorphism in $GL(n, \mathbb{Z})$ and gives rise to a factor that is conjugate to the initial action: one can set $S = \frac{1}{q}Id$ and obtain that $\rho_\alpha = \rho_{\alpha_{(1/q)\Gamma_0}}$. On the other hand, one can find, for any lattice $\Gamma_0 \subset \Gamma$, a positive integer q such that $\Gamma \subset \frac{1}{q}\Gamma_0$ (take q to be the least common multiple of the denominators of all coordinates for a basis of Γ). Thus $\alpha_{(1/q)\Gamma_0}$ appears as a factor of α_γ.

We summarize these considerations in the following proposition.

Proposition 2.2.2 *Let α and α' be $\mathbb{Z}^d$-actions by automorphisms of the torus $\mathbb{T}^n$. Then the following are equivalent:*

(i) ρ_α and $\rho_{\alpha'}$ are conjugate over $\mathbb{Q}$;

(ii) *there exists an action α'' such that both α and α' are isomorphic to finite algebraic factors of α''; and*

(iii) *α and α' are weakly algebraic isomorphic, that is, each of them is isomorphic to a finite algebraic factor of the other.*

2.2.2 The genuine higher rank condition

As we will see later, the absence of rank-one algebraic factors is one of the most general situations in which certain rigidity phenomena appear. The following equivalent characterization of this condition can be found in [164].

Proposition 2.2.3 *For an action α of $\mathbb{Z}^k$ by automorphisms of a torus the following two conditions are equivalent:*

(i) *The action α contains a subgroup isomorphic to $\mathbb{Z}^2$, for which all nontrivial elements are ergodic automorphisms of the torus.*

(ii) *The action α does not possess any non-trivial rank-one algebraic factor.*

Proof Let $\Sigma = \alpha(\mathbb{Z}^k) \subset SL(n, \mathbb{Z})$. Every element $g \in \Sigma$ admits a Jordan decomposition $g = g_s \times g_u$ into a semisimple and a unipotent part, and both g_s, g_u are in $SL(n, \mathbb{Q})$. Moreover, there exists $m \in \mathbb{N}$ such that $g^m = g_s^m \times g_u^m$ is a decomposition inside $SL(n, \mathbb{Z})$. So there exists $\Sigma' \subset \Sigma$ a subgroup of finite index such that any element in Σ' has a decomposition inside $SL(n, \mathbb{Z})$. We denote by Σ_s the semisimple part of Σ, and by Σ_u the unipotent part. Note that $\Sigma = \Sigma_s \times \Sigma_u$ and $\Sigma_s \cap \Sigma_u$ is finite.

If $g = g_s \times g_u$ is a Jordan decomposition and $g, g_s, g_u \in SL(n, \mathbb{Z})$, then it follows from Proposition 2.1.2 that g induces an ergodic automorphism of a torus if and only if g_s does. Consequently Σ contains a subgroup isomorphic to $\mathbb{Z}^2$ consisting of ergodic automorphisms if and only if Σ_s does.

We show now that if Σ_s admits a rank-one factor then Σ does. This will reduce the proof of the proposition to the case when Σ is semisimple.

Let $\mathbb{T} \subset \mathbb{T}^n$ be a Σ_s-invariant subtorus such that $\Sigma_s|_{\mathbb{T}^n/\mathbb{T}}$ is a finite extension of $\mathbb{Z}$. Then Σ_s contains a subgroup H such that Σ_s/H is a finite extension of $\mathbb{Z}$ and H acts trivially on $\mathbb{T}^n/\mathbb{T}$. Let $V = \mathrm{Fix}(H)$. It follows from the fact that H is semisimple that V has an H-invariant complement V'. Moreover, both V and V' are defined over $\mathbb{Q}$ and both are invariant under the centralizer of H in $SL(n, \mathbb{R})$. This implies that both V and V' are Σ and Σ_s invariant. If $\mathbb{T}' = V'/\mathbb{Z}^n \cap V'$, then the action of Σ_s on $\mathbb{T}^n/\mathbb{T}'$is of rank-one. Note now that the unipotent component of $\Sigma|_{\mathbb{T}^n/\mathbb{T}}$ has an invariant proper subspace W

defined over $\mathbb{Q}$ such that the induced action on $\mathbb{R}^n / W$ is trivial, and conclude that Σ has a rank-one factor.

If Σ has an ergodic element, one can show also that if Σ admits a rank-one factor then Σ_s does. Indeed, let $\mathbb{T} \subset \mathbb{T}^n$, $\Sigma(\mathbb{T}) = \mathbb{T}$ and $\Sigma|_{\mathbb{T}^n/\mathbb{T}}$ is a finite extension of $\mathbb{Z}$. Let H be the stabilizer of $\mathbb{T}^n/\mathbb{T}$ in Σ. Then Σ/H is a finite extension of $\mathbb{Z}$. Let T' be the maximal H_s-fixed subtorus. Clearly T' is non-trivial, and has an H_s-invariant complement T''. Then T'' is Σ_s-invariant and $\Sigma_s|_{\mathbb{T}^n/T''}$ contains an ergodic element and is included in a finite extension of $\mathbb{Z}$, so it has to be a finite extension of $\mathbb{Z}$.

Note now that if Σ is semisimple, as we will assume from now on, then the implication (1) $\Rightarrow$ (2) is obvious, so we need to show only (2) $\Rightarrow$ (1).

Observe that in any subgroup of $SL(n, \mathbb{Z})$ isomorphic to $\mathbb{Z}^k$ which has no proper invariant tori in $\mathbb{T}^n$, every (non-trivial) element is ergodic. Let $A \subset \Sigma$ ergodic, and let $\mathbb{T}^n = T_1 \times \cdots \times T_s$ be an almost (i.e., up to a finite extension) direct product of Σ-invariant irreducible subtori. From the observation above, for each i there exists $B_i \in \Sigma$ such that the restrictions of A and B_i to T_i generates a $\mathbb{Z}^2$-action with every (non-trivial) element ergodic. We want to find $B \in \Sigma$ such that A and B generate a $\mathbb{Z}^2$-action with every (non-trivial) element ergodic.

We will use the following fact that can be easily proved by induction:

Let $A, B, C \in SL(n, \mathbb{Z})$ semisimple commuting elements such that A, B generate a $\mathbb{Z}^2$ action on $\mathbb{T}^n$ with every (non-trivial) element ergodic. Then there exist at most a finite number of relatively prime triples (k, l, m) such that $A^k B^l C^m$ is not ergodic.

Suppose now that $A, B, C \in \Sigma$ such that $\mathbb{T}^n = T \times T'$, T, T' Σ-invariant, A, B generate a $\mathbb{Z}^2$-action on T, A, C generate a $\mathbb{Z}^2$-action on T', and both actions have each non-trivial element ergodic. Then the previous fact implies that there are only finitely many rational directions (k, l, m) in $\mathbb{R}^3$ with non-ergodic element $A^K B^l C^m$. So we can find a rational plane in $\mathbb{R}^3$ which consists of ergodic triples only, and the proposition follows now by induction. □

Either one of the conditions in Proposition 2.2.3 describes the most general "genuine higher rank" situation. Accordingly we will call such actions *genuinely higher rank*.

2.2.3 Rigidity of genuinely higher rank actions

Genuinely higher rank actions by automorphisms of a torus possess a number of strong rigidity properties:

- *Local differentiable rigidity*: Any smooth action whose generators are sufficiently close to those of a genuinely higher rank action α is differentiably conjugate to α [20, 22, 81].
- *Isomorphism rigidity*: Any action by automorphisms of a torus measurably isomorphic to a genuinely higher rank action by automorphisms is algebraically isomorphic to it [60, 70].
- *Measure rigidity (Anosov case)*: The only ergodic invariant measures for α such that some element has positive entropy are the Lebesgue measures on closed invariant subgroups [31].
- *Rigidity of measurable centralizer*: The centralizer of a genuinely higher rank action by automorphisms of a torus in the group of Lebesgue measure preserving transformations consists of affine transformations [60, 70].

Now we will consider various interesting classes and examples of genuinely higher rank actions by automorphisms of a torus. We begin with some preliminaries from algebraic number theory.

2.2.4 Irreducible actions and units in number fields

An important class of genuinely higher rank abelian actions are those irreducible over $\mathbb{Q}$.

Definition 2.2.4 A genuine higher rank $\mathbb{Z}^k$-action α on $\mathbb{T}^n$ is called *irreducible* if any non-trivial algebraic factor of α has finite fibers.

From Proposition 2.1.2 follows that:

Proposition 2.2.5 *Any irreducible action α over $\mathbb{Q}$ by automorphisms of a torus has all non-trivial elements acting ergodically.*

The following equivalent conditions for irreducibility can be found in [7].

Proposition 2.2.6 *Let α be a $\mathbb{Z}^k$-action on $\mathbb{T}^n$ by automorphisms of a torus. The following conditions are equivalent:*

(i) α is irreducible;
(ii) the image of α in $GL(n,\mathbb{Z})$ contains a matrix with characteristic polynomial irreducible over $\mathbb{Q}$; and
(iii) α does not have a non-trivial invariant rational subspace or, any α-invariant closed subgroup of $\mathbb{T}^n$ is finite.

There are close connections between irreducible actions on $\mathbb{T}^n$ and groups of units in number fields of degree n. In fact, algebraic number theory provides several important technical tools for the study of $\mathbb{Z}^k$-actions by automorphisms of a torus. We review below Section 3.3 in [70], which contains a discussion of these issues.

Let $A \in GL(n, \mathbb{Z})$ be a matrix with an irreducible characteristic polynomial f, hence with distinct eigenvalues. The centralizer of A in the group of matrices $M(n, \mathbb{Q})$ can be identified with the ring of all polynomials in A with rational coefficients modulo the principal ideal generated by the polynomial $f(A)$, and hence with the field $K = \mathbb{Q}(\lambda)$, where λ is an eigenvalue of A. The identification is given by the map

$$\mathcal{G} : p(A) \to p(\lambda), \tag{2.2.1}$$

with $p \in \mathbb{Q}[x]$. Notice that if $B = p(A)$ is an integer matrix then $\mathcal{G}(B)$ is an algebraic integer, and if $B \in GL(n, \mathbb{Z})$ then $\mathcal{G}(B)$ is an algebraic unit. The converse is not necessarily true.

Lemma 2.2.7 *The map $\mathcal{G}$ in* (2.2.1) *is injective.*

Proof If $\mathcal{G}(p(A)) = 1$ for $p(A) \neq Id$ then $p(A)$ has 1 an an eigenvalue and hence has a rational subspace consisting of all invariant vectors. This subspace must be invariant under A, which contradicts its irreducibility. □

Denote by $\mathcal{O}_K$ the ring of integers of K, by $\mathcal{U}_K$ the group of units of $\mathcal{O}_k$, by $C(A)$ the centralizer of A in $M(n, \mathbb{Z})$ and by $Z(A)$ the centralizer of A in the group $GL(n, \mathbb{Z})$.

Lemma 2.2.8 *$\mathcal{G}(C(A))$ is a ring in K such that $\mathbb{Z}[\lambda] \subset \mathcal{G}(C(A)) \subset \mathcal{O}_K$, and $\mathcal{G}(Z(A)) = \mathcal{U}_K \cap \mathcal{G}(C(A))$.*

Proof $\mathcal{G}(C(A))$ is a ring because $C(A)$ is a ring. As observed above, images of integer matrices are algebraic integers and images of matrices with determinants ± 1 are algebraic units. Hence $\mathcal{G}(C(A)) \subset \mathcal{O}_K$. Finally, for every polynomial p with integer coefficients, $p(A)$ is an integer matrix, hence $\mathbb{Z}[\lambda] \subset \mathcal{G}(C(A))$. □

Notice that $\mathbb{Z}[\lambda]$ is a finite index subring of $\mathcal{O}_K$. Hence $\mathcal{G}(C(A))$ has the same property.

Remark 2.2.9 The groups of units in two different rings, say $\mathcal{O}_1 \subset \mathcal{O}_2$, may coincide. Examples can be found in the table of totally real cubic fields [16].

Proposition 2.2.10 *$Z(A)$ is isomorphic to $\mathbb{Z}^{r_1+r_2-1} \times F$, where r_1 is the number of real embeddings, r_2 is the number of pairs of complex conjugate embeddings of the field K into $\mathbb{C}$, and is a finite cyclic group.*

Proof By Lemma 2.2.8 $Z(A)$ is isomorphic to the group of units in the ring $\mathcal{G}(C(A))$, so the statement follows from the classical Dirichlet unit theorem ([8], Chapter 2, Section 4.3). □

Note that since $r_1 + 2r_2 = n$, Proposition 2.2.10 gives an upper bound on the rank of an irreducible $\mathbb{Z}^k$-action on $\mathbb{T}^n$.

2.2.5 Cartan actions

Of particular interest are abelian groups of ergodic automorphisms of $\mathbb{T}^n$ of maximal possible rank $n-1$ in agreement with the real rank of the Lie group $SL(n, \mathbb{R})$. These are a particular class of irreducible actions.

Definition 2.2.11 An action of $\mathbb{Z}^{n-1}$ on $\mathbb{T}^n$ for $n \geq 3$ by ergodic automorphisms is called a *Cartan action*.

The following facts proved in [70, Proposition 4.1] are, essentially, consequences of the Dirichlet unit theorem.

Proposition 2.2.12 *Let α be a Cartan action on $\mathbb{T}^n$. Then:*

(i) any element of the action other than identity has real eigenvalues and is hyperbolic and thus Bernoulli;
(ii) α is irreducible; and
(iii) the centralizer of α is a finite extension of α.

Lemma 2.2.13 *Let A be a hyperbolic matrix in $SL(n, \mathbb{Z})$ with irreducible characteristic polynomial and distinct real eigenvalues. Then every element of the centralizer $Z(A)$ other than $\{\pm \mathrm{Id}\}$ is hyperbolic.*

Proof Assume that $B \in Z(A)$ is not hyperbolic. As B is simultaneously diagonalizable with A and has real eigenvalues, it has an eigenvalue $+1$ or -1. The corresponding eigenspace is rational and A-invariant. Since A is irreducible, this eigenspace has to coincide with the whole space and hence $B = \pm \mathrm{Id}$. □

Corollary 2.2.14 *Cartan actions are exactly the maximal rank irreducible actions corresponding to totally real number fields. The centralizer $Z(\alpha)$ for a Cartan action α is isomorphic to $\mathbb{Z}^{n-1} \times \{\pm \mathrm{Id}\}$. Lyapunov exponents for a Cartan action are simple and Lyapunov hyperplanes are in general position and are* completely irrational, *i.e., none of them contains an integer point.*

In addition to rigidity properties enumerated in Section 2.2.3 Cartan actions are *globally rigid*: any Anosov action homotopic to a Cartan action is differentiable conjugate to it [151]. There is also strong isomorphism rigidity for actions with *Cartan homotopy data*, i.e., actions whose elements are homotopic to those of a Cartan action [74].

The following example demonstrates highly non-trivial consequences of isomorphism rigidity in the case of Cartan actions:

Example 2.2.15 [70, Section 6.3] Consider two Cartan actions of $\mathbb{Z}^2$ on $\mathbb{T}^3$ generated by automorphisms F_A, F_B, and $F_{A'}$, $F_{B'}$ correspondingly, where

$$A = \begin{pmatrix} 0 & 1 & 0 \\ 0 & 0 & 1 \\ 1 & 8 & 2 \end{pmatrix} \quad B = \begin{pmatrix} 2 & 1 & 0 \\ 0 & 2 & 1 \\ 1 & 8 & 4 \end{pmatrix},$$

and

$$A' = \begin{pmatrix} -1 & 2 & 0 \\ -1 & 1 & 1 \\ -5 & 9 & 2 \end{pmatrix} \quad B' = \begin{pmatrix} 1 & 2 & 0 \\ -1 & 3 & 1 \\ -5 & 9 & 5 \end{pmatrix}.$$

These two actions are isomorphic over $\mathbb{Q}$ and hence are algebraic factors of each other with finite fibers. They are, however, not isomorphic over $\mathbb{Z}$; this remains true even if one is allowed to change generators within each group. By the isomorphism rigidity property (see Section 2.2.3), the actions are not measurably isomorphic. However, every element of the actions is Bernoulli, and hence has a huge measurable centralizer. This gives a remarkable example of rigid construction from soft elements.

2.2.6 Symplectic actions on $\mathbb{T}^4$

The $\mathbb{R}$-rank (see Section 2.3.3) of the group $Sp(4,\mathbb{R})$ of invertible symplectic 4×4 matrices is two. Accordingly, any maximal split Cartan subgroup of $Sp(4,\mathbb{R})$ may intersect the integer lattice $Sp(4,\mathbb{Z})$ by a group of rank at most two. In fact, such an intersection may have rank-two and be irreducible over $\mathbb{Q}$. Using it as the linear part one obtains an irreducible $\mathbb{Z}^2$ Anosov action on $\mathbb{T}^4$ by symplectic automorphisms. Let F_A and F_B be the generators of such an action. Each of the matrices A and B has two pairs of mutually inverse real eigenvalues. Hence, the four Lyapunov exponents of the action split into two pairs with exponents in each pair differing by sign. Thus there are only two Lyapunov hyperplanes (lines in this case). Geometrically, the picture of exponents and Weyl chambers is the same as for the product action generated by $C \times \mathrm{Id}$ and $\mathrm{Id} \times D$, where $C, D \in SL(2, \mathbb{Z})$ are hyperbolic matrices.

The difference between the product case and the irreducible case is in that the latter contains a subgroup isomorphic to $\mathbb{R}^2$ consisting of ergodic elements, while the former does not. Alternatively, one can explain this as follows. The Lyapunov lines in the irreducible case are irrational and in the product case they are simply coordinate axes. If one consider the suspension of the action in the irreducible case *every* one-parameter subgroup of $\mathbb{R}^2$ acts ergodically including those represented by the Lyapunov line. Each of those subgroups acts by isometries along one of the invariant one-dimensional Lyapunov foliations thus providing an essential geometric ingredient for rigidity properties.

Notice that since A is irreducible with real eigenvalues its centralizer has rank-three by Proposition 2.2.10. Thus the $\mathbb{Z}^2$ symplectic action is embedded into an Anosov action of $\mathbb{Z}^3$. A third generator of this action may be chosen that is not a symplectic matrix.

Example 2.2.16 We show now an explicit example of $\mathbb{Z}^2$ symplectic action on $\mathbb{T}^4$. Let

$$A = \begin{pmatrix} 2 & 2 & 2 & -1 \\ -1 & -2 & -1 & 1 \\ 0 & 5 & 1 & -3 \\ -7 & -4 & -6 & 1 \end{pmatrix}, \quad B = \begin{pmatrix} -41 & 100 & 40 & -25 \\ 25 & -61 & -25 & 15 \\ 40 & -95 & -36 & 25 \\ 55 & -130 & -50 & 34 \end{pmatrix}.$$

Both A and B are symplectic and hyperbolic. The eigenvalues of A are (approximately) $\lambda_1 = 5.2828, \lambda_2 = -3.1552, \lambda_3 = -0.3169, \lambda_4 = 0.1893$ and the eigenvalues of B are $\lambda_1 = -107.8924, \lambda_2 = 3.6259, \lambda_3 = 0.2758, \lambda_4 = -0.0093$. It is clear that A and B are not powers of the same element in $\mathrm{Sp}(4, \mathbb{R})$. The characteristic polynomial of A is $P_A = x^4 - 2x^3 - 17x^2 - 2x + 1$ and the characteristic polynomial of B is $P_B = x^4 + 104x^3 - 419x^2 + 104x + 1$. Both P_A and P_B are irreducible over $\mathbb{Q}$ and reciprocal. One can easily check that $AB = BA$. Thus F_A, F_B determine a genuine $\mathbb{Z}^2$-action on $\mathbb{T}^4$. The $\mathbb{Z}^2$-action can be embedded into a $\mathbb{Z}^3$-action. A possible choice for the extra generator is

$$C = \begin{pmatrix} -6 & 13 & 4 & -4 \\ 1 & -5 & -4 & 0 \\ 16 & -23 & 2 & 13 \\ 19 & -28 & 1 & 15 \end{pmatrix}. \tag{2.2.2}$$

The eigenvalues of C are $\lambda_1 = 18.5110$, $\lambda_2 = -12.6222$, $\lambda_3 = 0.1414$, $\lambda_4 = -0.0303$. The characteristic polynomial of C is $P_C = x^4 - 6x^3 - 233x^2 + 26x + 1$. P_C is irreducible over $\mathbb{Q}$. The matrix C is hyperbolic, but not symplectic because its characteristic polynomial is not reciprocal. So C does not belong to the abelian group generated by A and B. Nevertheless, C commutes to A and B.

2.2.7 Genuinely partially hyperbolic actions

We proceed to construct examples of genuinely higher rank partially hyperbolic actions on tori that are irreducible and not hyperbolic.

The first dimension in which there are partially hyperbolic ergodic automorphisms of a tori that are not hyperbolic is 4. If $N = 2$, since the purely hyperbolic case is excluded, the two commuting matrices have both eigenvalues real and of absolute value 1. Then they are roots of unity and the matrices cannot induce ergodic automorphisms. If $N = 3$, the matrix has two complex conjugate eigenvalues, and an additional one that is real. The real one has to be of 1. This is excluded by the ergodicity assumption.

The simplest reducible examples for $N = 4$ are given by products. For irreducible examples, one needs to find an irreducible matrix in $\mathrm{SL}(n, \mathbb{Z})$ that has two complex conjugate eigenvalues of absolute value 1, and two real eigenvalues λ and λ^{-1}, $0 < |\lambda| < 1$. To produce such a matrix start with an irreducible quadratic polynomial that has a root larger than 2 and one less than 2 in absolute value. Then the substitution $x \to x + \frac{1}{x}$ gives a quadric polynomial that is the characteristic polynomial of a matrix with the desired properties.

Example 2.2.17 For example, start with the quadratic polynomial $P = x^2 - 3x + 1$, which after substitution becomes $Q = x^4 - 3x^3 + 3x^2 - 3x + 1$, which is irreducible over $\mathbb{Q}$. Then the companion matrix of Q:

$$A = \begin{pmatrix} 0 & 1 & 0 & 0 \\ 0 & 0 & 1 & 0 \\ 0 & 0 & 0 & 1 \\ -1 & 3 & -3 & 3 \end{pmatrix}$$

has the desired properties. The eigenvalues of A are: $\lambda_1 = 0.4643, \lambda_2 = 2.1532, \lambda_3 = 0.1909 - 0.9816i, \lambda_4 = 0.1909 + 0.9816i$.

There is no reducible or irreducible matrix with the desired properties if $N = 5$. Since the matrix is not hyperbolic, there are two cases to consider: either there are two complex conjugate eigenvalues of absolute value 1 and three real eigenvalues, or there are two pairs of complex conjugate eigenvalues, exactly one of these of absolute value 1, and one remaining real eigenvalue.

In the first case, let the characteristic polynomial of the matrix be:

$$\begin{aligned} f(x) &= (x-\lambda_1)(x-\lambda_2)(x-\lambda_3)(x-\mu)(x-\bar{\mu}) \\ &= x^5+\alpha x^4+\beta x^3+\gamma x^2+\delta x-1, \end{aligned}$$

with $\lambda_1, \lambda_2, \lambda_3$ real, and $\mu, \bar{\mu}$ complex conjugate with $|\mu| = 1$. Then

$$f(x) = (x^3 - ax^2 + bx - 1)(x^2 + cx + 1),$$

where $a = \lambda_1 + \lambda_2 + \lambda_3, b = \lambda_1\lambda_2 + \lambda_2\lambda_3 + \lambda_1\lambda_3$, and $c = 2\mathrm{Re}\mu$. The coefficients of f are $\alpha = c - a, \beta = b + 1 - ac, \gamma = -a + bc - 1, \delta = b - c$. Since all are rational numbers, $\alpha + \delta = b - a$ and $\beta + \gamma = (b - a)(1 + c)$ are rational. If $a \neq b$ this implies c rational, which is in contradiction to ergodicity. If $a = b$, then $f(x)$ has $x - 1$ as a factor, so again we have a contradiction to ergodicity.

In the second case, let the characteristic polynomial of the matrix be:

$$\begin{aligned} f(x) &= (x-\lambda)(x-\nu)(x-\bar{\nu})(x-\mu)(x-\bar{\mu}) \\ &= (x-\lambda)(x^2-ax+1)(x^2-bx+\rho^2) \\ &= x^5+\alpha x^4+\beta x^3+\gamma x^2+\delta x-1, \end{aligned}$$

where $|\mu| = 1, |\nu| = \rho \neq 1, a = 2\mathrm{Re}\mu, b = 2\mathrm{Re}\nu = 2\rho\cos\theta$. If we denote $A = \lambda + b$ and $B = \lambda b + \rho^2$, then the coefficients of f are $\alpha = -A - a, \beta = Aa + 1 + B, \gamma = -A - 1 - aB, \delta = B + a$. Since all are rational, $\alpha + \delta = B - A$ and $\beta + \delta = (B - A)(1 - a)$ are rational. So either $B = A$, or a is rational. If $A = B$, then $\lambda\rho^2 = 1$ implies $|\rho| = 1$ or $1 + \rho^2 - 2\rho\cos\theta = 0$. The first situation implies $\lambda = 1$, which contradicts ergodicity. The second situation is equivalent to $\sin^2\theta + (\cos\theta - \rho)^2 = 0$, which implies $|\rho| = 1$, contradicting ergodicity. If a is rational, then μ has to be a root of unity, giving again a contradiction to ergodicity.

More generally, one can show that there are no irreducible partially hyperbolic and not hyperbolic automorphisms of a torus in any odd dimensions.

Let N be an odd integer, and let $f(x)$ be the irreducible characteristic polynomial of degree N of an integer matrix A. Let λ be an eigenvalue of A of absolute value 1, and let $\mu = \lambda + \frac{1}{\lambda}$. For the number fields $L = \mathbb{Q}(\lambda)$ and $K = \mathbb{Q}(\mu)$ one has:

$$|L : K| \cdot |K : \mathbb{Q}| = N.$$

Observe now that the field K is real, since $\mu = \lambda + \frac{1}{\lambda} = \lambda + \bar{\lambda} \in \mathbb{R}$, so $|L : K| > 2$. Since $\lambda^2 - \lambda\mu + 1 = 0$, we also have $|L : K| < 2$. This implies $|L : K| = 2$, so N is even.

We start investigating higher rank abelian partially hyperbolic automorphisms.

Definition 2.2.18 A genuinely higher rank action of $\mathbb{Z}^k$ is called *genuinely partially hyperbolic* if it has at least one zero Lyapunov exponent. In fact, multiplicity of the zero exponent for such an action is always even because the eigenvalues corresponding to the exponent are complex and hence come in conjugate pairs.

The following existence results appear in [22, Theorem 3].

Theorem 2.2.19 *Irreducible genuinely partially hyperbolic actions by automorphisms of a torus exist in any even dimension starting from six and not in any other dimension.*

Reducible genuinely partially hyperbolic actions exist in any odd dimension starting from nine.

No genuinely partially hyperbolic actions exist in dimensions strictly less than up to six.

Proof One shows first that there are no genuinely partially hyperbolic $\mathbb{Z}^2$ actions on a torus $\mathbb{T}^N$ if $N < 6$, or irreducible genuinely partially hyperbolic actions if N is odd. The case of an odd dimension and the cases $N = 2, 3, 5$ are discussed above. The only case remaining is $N = 4$.

If $N = 4$ it is possible to find a partially hyperbolic matrix which is genuinely partially hyperbolic and acts ergodically on $\mathbb{T}^4$. See Example 2.2.17. Nevertheless, it is impossible to find a genuinely partially hyperbolic $\mathbb{Z}^2$-action. However, assume that such an action exists and let A, B be its generators. Then A and B have a common neutral space. Otherwise a product of a power of A and a power of B is hyperbolic. Since the roots are either real or complex conjugate, the common neutral space has to be two dimensional, corresponding to a pair of complex conjugate roots. Without loss, one can assume that A has real roots $0 < \alpha < 1 < \alpha^{-1}$, and B has real roots $0 < \beta < 1 < \beta^{-1}$. Both α and β, as well as the complex conjugate roots of both A and B, have to be irrational. By choosing k, l large enough, one can assume that the eigenvalues of $A^l B^k$ are arbitrarily close to identity. So the matrix $A^l B^k$ is arbitrarily close to identity. But this is impossible because $A^l B^k$ is an integer matrix not equal to identity.

We continue by showing the existence of irreducible genuinely partially hyperbolic $\mathbb{Z}^2$ actions in any even dimension $N \geq 6$. We use here the Dirichlet units theorem.

Let Q be an irreducible integer polynomial of degree n with all roots real and a leading coefficient of 1. Assume that Q has s, $s \geq 2$, roots greater than 2 and at least one root less than 2 in absolute value. Let $R(x) = Q(x + \frac{1}{x})x^n$ be the reciprocal polynomial of degree $2n$ determined by Q. Let ψ be a root of R and $\theta = \psi + \frac{1}{\psi}$ a root of Q. Let $L = \mathbb{Q}(\theta)$ and $K = \mathbb{Q}(\psi)$ be the corresponding number fields. Let σ_i, $i = 1, \ldots, 2n$, be the embeddings of K into $\mathbf{C}$. Since all roots of R come in pairs $\{\lambda, \frac{1}{\lambda}\}$ we can assume that $\sigma_i(\psi)\sigma_{i+n}(\psi) = 1, i = 1, \ldots, n$. Let $\alpha = P(\psi) \in K$, where P is a polynomial of degree less than n, an arbitrary element in K. Then

$$\sigma_i(P(\psi^{-1})) = P(\sigma_i(\psi^{-1})) = P(\sigma_i(\psi)^{-1}) = P(\sigma_{n+1}(\psi)) = \sigma_{n+1}(P(\psi)).$$

So $P(\psi)$ and $P(\psi^{-1})$ have the same norm:

$$N(P(\psi)) = N(P(\psi^{-1})) = \sigma_1(\psi)\cdots\sigma_{2n}(\psi).$$

So $P(\psi)$ is a unit if and only if $P(\psi^{-1})$ is a unit.

Let U_L, U_K be the group of units of L, K respectively. Define a homomorphism $f : U_K \to U_L$ by $f(P(\psi)) = P(\psi)P(\psi^{-1})$. We show that $f(U_K) \subset U_L$. For any integer polynomial $P(x)$ one has $P(x)P(\frac{1}{x}) = S(x+\frac{1}{x})$, where S is a rational polynomial. This implies $f(P(\psi)) = S(\theta)$, so $f(U_K) \subset L$. Moreover, $P(\psi)$ unit implies $P(\psi^{-1})$ unit, so $S(\theta)$ is a unit in K. Since $\theta = \psi + \frac{1}{\psi}$ we have $|K : L| \leq 2$. Since $|K : K \cup \mathbb{R}| = 2$ and due to the fact that K is real, one has $K = K \cup \mathbb{R}$. But this implies that any unit in U_K must lie in U_L, so the image of f is in U_L.

Let A be an integer matrix in $\mathrm{SL}(2n, \mathbb{R})$ with characteristic polynomial R. For example, A can be the companion matrix of R. In order to obtain matrices that commute to A it is enough to show that the kernel of f contains at least two independent units. Indeed, let $v = P(\psi)$ be a unit such that $B = P(A)$ commutes to A, and A and B are independent under multiplication. If v is in the kernel of f then $\sigma_i(f(v)) = 1$ for all i, that is:

$$\begin{aligned}1 &= \sigma_i(P(\psi)P(\psi^{-1})) = \sigma_i(P(\psi))\sigma_i(P(\psi)^{-1})\\ &= P(\sigma_i(\psi))P(\sigma_i(\psi^{-1})) = P(\psi_i)P(\psi_i^{-1}).\end{aligned}$$

If ψ_i is a root of R of absolute value 1, then $\psi_i^{-1} = \bar{\psi}_i$, therefore

$$1 = P(\psi_i)P(\bar{\psi}_i) = P(\psi_i)\overline{P(\psi_i)} = |P(\psi_i)|.$$

Thus B has eigenvalues of absolute value 1 in the same directions as A, and the $\mathbb{Z}^2$-action generated by A and B is genuinely partially hyperbolic.

It remains to show that the kernel of f contains at least two independent units. We show that the kernel contains s independent units. The polynomial R has $2s$ real roots, and $n-s$ pairs of complex conjugate roots of absolute value 1. The structure of the groups U_L, U_K is given by the Dirichlet units theorem. Every unit in U_L can be written as $\rho u_1^{k_1} u_2^{k_2} \dots u_{s+n-1}^{k_{s+n-1}}$, where u_i are independent units. The images $f(u_i)$ are units in U_K. By the Dirichlet units theorem, U_K can have at most $n-1$ independent units. Therefore, in $\{f(u_i)\}_i$ there are at most $n-1$ independent units. Without loss, assume that these are $f(u_i), i = 1, \dots, n-1$. This implies the existence of at least s relations of type:

$$1 = f(u_1)^{l_{1,k}} f(u_2)^{l_{2,k}} \cdots f(u_{n-1})^{l_{n-1,k}} f(u_k)^{l_k},$$

where $k = n, n+1, \dots, n+s$. These relations implies the existence of s units in the kernel of f:

$$v_k = u_1^{l_{1,k}} u_2^{l_{2,k}} \cdots u_{n-1}^{l_{n-1,k}} u_k^{l_k}.$$

Moreover, u_i independent implies that v_k are independent.

Finally, the existence of reducible genuinely partially hyperbolic $\mathbb{Z}^2$-actions on tori of any odd dimension greater than 9 can be shown by taking products of partially hyperbolic $\mathbb{Z}^2$-actions on tori of even dimension with hyperbolic $\mathbb{Z}^2$-actions on tori of dimension 3. Examples of the latest are shown in Examples 2.2.16 and 2.2.15. □

Example 2.2.20 The following example of irreducible genuine partially hyperbolic $\mathbb{Z}^2$-action on $\mathbb{T}^6$ appears in [22, Section 6.2]. It was found by S. Katok using the program PARI [134].

We start with the irreducible degree three polynomial $Q = x^3 - 2x^2 - 8x + 1$, which has two real roots of absolute value larger than 2, and a real root of absolute value less than 2. Its reciprocal polynomial is

$$f(x) = x^6 - 2x^5 - 5x^4 - 3x^3 - 5x^2 - 2x + 1,$$

which is irreducible over $\mathbb{Q}$, has four real roots, and a pair of complex conjugate roots of absolute value 1. A system of fundamental units for $\mathbb{Q}(f)$ is given by $u_1 = x, u_2 = x+1, u_3 = x^4 - 2x^3 - 6x^2 - x + 1, u_4 = 2x^5 - 6x^4 - 3x^3 - 6x^2 - 6x$.

The $\mathbb{Z}^2$ partially hyperbolic action is generated by the companion matrix A of f and by $B = 2A^5 - 6A^4 - 3A^3 - 6A^2 - 6A$:

$$A = \begin{pmatrix} 0 & 1 & 0 & 0 & 0 & 0 \\ 0 & 0 & 1 & 0 & 0 & 0 \\ 0 & 0 & 0 & 1 & 0 & 0 \\ 0 & 0 & 0 & 0 & 1 & 0 \\ 0 & 0 & 0 & 0 & 0 & 1 \\ -1 & 2 & 5 & 3 & 5 & 2 \end{pmatrix}$$

and

$$B = \begin{pmatrix} 0 & -6 & -6 & -3 & -6 & 2 \\ -2 & 4 & 4 & 0 & 7 & -2 \\ 2 & -6 & -6 & -2 & -10 & 3 \\ -3 & 8 & 9 & 3 & 13 & -4 \\ 4 & -11 & -12 & -3 & -17 & 5 \\ -5 & 14 & 14 & 3 & 22 & -7 \end{pmatrix}.$$

The eigenvalues of A are $3.6863, -1.3236, 0.06076 + 0.9981i, 0.06076 - 0.9981i, -0.7555, 0.2712$. The eigenvalues of B are -22.1542, -2.1586, $0.9105 + 0.4133i, 0.9105 - 0.4133i, -0.4632, -0.04513$. Both A and B have four real eigenvalues and a pair of complex conjugate eigenvalues of absolute value 1. The matrices A and B commute and have a common two-dimensional neutral space.

Example 2.2.21 Here is an example of irreducible genuinely partially hyperbolic action of $\mathbb{Z}^2$ on $\mathbb{T}^8$. We start with the irreducible degree four polynomial $Q = x^4 - x^3 - 7x^2 - x + 1$. Q has two real roots of absolute value greater than 2, and two real roots of absolute value less than 2. The reciprocal polynomial of Q is $R = x^8 - x^7 - 4x^6 - 7x^4 - 4x^3 - 3x^2 - x + 1$, which has four real roots and two pairs of complex conjugate roots of absolute value 1. A set of fundamental units for R, obtained using the program PARI [134], is given by: $u_1 = x, u_2 = x^7 - x^6 - 4x^5 - 3x^4 - 3x^3 - x^2, u_3 = x^7 - 2x^6 - x^5 - 3x^4 - 4x^3 - 3x + 1, u_4 = 8x^7 - 17x^6 - 3x^5 - 32x^4 - 22x^3 - 15x^2 - 15x + 4$. The generators of the $\mathbb{Z}^2$ action are the companion matrix A of R and $B = 8A^7 - 17A^6 - 3A^5 - 32A^4 - 22A^3 - 15A^2 - 15A + 4$:

$$A = \begin{pmatrix} 0 & 1 & 0 & 0 & 0 & 0 & 0 & 0 \\ 0 & 0 & 1 & 0 & 0 & 0 & 0 & 0 \\ 0 & 0 & 0 & 1 & 0 & 0 & 0 & 0 \\ 0 & 0 & 0 & 0 & 1 & 0 & 0 & 0 \\ 0 & 0 & 0 & 0 & 0 & 1 & 0 & 0 \\ 0 & 0 & 0 & 0 & 0 & 0 & 1 & 0 \\ 0 & 0 & 0 & 0 & 0 & 0 & 0 & 1 \\ -1 & 1 & 3 & 4 & 7 & 4 & 3 & 1 \end{pmatrix}$$

$$B = \begin{pmatrix} 4 & -15 & -15 & -22 & -32 & -3 & -17 & 8 \\ -8 & 12 & 9 & 17 & 34 & 0 & 21 & -9 \\ 9 & -17 & -15 & -27 & -46 & -2 & -27 & 12 \\ -12 & 21 & 19 & 33 & 57 & 2 & 34 & -15 \\ 15 & -27 & -24 & -41 & -72 & -3 & -43 & 19 \\ -19 & 34 & 30 & 52 & 92 & 4 & 54 & -24 \\ 24 & -43 & -38 & -66 & -116 & -4 & -68 & 30 \\ -30 & 54 & 47 & 82 & 144 & 4 & 86 & -38 \end{pmatrix}.$$

The eigenvalues of A are 2.8854, -1.2628, $-0.2433+0.9699i$, $-0.2433-0.9699i$, $0.1547+0.9879i$, $0.1547-0.9879i$, -0.7918, 0.3465. The eigenvalues of B are -138.319, -4.4174, $0.7076+0.7065i$, $0.7076-0.7065i$, $0.7770+0.6294i$, $0.7770-0.6294i$, -0.2263, -0.0072. Both A and B have four real eigenvalues and two pairs of complex conjugate eigenvalues of absolute value 1. The matrices A and B commute and have a common four-dimensional neutral space; eight is the lowest dimension when the latter property is compatible with partial hyperbolicity for a higher rank action.

2.2.8 Affine actions of the torus without fixed points

Any affine map of the torus whose linear part A has no roots of unity has a fixed point and hence is isomorphic to the ergodic automorphism F_A via a translation which takes zero into a fixed point.

For several commuting affine maps the set of fixed points of any of them is invariant under the others. Thus any abelian group of affine maps of a torus which contains an element with linear part without roots of unity has a finite orbit and contains a subgroup of finite index which has a fixed point and is hence isomorphic to an action by automorphisms. However, the whole group may not have a fixed point even if all non-zero elements of the action are hyperbolic [54]. We describe such an example on the torus and the relevant theoretical background below.

Let Γ be a discrete finitely generated group. Let $\phi_0 : \Gamma \to \mathrm{Aut}(\mathbb{T}^n) \equiv GL(n, \mathbb{Z})$ be a representation. We recall in this particular setting some notions from Section 1.4. A map $\tau : \Gamma \to \mathbb{T}^n$ is a 1-cocycle over the action ϕ if for each $\gamma_1, \gamma_2 \in \Gamma$

$$\tau(\gamma_1\gamma_2) = \tau(\gamma_1) \cdot \phi_0(\gamma_1)(\tau(\gamma_2)). \tag{2.2.3}$$

A 1-cocycle is *trivial* if there exists a point $x_0 \in \mathbb{T}^n$ so that for all $\gamma \in \Gamma$

$$\tau(\gamma) = \phi_0(\gamma)(x_0)^{-1} \cdot x_0. \tag{2.2.4}$$

The cohomology group $H^1(\Gamma; \mathbb{T}^n_{\phi_0})$ is the quotient space of the 1-cocycles modulo the trivial 1-cocycles.

The following proposition relates the basic structure of affine structures to 1-cohomology.

Proposition 2.2.22 *Let Γ be a finitely generated group and $\phi : \Gamma \to GL(n, \mathbb{Z})$ be a representation.*

(i) There is a one-to-one correspondence between the affine actions ϕ of Γ on $\mathbb{T}^n$ with linear part ϕ_0 and 1-cocycles τ over the action ϕ_0. We will denote by τ_ϕ the 1-cocycle over ϕ_0 associated to an action α. The translational class associated to ϕ is the cohomology class $[\tau_\phi] \in H^1(\Gamma; \mathbb{T}^n_{\phi_0})$.
(ii) An affine action with linear part ϕ_0 has a fixed point if and only if $[\tau_\phi]$ is trivial.
(iii) An affine action with linear part ϕ_0 has a dense set of periodic points if and only if $[\tau_\phi]$ is torsion.

Proof (i) Let $\phi(\gamma)$ be an affine transformation. Define the translational part $\tau_\phi(\gamma) \in \mathbb{T}^n$ to be its action on the identity element, $\tau(\gamma)(0) = \phi(\gamma)(0)$. The group law for the action ϕ becomes the cocycle law (2.2.3). Vice versa, given a 1-cocycle $\tau : \Gamma \to \mathbb{T}^n$ over ϕ_0, define an affine action by the rule $\phi(\gamma)(x) = \tau(\gamma) \cdot \phi_0(\gamma)(x)$ for $x \in \mathbb{T}^n, \gamma \in \Gamma$.

(ii) Suppose that x_0 is a fixed-point for the action. For each $\gamma \in \Gamma$, $\tau(\gamma) \cdot \phi_0(\gamma)(x_0) = x_0$, so $\tau(\gamma) = \phi_0(\gamma)(x_0)^{-1} \cdot x_0$, and τ is a trivial cocycle. Vice versa, if x_0 satisfies equation (2.2.4), then

$$\phi_\tau(\gamma)(x_0) = \tau(\gamma) \cdot \phi_0(\gamma)(x_0) = \tau(\gamma) \cdot \tau(\gamma)^{-1}(x_0) = x_0.$$

(iii) Let ϕ be an affine action with a periodic orbit $x_0 \in \mathbb{T}^n$. The stabilizer of the orbit of x_0 is a normal subgroup $\Gamma' \subset \Gamma$ with finite index. Let $R : H^1(\Gamma; \mathbb{T}^n_{\phi_0}) \to H^1(\Gamma'; \mathbb{T}^n_{\phi_0})$ be the restriction map and $T : H^1(\Gamma'; \mathbb{T}^n_{\phi_0}) \to H^1(\Gamma; \mathbb{T}^n_{\phi_0})$ the transfer. The restriction class $R[\tau_\phi] \in H^1(\Gamma'; \mathbb{T}^n_{\phi_0})$ is 0 by statement (ii). The composition $T \circ R = [\gamma : \Gamma'] \cdot Id$ (by [13], Proposition 10.1), so $0 = T \circ R[\tau_\phi] = [\Gamma : \Gamma'][\tau_\phi]$, which implies that $[\tau_\phi]$ is a torsion class.

Conversely, let $p > 0$ so that $p \cdot \tau(\gamma) = \phi_0(\gamma)(x_0)^{-1} \cdot x_0$ for all $\gamma \in \Gamma$. Choose $y_o \in \mathbb{T}^n$ so that $p \cdot y_0 = x_0$ and calculate

$$p \cdot \left(\tau(\gamma) \cdot \phi + 0(\gamma)(y_0) \cdot y_0^{-1}\right) = e \in \mathbb{T}^n,$$

hence τ is cohomologous to a cocycle taking values in the finite group $(\frac{1}{p}) \cdot \mathbb{Z}/\mathbb{Z} \subset \mathbb{T}$. The statement from Proposition follows now if we show that an affine action ϕ, defined by a translational cocycle τ'_ϕ with values in a finite subgroup $G \subset \mathbb{T}$, has a dense set of periodic points.

The isotropy subgroup $\Gamma_G \subset \Gamma$ of G for the linear action ϕ_0 is a normal subgroup of finite index. The restriction of τ'_ϕ to Γ_G is a homomorphisms τ'_ϕ : $\Gamma_G \to \Gamma$ with kernel Γ' of finite index. It follows that every rational point of $\mathbb{T}^n$ is a periodic point for the action ϕ. □

The following result appears in [54].

Theorem 2.2.23 *Let Γ be a free abelian group of rank $r \geq 2$, and $\phi_0 : \Gamma \to SL(n, \mathbb{Z})$ a representation such that $\phi_0(\gamma_h)$ is hyperbolic for some $\gamma_h \in \Gamma$. Then there exists a subgroup $\Gamma' \subset \Gamma$ of finite index and an affine Anosov action ϕ of Γ' on $\mathbb{T}^n$ with linear part $\phi_0|\Gamma'$ and no fixed points.*

Proof The fixed-point set $\mathrm{Fix}(\phi(\gamma_h)) \subset \mathbb{T}^n$ for $\phi(\gamma_h))$ is finite, and as Γ is abelian, is invariant under the action of $\phi(\Gamma)$. Thus, the existence of a fixed point for the action ϕ is equivalent to the existence of a fixed point for the restricted action of Γ on $\mathrm{Fix}(\phi(\gamma_h))$. Introduce now the relative cohomology group $H^1(\Gamma, \gamma_h; \mathrm{Fix}(\phi_0(\gamma_h))_{\phi_0})$ spanned by the 1-cocycles $\tau : \Gamma \to \mathrm{Fix}(\phi_0(\gamma_h))$ over ϕ_0 which vanish on γ_h. Note that this defines a subcomplex as $\tau(\gamma_h^m) = \tau(\gamma_h)^m$ using that $\phi_0(\gamma_h)$ acts trivially on $\mathrm{Fix}(\phi_0(\gamma_h))$. So there is a natural map

$$H^1(\Gamma, \gamma_h; \mathrm{Fix}(\phi_0(\gamma_h))_{\phi_0}) \to H^1(\Gamma; \mathbb{T}^n_{\phi_0}). \tag{2.2.5}$$

We show now that this map is injective and that each affine action ϕ with the linear part ϕ_0 as in the theorem yields a 1-cohomology class in the image of this map. Together with Proposition 2.2.22 this implies that the set of affine actions of Γ with linear part ϕ_0 is indexed by the cohomology group $H^1(\Gamma, \gamma_h; \mathrm{Fix}(\phi_0(\gamma_h))_{\phi_0})$, and that each class in the previous group gives rise to an Anosov action without fixed points.

Assume that $\tau : \Gamma \to \mathrm{Fix}(\phi_0(\gamma_h))$ is a 1-cocycle over ϕ_0 which vanishes on γ_h and is a coboundary as a map into $\mathbb{T}^n$. Then the corresponding affine action ϕ_τ admits a fixed point, so is conjugate to the linear action ϕ_0 via translation by some $x_0 \in \mathbb{T}^n$. Translation by x_0 maps the fixed point set of ϕ to that of ϕ_0. The hypothesis $\tau(\gamma_h) = 0$ implies that $\mathrm{Fix}(\phi(\gamma_h))$ contains 0, so $x_0 \in \mathrm{Fix}(\phi_0(\gamma_h))_{\phi_0}$ and thus τ also defines the zero class in $H^1(\gamma; \mathrm{Fix}(\phi_0(\gamma_h))_{\phi_0})$. A corollary of this argument is that $\mathrm{Fix}(\phi(\gamma_h)) = \mathrm{Fix}(\phi_0(\gamma_h))$ whenever $0 \in \mathrm{Fix}(\phi(\gamma_h))$.

Assume now that ϕ is an affine action as in the theorem. We can conjugate the action by a translation so that $0 \in \mathrm{Fix}(\phi(\gamma_h))$. Let $\tau_\phi : \Gamma \to \mathbb{T}^n$ denote the corresponding 1-cocycle. The set $\mathrm{Fix}(\phi(\gamma_h))$ is invariant under the action $\phi(\Gamma)$, so $\tau_\phi(\gamma) = \phi(\gamma_i)(0) \in \mathrm{Fix}(\phi(\gamma_h)) = \mathrm{Fix}(\phi_0(\gamma_h))$, and so the class $[\tau_h]$ is in the image of the homomorphisms (2.2.5).

If $\Gamma' \subset \Gamma$ is the subgroup of finite index consisting of the elements which act trivially when restricted to $\mathrm{Fix}(\phi(\gamma_h))$, and $\mathcal{A}$ an abelian group, let $\mathrm{Hom}(\Gamma', \gamma_h; \mathcal{A})$ denote the group homomorphisms that maps γ_h into the trivial element. The theorem now follows by observing that

$$\mathrm{Hom}(\Gamma', \gamma_h; \mathrm{Fix}(\phi_0(\gamma_h))) \subset H^1(\Gamma', \gamma_0; \mathrm{Fix}(\phi_0(\gamma_h))_{\phi_0}), \tag{2.2.6}$$

and using Proposition 2.2.22. Note that for the cohomology that appears on the right-hand side of (2.2.6), 1-cocycles are representations and there is only one trivial 1-cocycle, i.e., the trivial one. □

We show now a concrete example of higher rank abelian action on tori without fixed points.

Example 2.2.24 Consider the action $\phi_0 : \mathbb{Z}^2 \times \mathbb{T}^3 \to \mathbb{T}^3$ on the three-dimensional torus induced by F_A, F_B, where

$$A = \begin{pmatrix} 0 & 1 & 0 \\ 0 & 0 & 1 \\ 1 & 8 & 2 \end{pmatrix} \quad B = \begin{pmatrix} 2 & 1 & 0 \\ 0 & 2 & 1 \\ 1 & 8 & 4 \end{pmatrix}.$$

One can easily check that both A and B are hyperbolic, not powers of each other, and that $AB = BA$. We take $A = \gamma_h$, where γ_h is as above. Let

$$\mathrm{Fix}(A) = \left\{ \left(\frac{i}{10}, \frac{i}{10}, \frac{i}{10} \right) \middle| 0 \leq 1 \leq 1 \right\}$$

be the set of fixed elements of F_A. $\mathrm{Fix}(A)$ is an abelian group under addition isomorphic to $\mathbb{Z}_{10}$. The action of B on $\mathrm{Fix}(A)$ is given by multiplication by 3, so B^{10} is the first power of B that acts trivially on $\mathrm{Fix}(A)$. The subgroup $\Gamma' \subset \mathbb{Z}^2$ that acts trivially on $\mathrm{Fix}(A)$ is the subgroup generated by A and B^{10}. Let $\tau \in \mathrm{Hom}(\Gamma', \gamma_h; \mathrm{Fix}(\phi_0(\gamma_h)))$ be defined by $\tau(A) = \hat{0}$ and $\tau(B^{10}) = \hat{1}$. Since the action of Γ' on $\mathrm{Fix}(A)$, τ is also a non-trivial 1-cocycle, and hence induces a nontrivial element in $H^1(\Gamma', \gamma_0; \mathrm{Fix}(\phi_0(\gamma_h))_{\phi_0})$. Moreover, $\tau : \Gamma' \to \mathbb{T}^3$ is a 1-cocycle that is not trivial.

One defines now the affine action $\phi : \Gamma' \times \mathbb{T}^3 \to \mathbb{T}^3$ by $\phi(\gamma)(x) = \tau(\gamma)\phi_0(\gamma)(x)$ for all $\gamma \in \Gamma', x \in \mathbb{T}^3$. The action ϕ does not have any fixed point.

2.2.9 Higher rank abelian actions on nilmanifolds

Examples of higher rank abelian actions on nilmanifolds are constructed by Qian [145], developing the work of Auslander and Scheuneman [4].

Let V be an n-dimensional vector space over $\mathbb{R}$. Let $\{x_1, x_2, \dots, x_n\}$ be a basis. We denote by $N(V)$ the free Lie algebra associated to V, that is, the Lie algebra generated by all non-associative words

$$x_1, x_2, \dots, x_n, [x_1, x_2], \dots, [[x_1, x_2], x_3], \dots, \text{ etc.} \tag{2.2.7}$$

subject only to the Jacobi and skew-symmetry relations. The free k-step nilpotent Lie algebra, denoted by $N_k(V)$, is obtained by imposing the condition that every word of length $k+1$ is zero.

To find a lattice in $N_k(V)$, consider C_0 to be the basis in $N_k(V)$ consisting of all words in (2.2.7) of length at most k, and let $C = \mathbb{Z}$-span of C_0. Then there exists an integer $m \in \mathbb{Z}$ such that mC is a uniform lattice in $N_K(V)$. For example, using Campbell–Baker–Hausdorff formula it is easy to see that for $k = 1, 2, 3, 4$ one may take $m = 1, 2, 12, 48$ respectively.

We denote the lattice $\Lambda = \exp(mC)$, the corresponding nilmanifold $N(n,k) := N_k(V)/\Lambda$, and the group of automorphisms of $N(n,k)$ by $\mathrm{Aut}(n,k)$. Recall that an element $A \in \mathrm{Aut}(n,k)$ is a linear automorphism $A : N(n,k) \to N(n,k)$ that preserves the bracket operation and satisfies $A(\Lambda) = \Lambda$.

The standard basis in $N_k(V)$, consisting of words of length at most k in (2.2.7), determines a decomposition $N_k(V) = V^{(1)} \oplus V^{(2)} \oplus \dots \oplus V^{(k)}$, where $V^{(i)}$ is the subspace generated by the words of length exactly i. The decomposition is kept invariant by any automorphism $A \in N(n,k)$. One can check, using the fact that A preserves the brackets, that the matrix of A with respect to the standard basis is of type

$$A = \begin{pmatrix} A_0^{(0)} & 0 & 0 & \cdots & 0 \\ A_1^{(0)} & A_0^{(1)} & 0 & \cdots & 0 \\ A_2^{(0)} & A_1^{(1)} & A_0^{(2)} & \cdots & 0 \\ \cdots & \cdots & \cdots & \cdots & \cdots \\ A_{k-1}^{(0)} & A_{k-2}^{(1)} & A_{k-3}^{(2)} & \cdots & A_0^{(k-1)} \end{pmatrix}, \tag{2.2.8}$$

where $A_0^{(0)} \in SL(n, \mathbb{Z})$, $A_i^{(0)} \in Mat(n_i \times n, \mathbb{Z})$, $n_i = \dim(V^{(i)})$, $A_i^{(0)}$ are arbitrary, and the matrices $A_i^{(l)}$ are determined inductively by $A_j^{(0)}$, $j < i$.

Example 2.2.25 Consider the free Lie algebra $N_2(2)$, which is generated by $\{x_1, x_2, [x_1, x_2]\}$. An automorphism $A \in \mathrm{Aut}(2,2)$ can be represented as a matrix:

$$A = \begin{pmatrix} a_{11} & a_{12} & 0 \\ a_{21} & a_{22} & 0 \\ b_1 & b_2 & c \end{pmatrix} \tag{2.2.9}$$

where $\begin{pmatrix} a_{11} & a_{12} \\ a_{21} & a_{22} \end{pmatrix} \in SL(2, \mathbb{Z}), b_1, b_2 \in \mathbb{Z}$ arbitrary, and $c = a_{11}a_{22} - a_{12}a_{21} = 1$. In particular, $N_2(2)$ does not have any Anosov automorphism.

Note that by choosing in (2.2.8) the matrices $A_1^{(0)}, A_2^{(0)}, \ldots, A_{k-1}^{(0)}$ to be trivial, one obtains a representation of $SL(n, \mathbb{Z})$ into $\mathrm{Aut}(n, k)$. We will refer to this representation as the *standard representation* of $SL(n, \mathbb{Z})$ into $\mathrm{Aut}(n, k)$. Note that the standard representation is block diagonal. In this case the only non-trivial entries in (2.2.8) are the diagonal blocks $A_0^{(i)}, 0 \leq i \leq k$.

One can generalize the notion of Cartan action, introduced in Section 2.2.5, to arbitrary manifolds and, in particular, to (infra)nilmanifolds. The definition below was introduced by Hurder in [55]. It was frequently used in the rigidity theory of higher rank lattice actions on compact manifolds.

Definition 2.2.26 Let $\phi : \mathbb{Z}^k \times M \to M$ be a C^∞ action on a compact manifold M. The action ϕ is called Cartan if there exists a set of generators $\{\gamma_1, \ldots, \gamma_n\}$ of $\mathbb{Z}^k$ such that:

(i) each $\phi(\gamma_i)$ is hyperbolic and has a one-dimensional, strongest stable foliation $\mathcal{F}^{ss}$;
(ii) the tangential distributions are pair-wise transverse with their direct sum $E_1^{ss} \oplus \cdots \oplus E_n^{ss} \cong TM$.

More general, if Γ is a discrete group and $\phi : \Gamma \times M \to M$ is a C^∞ action, then ϕ is called Cartan if there exists a set of commuting elements $\{\gamma_1, \ldots, \gamma_n\}$, which generate an abelian group $\mathcal{A}$, such that the restriction of ϕ to $\mathcal{A}$ is a Cartan action on M.

Remark 2.2.27 Note that from a result of Franks [39] it follows that if a manifold admits an Anosov diffeomorphism with a one-dimensional stable foliation, then the manifold is a torus. Some of the following examples are on nilmanifolds. In these cases, the one-dimensional foliations $\mathcal{F}^{ss}$ that appear in the Cartan actions are always strict subfoliations of some stable foliations.

Remark 2.2.28 It is immediate that a $\mathbb{Z}^k$ Cartan action generated by hyperbolic automorphisms of an infranilmanifold is a genuine higher rank $\mathbb{Z}^k$-action.

The following lemma shows the existence of many Cartan actions on tori. It allows to build hyperbolic matrices in $SL(n, \mathbb{Z})$ with specific eigenvalue structure, and it is used to build examples of higher rank abelian actions on nilmanifolds. Its proof is based on more general results from [142].

Lemma 2.2.29 *Let $\Gamma = SL(n, \mathbb{Z})$, $n \geq 2$, or, more generally, a subgroup of finite index in $SL(n, \mathbb{Z})$. Then there exists a Cartan subgroup H of $SL(n, \mathbb{R})$ such that the quotient $H/(H \cup \Gamma)$ is compact. In particular, there exists a subgroup $\mathcal{A} \subset \Gamma$ such that:*

(i) the elements of $\mathcal{A}$ are simultaneously diagonalizable over $\mathbb{R}$; and
(ii) $\mathcal{A}$ is isomorphic to a free abelian group of rank $n - 1$.

Let $v_1, \ldots, v_n \in \mathbb{R}^n$ be a basis of simultaneously eigenvectors for the group $\mathcal{A}$, and $\lambda_i : \mathcal{A} \to \mathbb{R}-\{0\}$ the character of $\mathcal{A}$ defined by $Av_i = \lambda_i(A)v_i$, $A \in \mathcal{A}$. One can pass to a subgroup of finite index and assume that each λ_i takes values in $\mathbb{R}^+$. If H^0 is the connected component of the identity in the the Cartan group in Lemma 2.2.29 then the λ_is extend to H^0:

$$\lambda_i : H^0 \to \{(x_1 \ldots x_n) \in (\mathbb{R}^+)^n | x_1 \ldots x_n = 1\} \tag{2.2.10}$$

as an isomorphism of analytic groups. Note that $H^0/\mathcal{A}$ is compact. As a consequence of the compactness of $\mathcal{A}$ in H^0, one has that for each i, $1 \leq i \leq n$, there exists $A_i \in \mathcal{A}$ such that $\lambda_i(A_i) < 1$ and $\lambda_j(A_i) > 1$ for each $j \neq i$. Moreover, for each $1 \leq i < j \leq n$ there exists $B_{ij} \in \mathcal{A}$ such that $\lambda_i(B_{ij}) < 1, \lambda_j(B_{ij}) < 1$, and $\lambda_k(B_{ij}) > 1$ for all $k \neq i, k \neq j$. Note that by standard results the torus H^0 is $\mathbb{Q}$-anisotropic, that is, none of the eigenspaces $\mathbb{R}v_i$ is a rational line and its image under the standard projection is dense in $\mathbb{T}^n$. As a consequence, if $A \in \mathcal{A}$, $A \neq 1$, then $\lambda_i(A) \neq 1$ for each $1 \leq i \leq n$.

Theorem 2.2.30 *If $n \geq 3$, then the action induced on $N(n, 2)$ by the standard representation of $SL(n, \mathbb{Z})$ is a Cartan action.*

Proof Let $\mathcal{A} \cong \mathbb{Z}^{n-1}$ be the diagonalizable subgroup of $SL(n, \mathbb{Z})$ whose existence is guaranteed by Lemma 2.2.29. Let $\{v_1, \ldots, v_n\}$ be the basis in $\mathbb{R}^n$ in which $\mathcal{A}$ is diagonalizable. Then there exist hyperbolic matrices $A_k \in \mathcal{A}$, $1 \leq k \leq n$ such that A_k has all eigenvalues positive numbers, with only one eigenvalue $\lambda_k^{(k)}$ in position k greater than one, and all the other eigenvalues less than one, and there exist hyperbolic matrices B_{ij}, $1 \leq i < j \leq n$ such that B_{ij} has all eigenvalues positive numbers, with only two eigenvalues, $\lambda_i^{(ij)}, \lambda_j^{(ij)}$, in positions i, j, greater than one, and all other eigenvalues are less than one.

Then the set

$$\mathcal{B} = \{v_k | 1 \leq k \leq n\} \cup \{[v_i, v_j] | 1 \leq i < j \leq n\} \tag{2.2.11}$$

is a basis for $N(2, n)$ in which the linear action induced by $\mathcal{A}$ on $N(2, n)$ is diagonalizable. It is clear now that the action induced by the matrix A_k on $N(2, n)$ is hyperbolic and has a strong stable one-dimensional foliation parallel to the vector v_k, and the action induced by the matrix B_{ij} is hyperbolic and has a strong stable one-dimensional foliation parallel to the vector $[v_i, v_j]$. □

Remark 2.2.31 One can show that for any $k > 2$ the standard representation of $SL(n, \mathbb{Z})$ into $\mathrm{Aut}(n, k)$ is not a Cartan action. In the case $n = 3$ this follows from the fact that two different eigenvectors v_i, v_j, v_k belonging to a basis in $\mathbb{R}^n$ generate two different words of length 3 in $N(n, 3)$, namely $[v_i, [v_j, v_k]]$ and $[v_k, [v_i, v_j]]$.

Nevertheless, the standard action still has many Anosov diffeomorphisms, and, as follows from the theorem below, they actually generate the action.

First we prove an auxiliary lemma.

Lemma 2.2.32 *Let* $A \in SL(n, \mathbb{R})$. *Let* $\lambda_1, \dots, \lambda_n \in \mathbb{R}^+, \lambda_i \neq 1$ *for all* $1 \leq i \leq n$. *Then there exists k positive integer such that the matrix* $ADiag\{\lambda_1^k, \dots, \lambda_n^k\}$ *is hyperbolic.*

Proof It is enough to show that for any non-zero unit vector v and for k positive integers sufficiently large one has $\|A\mathrm{Diag}\{\lambda_1^k, \dots, \lambda_n^k\}v\| \neq \|v\|$. We proceed by contradiction. Let $v = (v_1, v_2, \dots, v_n)$ be a unit vector such that

$$\|A\mathrm{Diag}\{\lambda_1^k, \dots, \lambda_n^k\}v\| = \|v\|. \tag{2.2.12}$$

Since all λ_is are positive we can assume without loss that

$$\lambda_1 = \dots = \lambda_l > \lambda_{l+1} \geq \dots \geq \lambda_n$$

for some $2 \leq l < n$. We rewrite (2.2.12) as

$$\lambda_1^k \left\| A\mathrm{Diag}\{v_1, \dots, v_l, \left(\frac{\lambda_{l+1}}{\lambda_1}\right)^k v_{l+1}, \dots, \left(\frac{\lambda_n}{\lambda_1}\right)^k v_n\} \right\| = \|v\|. \tag{2.2.13}$$

Observe now that $A\mathrm{Diag}\{v_1, \dots, v_l, 0, 0, \dots, 0\} \neq 0$ and k large enough gives a contradiction in (2.2.13). Thus the first l components of v are zero. In a similar way one can show that all components of v are zero, so $v = 0$. But this is in contradiction with v of length 1. □

Theorem 2.2.33 *Consider the standard representation of* $SL(n, \mathbb{Z})$ *into* $Aut(n, k)$. *Then its image is generated by Anosov diffeomorphisms if* $n \geq k+1$.

Proof It follows from Lemma 2.2.29 that $SL(n, \mathbb{Z})$ contains a hyperbolic element A that has all eigenvalues λ_i, $1 \leq i \leq n$, real eigenvalues, and, moreover, any product of λ_is of length strictly less than n is different from 1. Choose now a basis $\mathcal{B} = \{v_1, v_2, \ldots, v_n\}$ of V consisting of eigenvectors of A. Then the set of words of length less then $k + 1(\leq n)$ in letters from $\mathcal{B}$ gives a basis in $N(n, k)$ which consists of eigenvectors for the action of A on $N(n, k)$. The corresponding eigenvalues are products of distinct eigenvalues λ_i, $1 \leq i \leq n$, of length less than n, so are different from 1 and positive. This implies that A, viewed as an element in $\mathrm{Aut}(n, k)$, has all eigenvalues different from 1. For any $M > 0$, by taking higher powers of A we can assume that the eigenvalues of A on $N(n, k)$ are such that $\lambda > M$ or $\lambda^{-1} > M$.

It is well known that $SL(n, \mathbb{Z})$ is generated by the elementary matrices $E_{i,j} := \mathrm{Id} + e_{i,j}$, where $e_{i,j}$ is the matrix with 1 in the (i, j) position and 0 elsewhere. Since each $E_{i,j}$ can be written as

$$E_{i,j} = (E_{i,j} A^{-1} E_{i,j}^{-1})(E_{i,j} A),$$

to finish the proof of the theorem it is enough to show that the action of the matrix $E_{i,j}A$ on $N(n, k)$ is hyperbolic. But this follows from Lemma 2.2.32. □

Remark 2.2.34 Using Theorem 2.2.33 one can show that the whole group $\mathrm{Aut}(n, k)$ is generated by Anosov diffeomorphisms.

Example 2.2.35 We describe now another example of a $\mathbb{Z}^3$-action on a nilmanifold. Let $\mathfrak{n}$ be the 2-step nilpotent Lie algebra generated by $\{e_i; 1 \leq i \leq 10\}$, with the following relations:

$$[e_1, e_2] = e_5, [e_1, e_3] = e_6, [e_1, e_4] = e_7,$$
$$[e_2, e_3] = e_8, [e_2, e_4] = e_9, [e_3, e_4] = e_{10},$$

and all other brackets between the generators are zero. Define $C = \mathrm{span}_{\mathbb{Z}}\{e_i\}$. Denote $N = \exp(\mathfrak{n})$ and $\Gamma = \exp(2C)$. Then N is a connected, simply connected nilpotent Lie group, and Γ is a co-compact lattice in N.

Consider the standard representation of $SL(4, \mathbb{Z})$ on $\mathrm{span}\{e_i; 1 \leq i \leq 4\}$. Then, using the relations between the e_is, we find a representation of $SL(4, \mathbb{Z})$ on $\mathrm{span}\{e_i; 5 \leq i \leq 10\}$. So we have a representation of $SL(4, \mathbb{Z})$ on $\mathfrak{n}$, and therefore an action on N, which preserves Γ. One can find an abelian subgroup generated by three hyperbolic matrices in $SL(4, \mathbb{Z})$ that gives a $\mathbb{Z}^3$-action on the nilmanifold N/Γ.

2.3 Higher rank $\mathbb{R}^k$-actions

2.3.1 Summary

Starting with Anosov flows and doing standard constructions as products, quotients, and covers, or starting with Anosov diffeomorphisms and doing suspensions, provide a large class of hyperbolic $\mathbb{R}^k$-actions. These examples do not exhibit many rigidity properties.

An interesting class of examples of higher rank $\mathbb{R}^k$-actions, $k \geq 2$, which exhibit many rigidity properties, comes from the following unified algebraic construction. None of the examples below have a finite cover with a smooth factor on which the action is not faithful, not transitive or is generated by a rank-one group.

Let G be a connected Lie group, $A \subset G$ a closed abelian subgroup which is isomorphic with $\mathbb{R}^k$, M a compact subgroup of the centralizer $Z(A)$ of A, and Γ a co-compact lattice in G. Then A acts by left translation on the compact space $M \setminus G/\Gamma$.

We will discuss in this section the following specific types corresponding to the general construction:

(i) For *suspensions of actions by automorphisms of tori and nilmanifolds* take $G = \mathbb{R}^k \rtimes \mathbb{R}^m$ or $G = \mathbb{R}^k \rtimes N$, the semidirect product of $\mathbb{R}^k$ with $\mathbb{R}^m$ or a simply connected nilpotent Lie group N.

(ii) For the *symmetric space examples* take G a semisimple Lie group of the non-compact type.

(iii) For the *twisted symmetric space examples* take $G = H \rtimes \mathbb{R}^m$ or $G = H \rtimes N$, a semidirect product of a reductive Lie group H with semisimple factor of the non-compact type with $\mathbb{R}^m$ or a simply connected nilpotent group N.

Those examples are partially hyperbolic and many among them are Anosov. Further interesting partially hyperbolic examples are obtained by taking restrictions of those actions to higher rank subgroups of A, i.e., subgroups which contain a discrete $\mathbb{Z}^2$ subgroup.

2.3.2 Suspensions of automorphisms of tori and nilmanifolds

Consider a genuine higher rank $\mathbb{Z}^k$-action on $\mathbb{T}^n$ by automorphisms of a torus. Recall that such an action contains a $\mathbb{Z}^2$-action such that every non-trivial element of $\mathbb{Z}^2$ acts ergodically on $\mathbb{T}^n$ with respect to the Haar measure. Embed $\mathbb{Z}^k$ as a lattice in $\mathbb{R}^k$. Let $\mathbb{Z}^k$ acts on $\mathbb{R}^k \times \mathbb{T}^n$ by $z(x, m) = (x - z, zm)$, and let $M = (\mathbb{R}^k \times \mathbb{T}^n)/\mathbb{Z}^k$ be the orbit space of the action.

Observe now that the group $\mathbb{R}^k$ acts naturally on $\mathbb{R}^k \times \mathbb{T}^n$ by $x(y, m) = (x + y, m)$ and the $\mathbb{R}^k$-action commutes with the previous $\mathbb{Z}^k$-action. Hence the $\mathbb{R}^k$-action descends to M. The induced $\mathbb{R}^k$-action is called the *suspension* of the $\mathbb{Z}^k$-action.

These examples generalize to irreducible Anosov $\mathbb{Z}^k$-actions by automorphisms of nilmanifolds. See Section 2.2.9 for examples of such actions.

2.3.3 Symmetric spaces and Weyl chamber flows

Now we consider the leading class of algebraic Anosov and partially hyperbolic higher rank $\mathbb{R}^k$-actions. We start with a review of relevant facts from Lie group theory. A good reference for this material is the book of Helgason [48].

Summary of Lie group theory

All our Lie groups are considered to be real, that is Lie groups over $\mathbb{R}$. Let G be a Lie group. The maximal connected solvable normal subgroup $R \subset G$ is said to be the *radical* of G. A connected Lie group is said to be *semisimple* if its radical is trivial. G is said to be *simple* if it has no non-trivial proper normal connected subgroups. Every connected semisimple Lie group G can be uniquely decomposed into an *almost direct product* $G = G_1 \cdots G_n$ of its normal simple subgroups, called the *simple factors* of G. In the previous product, G_i, G_j commute and the intersection $G_i \cap G_j$ is discrete if $i \neq j$. If, in addition, G is simply connected and center free, then G is the direct product $G_1 \times \cdots \times G_n$. Note that if G is semisimple, its center $Z(G)$ is discrete, and, moreover, it coincides with the center of the adjoint representation if G is connected.

Let G be a connected semisimple Lie group. Then G is equal to the almost direct product $G = KS$ of its compact and totally non-compact parts, where $K \subset G$ is the product of all compact simple components of G, and S is the product of all non-compact simple components. G is called *totally non-compact* if K is trivial. A subgroup $H \subset G$ is said to be *Cartan* if H is a maximal connected abelian subgroup consisting of semisimple elements. Any Cartan subgroup has a unique decomposition $H = T \times A$ into a direct product of a compact torus T and an $\mathbb{R}$-diagonalizable subgroup A.

All maximal connected $\mathbb{R}$ diagonalizable subgroups of G are conjugate and their common dimension is called the $\mathbb{R}$-*rank* of G. If maximal $\mathbb{R}$-diagonalizable subgroups are Cartan subgroups then G is said to be an $\mathbb{R}$-*split* group.

If $\mathfrak{g}$ is a finite dimensional Lie algebra, then there exists a uniquely solvable ideal in $\mathfrak{g}$ containing all solvable ideals of $\mathfrak{g}$. This ideal is called the *radical* of

$\mathfrak{g}$ and is denoted rad $\mathfrak{g}$. A finite dimensional Lie algebra $\mathfrak{g}$ is called *simple* if it is non-abelian and has no proper non-zero ideals. It is called *semisimple* if rad $\mathfrak{g} = 0$, that is, $\mathfrak{g}$ has no non-trivial solvable ideals. The Lie algebra of a semisimple Lie group is semisimple. If $\mathfrak{g}$ is semisimple then $[\mathfrak{g}, \mathfrak{g}] = \mathfrak{g}$, and if $\mathfrak{g}$ is an arbitrary finite-dimensional Lie algebra, then $\mathfrak{g}/\text{rad}\ \mathfrak{g}$ is semisimple.

Let $\mathfrak{g}$ be a Lie algebra. The bilinear form $B(X, Y) = Tr(adXadY)$ on $\mathfrak{g} \times \mathfrak{g}$ is called the *Killing form* of $\mathfrak{g}$.

An automorphism Θ of a Lie algebra is called *involutive* if $\Theta^2 = \text{Id}_{\mathfrak{g}}$. An involutive automorphism Θ of a semisimple Lie algebra $\mathfrak{g}$ is called *Cartan involution* if the bilinear form $B_\Theta(X, Y) = -B(X, \Theta Y)$ is strictly positive definite.

Let G be a semisimple Lie group of non-compact type with $\mathfrak{g}$ its semisimple Lie algebra. Let B be the Killing form and let Θ be any Cartan involution of $\mathfrak{g}$. Let $\mathfrak{g} = \mathfrak{k} + \mathfrak{p}$ be the corresponding *Cartan decomposition*, that is, the decomposition in eigenspaces corresponding to eigenvalues 1 and -1 of Θ. Note that $\mathfrak{k}$ is the fixed point set of Θ. It also coincides with the Lie algebra of a maximal compact subgroup $K \subset G$. Let $\mathfrak{a} \subset \mathfrak{p}$ be any maximal abelian subspace. All such subspaces have the same dimension. Let $A = \exp \mathfrak{a} \subset G$ the corresponding subgroup. Then A is the connected component of identity of a split Cartan subgroup of G. We denote by $\log : A \to \mathfrak{a}$ the inverse of the exponential map.

The centralizer $Z(A)$ of A splits as a product $Z(A) = M A$, where M is compact. M coincides with the centralizer of $\mathfrak{a}$ in K. If $\mathfrak{m}$ is the centralizer of $\mathfrak{a}$ in $\mathfrak{t}$, then $\mathfrak{m}$ is the Lie algebra of M.

For each λ in the dual space $\mathfrak{a}^*$ of $\mathfrak{a}$ let

$$\mathfrak{g}_\lambda = \{X \in \mathfrak{g} | [H, X] = \lambda(H)X \text{ for } H \in a\}.$$

Then λ is called a *restricted root* if $\lambda \neq 0$ and $\mathfrak{g}_\lambda \neq 0$.

The simultaneous diagonalization of $ad_{\mathfrak{g}}(\mathfrak{a})$ gives the decomposition

$$\mathfrak{g} = \mathfrak{g}_o + \sum_{\lambda \in \Lambda} \mathfrak{g}_\lambda, \ \ \mathfrak{g}_0 = \mathfrak{a} + \mathfrak{m},$$

where Λ is the set of restricted roots. The spaces $\mathfrak{g}_\lambda$ are called *root spaces*. A point $H \in \mathfrak{a}$ is called *regular* if $\lambda(H) \neq 0$ for all $\lambda \in \Lambda$. Otherwise it is called *singular*. The set of regular elements consists of the complement of finitely many hyperplanes, and its components are called *Weyl chambers*.

Definition of Weyl chamber flow and hyperbolicity

Let G be a semisimple connected real Lie group of the non-compact type, with Lie algebra $\mathfrak{g}$. Let $K \subset G$ be a maximal compact subgroup that gives a

Cartan decomposition $\mathfrak{g} = \mathfrak{k}+\mathfrak{p}$, where $\mathfrak{k}$ is the Lie algebra of K and $\mathfrak{p}$ is the orthogonal complement of $\mathfrak{k}$ with respect to the Killing form of $\mathfrak{g}$. Let $\mathfrak{a} \subset \mathfrak{p}$ be a maximal abelian subalgebra and $A = \exp \mathfrak{a}$ be the corresponding subgroup. Let M be the centralizer of A in K. Suppose Γ is an irreducible torsion-free co-compact lattice in G. Since A commutes with M, the action of A by left translations on G/Γ descends to an A-action on $N \stackrel{\text{def}}{=} M \setminus G/\Gamma$. We call this action the *Weyl chamber flow* of A. Notice that the rank of the acting group is equal to the $\mathbb{R}$-*rank* of the group G. From the dynamical point of view there is a great difference between the cases of rank-one and higher rank. Usually the name Weyl chamber flow is applied only to the higher rank case. The rank-one case corresponds to the geodesic flow on the corresponding locally symmetric space $C\setminus G/\Gamma$, where C is a maximal compact subgroup of G. It is an Anosov flow and does not have any rigidity properties beyond structural stability.

The following result appears in [57].

Proposition 2.3.1 *Any Weyl chamber flow $\alpha : A \times N \to N$,*

$$\alpha(a, \widehat{Mg}) = \alpha_a(\widehat{Mg}) = \widehat{Mag},$$

where $\widehat{Mg}$ is the class of Mg in $M\setminus(G/\Gamma)$, is an Anosov action. If the real rank of G is higher than 2, then the action α is a higher rank hyperbolic $\mathbb{R}^k$-action.

Proof We shall prove that all regular elements of A are Anosov elements for α. It is enough to prove this for the lifted action on $M \setminus G$, which for simplicity we denote by α as well. We need to compute the differential of $\alpha_a : M \setminus G \to M \setminus G$. Since the tangent spaces to $M \setminus G$ can be canonically identified with $\mathfrak{m} \setminus \mathfrak{g}$ using right translations with elements in G, it is enough to compute the differential $d\alpha_a : \mathfrak{m} \setminus \mathfrak{g} \to \mathfrak{m} \setminus \mathfrak{g}$. It is clear that $d\alpha_a = \mathrm{Ad}(a)$, where $\mathrm{Ad}(a)$ is the projection on $\mathfrak{m} \setminus \mathfrak{g}$ of $\mathrm{Ad}(a) : \mathfrak{m} \setminus \mathfrak{g}$.

Let Λ denote the restricted root system of G. Then the Lie algebra $\mathfrak{g}$ of G has the root space decomposition

$$\mathfrak{g} = \mathfrak{m} + \mathfrak{a} + \sum_{\alpha\in\Lambda} \mathfrak{g}^\alpha,$$

where $\mathfrak{g}^\alpha$ is the root space of α and $\mathfrak{m}$ and $\mathfrak{a}$ are the Lie algebras of M and A. Then $\mathfrak{m} \setminus \mathfrak{g} \cong \mathfrak{a} + \sum_{\alpha\in\Lambda} \mathfrak{g}^\alpha$ and any element $\xi \in \mathfrak{m} \setminus \mathfrak{g}$ can be written as $\xi = \xi_0 + \sum_{\lambda\in\Lambda} \xi_\lambda$, where $\xi_0 \in \mathfrak{a}$ and $\xi_\lambda \in \mathfrak{g}^\alpha$.

The identity $\mathrm{Ad}(a) = \exp(\mathrm{ad}(\log a))$ implies that for ξ as above one has

$$\mathrm{Ad}(a)(\xi) = \xi_0 + \sum_{\lambda\in\Lambda} e^{\lambda(\log a)}\xi_\lambda. \tag{2.3.1}$$

Assume now that $a \in A$ is a regular element. The set Λ splits into two subsets,

$$\Lambda^+ = \{\lambda \in \Lambda | \lambda(\log a) > 0\}, \quad \Lambda^- = \{\lambda \in \Lambda | \lambda(\log a) < 0\},$$

and $\mathfrak{m} \setminus \mathfrak{g}$ splits into a direct sum invariant under $d\alpha_a$,

$$\mathfrak{m} \setminus \mathfrak{g} = \mathfrak{a} + \mathfrak{n}^+ + \mathfrak{n}^-,$$

where

$$\mathfrak{n}^+ = \sum_{\alpha \in \Lambda^+} \mathfrak{g}^\alpha, \quad \mathfrak{n}^- = \sum_{\alpha \in \Lambda^-} \mathfrak{g}^\alpha.$$

One has

$$d\alpha_a(\xi) = \xi, \text{ if } \xi \in \mathfrak{a},$$
$$d\alpha_a(\xi) = \sum_{\lambda \in \Lambda^+} e^{\lambda(\log a)} \xi_\lambda, \text{ if } \xi \in \mathfrak{n}^+,$$
$$d\alpha_a(\xi) = \sum_{\lambda \in \Lambda^-} e^{\lambda(\log a)} \xi_\lambda, \text{ if } \xi \in \mathfrak{n}^-.$$

This decomposition, extended to the tangent space $T(M \setminus G)$, gives a hyperbolic decomposition. Indeed, inverting the sign of the Killing form on $\mathfrak{k}$ and the extending the scalar product on $\mathfrak{g}$ by left translations to the $M \setminus G$, one obtains a Riemannian metric on $M \setminus G$. The root spaces are orthogonal in this metric, and, moreover, one has

$$\|d\alpha_a(\xi)\| = \|\xi\|, \text{ for } \xi \in \mathfrak{a},$$
$$\|d\alpha_a(\xi)\| \le e^{-k}\|\xi\|, \text{ for } \xi \in \mathfrak{n}^+,$$
$$\|d\alpha_a(\xi)\| \le e^{k}\|\xi\| \text{ for } \xi \in \mathfrak{n}^-,$$

where $k = \min\{\lambda(\log a) | \lambda \in \Lambda^+\}$ is a positive constant depending only on $\mathfrak{a}$. □

Notice that the Weyl chambers in our sense are the same as those in the classical theory of simple Lie groups.

Remark 2.3.2 One needs Γ to be torsion-free only to assure that N is a manifold. Treating the orbifold case can be done in a similar way.

If the group G is $\mathbb{R}$-*split*, i.e., its real rank equals its complex rank, then $M = \{\text{Id}\}$. In this case the Weyl chamber flow acts on G/Γ.

In the non-split case the action of A on the whole group G is a compact group extension of the Weyl chamber flow and hence is partially hyperbolic with the zero Lyapunov exponent of extra multiplicity $\dim M$.

2.3.4 Examples of Weyl chamber flows

Weyl chamber flow on $SL(n, \mathbb{R})$

The Lie algebra $\mathfrak{sl}(n, \mathbb{R})$ of $SL(n, \mathbb{R})$ can be identified with the set of $n \times n$ matrices of trace zero. The subgroup $D_n^+ \subset SL(n, \mathbb{R})$ of matrices with positive diagonals is the connected component of the identity in a maximal Cartan subgroup H of $SL(n, \mathbb{R})$. The diagonal entries of $d \in D_n^+$ can be written as exponentials e^{t_i}, $i = 1, \dots, n$, where $t_1 + \dots + t_n = 0$. Thus it is convenient to parameterize elements A in the Lie algebra $\mathfrak{h}$ of D_n^+ by coordinates $t_1, \dots, t_n$ satisfying the relation $t_1 + \dots + t_n = 0$. The dimension of $\mathfrak{h}$ is $n - 1$.

Let $\Gamma \subset \mathrm{SL}(n, \mathbb{R})$ be a co-compact lattice. We describe the invariant foliations for the Weyl chamber flow. Note that the derivative of the right multiplication by elements from H on $\Gamma \setminus M$ coincides with the inverse on H compose to the adjoint representation. To find the Lyapunov exponents it is enough to find the eigenvalues of this map. The invariant foliations can be obtained as before by taking the exponential of the subspaces generated by eigenvectors.

The following basis in $\mathfrak{sl}(n, \mathbb{R})$ consists of eigenvectors for $\mathrm{ad}(A)$, $A = \mathrm{diag}(t_1, \dots, t_n)$:

$$N_i = v_{i+1,i+1} - v_{i,i}, 1 \le i \le n - 1,$$
$$C_{i,j} = v_{i,j}, 1 \le i, j \le n, i \ne j.$$

A direct computation shows that

$$\mathrm{ad}(A)(N_i) = AN_i - N_i A = 0,$$
$$\mathrm{ad}(A)(C_{i,j}) = AC_{i,j} - C_{i,j} A$$
$$= \mathrm{diag}(t_1, \dots, t_n) v_{i,j} - v_{i,j} \mathrm{diag}(t_1, \dots, t_n) = (t_i - t_j) C_{i,j}.$$

Thus we can give the following description for the Lyapunov exponents.

Proposition 2.3.3 *The non-zero Lyapunov exponents for an element $a = \mathrm{diag}(e^{t_1}, \dots, e^{t_n})$ of the Weyl chamber flow on $SL(n, \mathbb{R})/\Gamma$ are $t_i - t_j$, where $i \ne j$ and $1 \le i, j \le n$. The zero Lyapunov exponent comes only from the orbit foliation and hence has multiplicity $n - 1$. Consequently any matrix $d \in D_n^+$ whose elements are pairwise different acts normally hyperbolically on $SL(n, \mathbb{R})/\Gamma$ and hence is regular.*

For every $i \ne j$ the equation $t_i = t_j$ defines a Lyapunov hyperplane $H_{i,j} \subset D_n^+$. The connected components of

$$D_n^+ \setminus \bigcup_{i \ne j} H_{i,j}$$

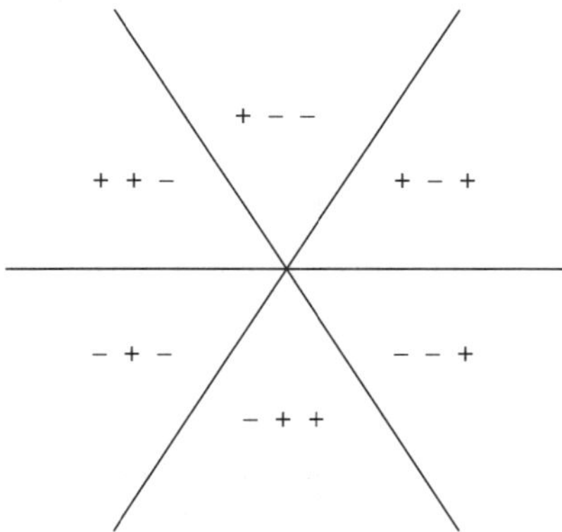

Figure 2.1 Weyl chambers for $SL(3, \mathbb{R})$.

are the Weyl chambers of the flow α. We recall that any element belonging to a Weyl chamber is regular element. The picture of the Weyl chambers for $n = 3$ is shown in Figure 2.1. The signs that appear in each chamber are the signs of half of the Lyapunov exponents of a regular element from the chamber with respect to a certain fixed basis. For this action, the Lyapunov exponents appear in pairs of opposite signs.

One should note that the Weyl chamber flows on certain factors of $SL(n, \mathbb{R})$, especially on $SL(n, \mathbb{Z})/SL(n, \mathbb{R})$, which is identified with the space of all unimodular lattices in $\mathbb{R}^n$, appear in several problems in number theory. The former space, however, is not compact.

Notice the highly resonant character of Lyapunov exponents, in contrast to the case of Cartan actions on the torus. Out of $n(n-1)$ exponents there are only $n-1$ independent ones, say $t_i - t_{i+1},\ i = 1, \ldots, n-1$. The most important resonances geometrically are those which bring pairs of exponents differing by sign. Thus every Lyapunov hyperplane is the kernel of two exponents. This is in fact a general feature of all Weyl chamber flows.

Weyl chamber flow on Sp$(n, \mathbb{R})$

The symplectic group $\mathrm{Sp}(n, \mathbb{R})$ is the group of matrices that leaves invariant the exterior form:

$$x_1 \wedge x_{n+1} + x_2 \wedge x_{n+2} + \cdots + x_n \wedge x_{2n}.$$

Equivalently, it is the set of $2n \times 2n$ matrices g with real entries that satisfy $g^t J_n g = J_n$, where $J_n = \begin{pmatrix} 0 & I_n \\ -I_n & 0 \end{pmatrix}$ and $I_n \in \mathrm{GL}(n, \mathbb{R})$ is the identity.

$\mathrm{Sp}(n,\mathbb{R})$ has a natural embedding in $SL(n,\mathbb{R})$ and its Lie algebra $\mathfrak{sp}(n,\mathbb{R})$ can be identified with a Lie subalgebra of $\mathfrak{sl}(n,\mathbb{R})$:

$$\mathfrak{sp}(n,\mathbb{R}) = \left\{ \begin{pmatrix} X_1 & X_2 \\ X_3 & -X_1^t \end{pmatrix} \right\},$$

where X_1, X_2, X_3 are $n \times n$ matrices with real entries, and in addition X_2, X_3 are symmetric matrices, that is, $X_2^t = X_2$, $X_3^t = X_3$, where A^t is the transpose of the matrix A. The dimension of $\mathrm{Sp}(n,\mathbb{R})$ is $\frac{3n^2+n}{2}$.

Let H be the maximal $\mathbb{R}$-split torus in $\mathrm{Sp}(n,\mathbb{R})$ that is the exponential of the n dimensional Cartan subalgebra $\mathfrak{h}$ which has a basis consisting of the elements

$$v_{1,1} - v_{n+1,n+1}, v_{2,2} - v_{n+2,n+2}, \dots, v_{n,n} - v_{2n,2n}.$$

The elements $A \in \mathfrak{h}$ and $\exp(A) \in H$ are diagonal matrices

$$\begin{aligned} A &= \mathrm{diag}(t_1, t_2, \dots, t_n, -t_1, \dots, -t_n), \\ \exp(A) &= \mathrm{diag}(e^{t_1}, \dots, e^{t_n}, e^{-t_1}, \dots, e^{-t_n}), \end{aligned} \tag{2.3.2}$$

with $t_1, \dots, t_n$ real numbers.

Let $\Gamma \subset \mathrm{Sp}(n,\mathbb{R})$ be a co-compact lattice. We describe the invariant foliations for the Weyl chamber flow. We use the same procedure as before. The following basis in $\mathfrak{sp}(n,\mathbb{R})$ consists of eigenvectors for $\mathrm{ad}(A)$, $a \in \mathfrak{h}$ given by formula (2.3.2):

$$\begin{aligned} N_i &= v_{i,i} - v_{n+i,n+i}, \ 1 \le i \le n, \\ C_{i,j}^1 &= v_{i,j} - v_{n+j,n+i}, \ 1 \le i, j \le n, i \ne j, \\ C_i^2 &= v_{i,n+i}, \ 1 \le i \le n, \\ C_{i,j}^3 &= v_{i,n+j} + v_{j,n+i}, \ 1 \le i, j \le n, i > j, \\ C_i^4 &= v_{i+n,i}, \ 1 \le i \le n, \\ C_{i,j}^5 &= v_{i,j+n} + v_{j,i+n}, \ 1 \le i, j \le n, i > j. \end{aligned}$$

An easy computation shows that

$$\begin{aligned} \mathrm{ad}(A)(N_i) &= 0, \\ \mathrm{ad}(A)(C_{i,j}^1) &= (t_i - t_j)C_{i,j}^1, \\ \mathrm{ad}(A)(C_i^2) &= (2t_i)C_i^2, \\ \mathrm{ad}(A)(C_{i,j}^3) &= (t_i + t_j)C_{i,j}^3, \\ \mathrm{ad}(A)(C_i^4) &= (-2t_i)C_i^4, \\ \mathrm{ad}(A)(C_{i,j}^5) &= (-t_i - t_j)C_{i,j}^5. \end{aligned}$$

Thus we can give the following description for the Lyapunov exponents.

Proposition 2.3.4 *The non-zero Lyapunov exponents for a transformation $a = \operatorname{diag}(e^{t_1}, \ldots, e^{t_n}, e^{-t_1}, \ldots, e^{-t_n}) \in H$ of the Weyl chamber flow on $\Gamma \setminus Sp(n, \mathbb{R})$ are $t_i - t_j, t_j - t_i, 2t_i, -2t_i, t_i + t_j, -t_i - t_j$, where $i > j$ and $1 \le i, j \le n$. The zero Lyapunov exponent comes only from the orbit foliation and hence has multiplicity n. Consequently any matrix $a \in H$ whose elements are pairwise different in absolute value and different from 1 acts normally hyperbolically on $\Gamma \setminus Sp(n, \mathbb{R})$ and hence is regular.*

Weyl chamber flow on $\mathbf{SO}(n, n, \mathbb{R})^\circ$

The orthogonal group $\mathrm{SO}(n, n, \mathbb{R})$ has a natural embedding in $\mathrm{SL}(2n, \mathbb{R})$ as the group of matrices that keep invariant the quadratic form

$$-x_1^2 - \cdots - x_n^2 + x_{n+1}^2 + \cdots + x_{2n}^2.$$

Equivalently, it is the set of $2n \times 2n$ matrices g with real entries that satisfy $g^t I_{n,n} g = I_{n,n}$, where $I_{n,n} = \begin{pmatrix} -I_n & 0 \\ 0 & I_n \end{pmatrix}$ and $I_n \in \mathrm{GL}(n, \mathbb{R})$ is the identity. Its Lie algebra is denoted $\mathfrak{so}(n, n, \mathbb{R})$ and can be viewed as a subalgebra of $\mathfrak{sl}(2n, \mathbb{R})$:

$$\mathfrak{so}(n, n, \mathbb{R}) = \left\{ \begin{pmatrix} X_1 & X_2 \\ X_2^t & X_3 \end{pmatrix} \right\},$$

where X_1, X_2, X_3 are $n \times n$ matrices with real entries, and in addition X_1, X_3 are skew symmetric, that is, $X_1^t = -X_1, X_3^t = -X_3$, where A^t is the transpose of the matrix A. The dimension of $\mathfrak{so}(n, n, \mathbb{R})$ is $\frac{n^2-n}{2}$.

Let H be the maximal $\mathbb{R}$-split torus in $\mathrm{SO}(n, n, \mathbb{R})^\circ$ that is the exponential of the n dimensional Cartan subalgebra $\mathfrak{h}$ which has a basis consisting of the elements

$$v_{1,n+1} + v_{n+1,1}, v_{2,n+2} + v_{n+2,2}, \ldots, v_{n,2n} + v_{2n,n}.$$

Let $\Gamma \subset \mathrm{SO}(n, n, \mathbb{R})^\circ$ be a co-compact lattice. We describe the invariant foliations for the Weyl chamber flow. We again use the same procedure as before.

Consider now the following vectors in $\mathfrak{so}(n, n, \mathbb{R})$:

$$C_{i,j}^1 = v_{i,j} - v_{j,i}, 1 \le i, j \le n, i < j,$$
$$C_{i,j}^2 = v_{i+n,j} - v_{j+n,i}, 1 \le i, j \le n, i < j,$$
$$C_{i,j}^3 = v_{j+n,i} + v_{i,j+n}, 1 \le i, j \le n, i < j,$$
$$C_{i,j}^4 = v_{i+n,j} + v_{j,i+n}, 1 \le i, j \le n, i < j.$$

The following basis in $\mathfrak{so}(n, n, \mathbb{R})$ consists of eigenvectors for $\mathrm{ad}(A)$, $A \in \mathfrak{h}$, $A = \sum_{i=1}^{n} t_i N_i$:

$$\begin{aligned} N_i &= v_{i,n+i} + v_{n+i,i},\ 1 \le i \le n, \\ D^1_{i,j} &= C^1_{i,j} + C^2_{i,j} + C^3_{i,j} + C^4_{i,j}, \\ D^2_{i,j} &= C^1_{i,j} - C^2_{i,j} - C^3_{i,j} + C^4_{i,j}, \\ D^3_{i,j} &= C^1_{i,j} + C^2_{i,j} - C^3_{i,j} - C^4_{i,j}, \\ D^4_{i,j} &= C^1_{i,j} - C^2_{i,j} + C^3_{i,j} - C^4_{i,j}. \end{aligned}$$

An easy computation shows that

$$\begin{aligned} \mathrm{ad}(A)N_i &= 0, \\ \mathrm{ad}(A)D^1_{i,j} &= (t_i - t_j)D^1_{i,j}, \\ \mathrm{ad}(A)D^2_{i,j} &= (t_i + t_j)D^2_{i,j}, \\ \mathrm{ad}(A)D^3_{i,j} &= (-t_i + t_j)D^3_{i,j}, \\ \mathrm{ad}(A)D^4_{i,j} &= (-t_i - t_j)D^4_{i,j}. \end{aligned}$$

We can give now the following description for the Lyapunov exponents.

Proposition 2.3.5 *The non-zero Lyapunov exponents for a transformation* $a = \exp(A)$, $A \in H$, $A = \sum_{i=1}^{n} t_i N_i$ *of the Weyl chamber flow on* $\Gamma \setminus SO(n, n, \mathbb{R})^\circ$ *are* $t_i - t_j, t_j - t_i, t_i + t_j, -t_i - t_j$, *where* $i \ne j$ *and* $1 \le i, j \le n$. *The zero Lyapunov exponent comes only from the orbit foliation and hence has multiplicity* n. *Consequently any matrix* $a \in H$ *whose eigenvalues are pairwise different in absolute value and different from* 1 *acts normally hyperbolically on* $\Gamma \setminus SO(n, n, \mathbb{R})^\circ$ *and hence is regular.*

Weyl chamber flow on $SU(m, n)$

Let $m \ge n$ positive integers. The unitary group $\mathrm{SU}(m, n)$ has a natural embedding in $\mathrm{SL}(m+n, \mathbb{C})$ as the group of matrices of determinant 1 that keep invariant the Hermitian form H:

$$-x_1\bar{x}_1 - \cdots - x_m\bar{x}_m + x_{m+1}\bar{x}_{m+1} + \cdots + x_{m+n}\bar{x}_{m+n}.$$

Equivalently, it is the set of $(m+n) \times (m+n)$ matrices g with complex entries that satisfy $g^t I_{m,n}\bar{g} = I_{m,n}$, where $I_{m,n} = \begin{pmatrix} -I_m & 0 \\ 0 & I_n \end{pmatrix}$ and $I_n \in \mathrm{GL}(n, \mathbb{R})$ is the identity.

In what follows we denote by $\bar{A}^t$ the complex conjugate transpose of the matrix A. The Lie algebra $\mathfrak{su}(m, n)$ of $\mathrm{SU}(m, n)$ can be expressed as the set of $2(m+n) \times 2(m+n)$ matrices:

$$\begin{pmatrix} X_1 & X_2 \\ \bar{X}_2^t & X_3 \end{pmatrix},$$

where X_1, X_2, X_3, X_4 are complex matrices, X_2 is arbitrary of order $m \times n$, X_1 of order $m \times m$, and $\bar{X}_1^t = X_1$, X_3 of order $n \times m$ and $\bar{X}_3^t = X_3$, and in addition $\mathrm{Tr}(X_1) + \mathrm{Tr}(X_3) = 0$.

Alternatively, and this is the approach we use in the sequel, one can choose a base $(e_i)_{i=1}^{m+n}$ in $\mathbb{C}^{m+n}$ (under a linear transformation that has real coefficients) for which

$$\begin{aligned} &H(e_i, e_{i+n}) = 1, && 1 \le i \le n, \\ &H(e_j, e_j) = 1, && 2n+1 \le j \le m+n, \\ &H(e_i, e_j) = 0, && \text{otherwise.} \end{aligned}$$

Using this base, the Lie algebra $\mathfrak{su}(m, n)$ can be expressed as the set of $(m+n) \times (m+n)$ matrices

$$\begin{pmatrix} X_1 & X_2 & Y_1 \\ X_3 & X_4 & Y_2 \\ Z_1 & Z_2 & W \end{pmatrix},$$

where all entries are complex matrices, X_1, X_2, X_3, X_4 are $n \times n$ matrices, Y_1, Y_2 are $n \times (m-n)$ matrices, Z_1, Z_2 are $(m-n) \times n$ matrices, and W is a $(m-n) \times (m-n)$ matrix. In addition

$$\begin{aligned} &X_1 = -\bar{X}_4^t, \quad \bar{X}_2^t = -X_2, \quad \bar{X}_3^t = -X_3, \\ &\bar{W}^t = -W, \quad Y_1 = -\bar{Z}_2^t, \quad Y_2 = -\bar{Z}_1^t, \\ &\mathrm{Tr}(X_1) + \mathrm{Tr}(X_4) + \mathrm{Tr}(W) = 0. \end{aligned}$$

The $\mathbb{R}$-dimension of $\mathfrak{su}(m, n)$ is $(m+n)^2 - 1$.

We denote by $v_{i,j}$ the $(m+n) \times (m+n)$ matrix that has the entry (i, j) equal to 1 and all other entries equal to zero. Let H be the maximal $\mathbb{R}$-split torus in $\mathrm{SU}(m, n)$ that is the exponential of the n dimensional Cartan subalgebra $\mathfrak{h}$ which has a basis consisting of the elements

$$N_i = v_{i,i} - v_{n+i,n+i}, \; 1 \le i \le n. \tag{2.3.3}$$

Let $\Gamma \subset \mathrm{SU}(m, n)$ be a co-compact lattice. We describe now the foliations for the Weyl chamber flow induced by H on $\Gamma \setminus \mathrm{SU}(m, n)$. We use the same approach as for the other Weyl chamber flows.

The following elements in $su(m,n)$ are eigenvectors for $\mathrm{ad}(A)$, $A \in \mathfrak{h}$, $A = \sum_{i=1}^{n} t_i N_i$:

$$\begin{aligned}
A^1_{i,j} &= v_{i,j+n} - v_{j,i+n}, 1 \le i, j \le n, i < j,\\
A^2_{i,j} &= i(v_{i,j+n} + v_{j,i+n}), 1 \le i, j \le n, i < j,\\
B^1_{i,j} &= v_{i,j} - v_{j+n,i+n}, 1 \le i, j \le n, i \ne j,\\
B^2_{i,j} &= i(v_{i,j} + v_{j+n,i+n}), 1 \le i, j \le n, i \ne j,\\
C^1_{i,j} &= v_{j+n,i} - v_{i+n,j}, 1 \le i, j \le n, i < j,\\
C^2_{i,j} &= i(v_{j+n,i} + v_{i+n,j}), 1 \le i, j \le n, i < j,\\
D^1_{i,l} &= v_{i,2n+l} - v_{2n+l,i+n}, 1 \le i \le n, l \le m-n,\\
D^2_{i,l} &= i(v_{i,2n+l} + v_{2n+l,i+n}), 1 \le i \le n, l \le m-n,\\
E^1_{i,l} &= v_{i+n,2n+l} - v_{2n+l,i+n}, 1 \le i \le n, l \le m-n,\\
E^2_{i,l} &= i(v_{i+n,2n+l} + v_{2n+l,i+n}), 1 \le i \le n, l \le m-n,\\
F_i &= i v_{i+n,i}, 1 \le i \le n,\\
G_i &= i v_{i,i+n}, 1 \le i \le n.
\end{aligned} \tag{2.3.4}$$

One can complete the set of vectors given by (2.3.3) and (2.3.4) to a base for $\mathfrak{su}(m,n)$ by adding vectors in a base for the compact part of the centralizer H.

An easy computation shows that

$$\begin{aligned}
\mathrm{ad}(A)N_i &= 0,\\
\mathrm{ad}(A)A^1_{i,j} &= (t_i + t_j)A^1_{i,j},\\
\mathrm{ad}(A)A^2_{i,j} &= (t_i + t_j)A^2_{i,j},\\
\mathrm{ad}(A)B^1_{i,j} &= (t_i - t_j)B^1_{i,j},\\
\mathrm{ad}(A)B^2_{i,j} &= (t_i - t_j)B^2_{i,j},\\
\mathrm{ad}(A)C^1_{i,j} &= (-t_i - t_j)C^1_{i,j},\\
\mathrm{ad}(A)C^2_{i,j} &= (-t_i - t_j)C^2_{i,j},\\
\mathrm{ad}(A)D^1_{i,l} &= t_i D^1_{i,l},\\
\mathrm{ad}(A)D^2_{i,l} &= t_i D^2_{i,l},\\
\mathrm{ad}(A)E^1_{i,l} &= -t_i E^1_{i,l},\\
\mathrm{ad}(A)E^2_{i,l} &= -t_i E^2_{i,l},\\
\mathrm{ad}(A)F_i &= -2t_i F_i,\\
\mathrm{ad}(A)G_i &= 2t_i G_i.
\end{aligned}$$

We give the description for the Lyapunov exponents of the Weyl chamber flow.

Proposition 2.3.6 *The non-zero Lyapunov exponents for a transformation $a = \exp(A)$, $A \in H$, $A = \sum_{i=1}^{n} t_i N_i$ of the Weyl chamber flow on $\Gamma \setminus SU(m,n)$ are $t_i + t_j, t_i - t_j, t_j - t_i, -t_i - t_j, t_i, -t_i$, all appearing with multiplicity 2, and $2t_i, -2t_i$, appearing with multiplicity 1, where $i \neq j$ and $1 \leq i, j \leq n$. If $m = n$ the exponents $t_i, -t_i$ do not appear. The zero Lyapunov exponent comes from the orbit foliation and from the compact part of the centralizer of the maximal $\mathbb{R}$-split Cartan subgroup, and has multiplicity $2n - 1 + (m-n)^2$. Consequently, any matrix $a \in H$ whose eigenvalues are pairwise different in absolute value and different from 1 acts only partially hyperbolically on $\Gamma \setminus SU(m,n)$.*

2.3.5 Twisted Weyl chamber flows and further extensions

Now we describe another class of algebraic Anosov actions of $\mathbb{R}^n$, which is obtained from the Weyl chamber flows by a very special extension procedure. This example appears in Example 2.7 in [78].

Let G, Γ, A, and M be as in 'Definition of Weyl Chamber flow and hyperbolicity', on p. 79. Let $\rho : \Gamma \to SL(n, \mathbb{Z})$ be a representation of Γ which is irreducible over $\mathbb{Q}$. Then Γ acts on the n-torus $\mathbb{T}^n$ via ρ and hence on the product space $(M \setminus G) \times \mathbb{T}^n$ via

$$\gamma(x,t) = (x\gamma^{-1}, \rho(\gamma)(t)).$$

Let $N = (M \setminus G \times \mathbb{T}^n)/\Gamma$ be the orbit space of this action. As the action of A on $M \setminus G \times \mathbb{T}^n$ given by $a(x,t) = (ax, t)$ commutes with the Γ-action, it induces an action of A on N.

Assume that $\rho(\gamma)$ for some element $\gamma \in \Gamma$ is an Anosov diffeomorphism on $\mathbb{T}^n$. The image under ρ of the center of Γ is finite by the Schur lemma. Hence Γ may be assumed to be a lattice in a semisimple Lie group with finite center. By Margulis' superrigidity theory [110], semisimplicity of the algebraic hull H of $\rho(\Gamma)$, and the existence of a hyperbolic element $\rho(\gamma)$, which guarantee that the image is not compact, the representation ρ of Γ extends to a homomorphism $G \to H_{\text{ad}}$, where H_{ad} is the adjoint group of H. Assume now that $\rho(\Gamma)$ has a trivial center (otherwise one can consider the case of an orbifold). Since Γ is co-compact, γ is a semisimple element of G. Let $\gamma = k_\gamma s_\gamma$ be the decomposition of γ into compact and split semisimple parts. Then s_γ is conjugate to an element $a \in A$. As ρ extends to G, it follows that $\rho(s_\gamma)$ and $\rho(a)$ have no eigenvalues of absolute value 1. Moreover, one can pick a such that

$\log a$ belongs to an open Weyl chamber of the Lie algebra $\mathfrak{g}$. Then it follows from Theorem 2.3.1 that a acts normally hyperbolic on $M \setminus G/\Gamma$.

One shows now that the action of a on N is hyperbolic as well. Let $(x, t) \in N$. Since Γ is co-compact, there is a uniformly bounded sequence of elements $u_n \in G$ such that $x^{-1}a^n x = u_n(x)\gamma_n(x)$, for some $\gamma_n(x) \in \Gamma$. Since $u_n(x)$ is uniformly bounded in x and n, the stable tangent vectors for $x^{-1}ax$ are exponentially contracted by $\gamma(x_n)$ with estimates uniform in x. The same conclusion applies to unstable vectors. Thus

$$a^n(x, t) = (x(x^{-1}a^n x), t) = (xu_n, \rho(\gamma_n)t),$$

and since a acts normally hyperbolic on $M \setminus G/\Gamma$, it follows that a is normally hyperbolic with respect to the orbit foliation of A.

The above construction can be generalized by considering extensions by automorphisms of a torus of other higher rank actions for which one of the monodromy elements is Anosov. For example, using a twisted Weyl chamber flow as above as the base we obtain nilmanifold extensions of the Weyl chamber flow. Starkov [78] pointed out that one can also start with the product of a Weyl chamber flow with a transitive action of some $\mathbb{R}^l$ on a torus and produce a toral extension which is Anosov and no finite cover splits as a product. These two extension constructions can be combined and iterated.

2.3.6 Reduction of scalars

For completeness we give a brief presentation of the available methods of constructing lattices in a semisimple Lie group G. More details can be found in [110, 178, 179]. In particular, this discussion will allow us to show an explicit example of a twisted Weyl chamber flow.

If G is a semisimple Lie group, and, in addition, a matrix group, the simplest way to construct a lattice is to take the integer points in G. For example take $SL(n, \mathbb{Z})$ in $SL(n, \mathbb{R})$. This construction always gives non-compact lattices because of the existence of unipotent elements in the lattice. To obtain co-compact lattices, one can use a standard construction called in the literature "restriction of scalars" [176].

Recall that an algebraic variety over a closed field K is the set of zeroes in K^l of a finite family of polynomials in $K[x_1, \ldots, x_l]$. If V is an algebraic variety in K^l, and A is a subring of K, then we denote by V_A the set of A-points in V, that is, $V \cap A^l$. If $k \subset K$ is a subfield, an algebraic variety V over K is called k-variety if the polynomials defining the variety have all coefficients in k. Suppose now that k is an algebraic number field, that is, $\mathbb{Q} \subset k \subset \mathbb{C}$ and $d = [k : \mathbb{Q}] < \infty$, and G is an algebraic k-group. Then there are d distinct field embeddings σ_1 = identity, $\sigma_2, \ldots, \sigma_d$ of k into $\mathbb{C}$ which are linearly independent in the $\mathbb{C}$-vector space of all functions $k \to \mathbb{C}$, and such that, if $\alpha \in k$, then $\sigma_i(\alpha) = \alpha$ for all i if and only if $\alpha \in \mathbb{Q}$. Any embedding $\sigma : k \to \mathbb{C}$ can be extended to an automorphism of $\mathbb{C}$, and, moreover, defines a $\mathbb{C}$-isomorphism from $\mathbb{C}^l$ into itself. We denote by G^σ the image of G under σ, which is an algebraic $\sigma(k)$-group.

For $k, \sigma_1, \ldots, \sigma_d$ as above, let $R_{k/\mathbb{Q}}(G) = \prod_{i=1}^d G^{\sigma_i}$, and for $g \in G_k$ let $g' = (\sigma_1(g), \ldots, \sigma_d(g))$. Let $(G_k)' = \{g' | g \in G_k\}$. Then $R_{k/\mathbb{Q}}(G)$ is isomorphic to an algebraic $\mathbb{Q}$-group such that $(R_{k/\mathbb{Q}}(G))_{\mathbb{Q}} = (G_k)'$ and $(R_{k/\mathbb{Q}}(G))_{\mathbb{Z}} = (G_{\mathcal{O}})'$, where $\mathcal{O} \subset k$ is the subring of algebraic integers in k. The projection map $R_{k/\mathbb{Q}}(G) \to G$ onto the first factor is defined over k and defines bijections $(R_{k/\mathbb{Q}}(G))_{\mathbb{Q}} \to G_k$ and $(R_{k/\mathbb{Q}}(G))_{\mathbb{Z}} \to G_{\mathcal{O}}$. In particular, $G_{\mathcal{O}}$ is isomorphic to a lattice in $(R_{k/\mathbb{Q}}(G))_{\mathbb{R}}$.

We illustrate the restrictions of scalars by the following example.

Example 2.3.7 Let $G = SO(4, 2)$, which is viewed as embedded naturally in $SL(6, \mathbb{R})$ as the group of transformations that preserves the bilinear form

$$< x, y >= -\sqrt{2}(x_1 y_1 + x_2 y_2 + x_3 y_3 + x_4 y_4) + x_5 y_5 + x_6 y_6.$$

This form and therefore G are defined over the field $\mathbb{Q}(\sqrt{2})$, which has a conjugation σ defined by $\sigma(\sqrt{2}) = -\sqrt{2}$. Define $\Gamma = G(\mathbb{Z}[\sqrt{2}])$. Then there exists an embedding of Γ into $H = SO(4, 2) \times SO(6)$ given by $\gamma \in \Gamma$ to $(\gamma, \sigma(\gamma))$. It is easy to check that this embedding is discrete. Moreover, Γ is embedded as integral points for the rational structure on H with rational points $(m, \sigma(m))$, where $m \in G(\mathbb{Q}(\sqrt{2}))$. This implies that Γ is a lattice in H. Moreover, Γ projects to a lattice in G, since G is co-compact in H, and it follows from the general theory in [110] that this lattice is actually co-compact in H.

Observe now that $SO(4, 2)$ has real rank 2 and two connected components. Choose the maximal abelian Lie subalgebra $\mathfrak{h} \subset so(4, 2)$ generated by

$$v_{1,6} + v_{6,1}, v_{2,5} + v_{5,2},$$

with the corresponding abelian subgroup $H = \exp(\mathfrak{h})$, and let $M = SO(2)$ be embedded in $SO(4, 2)$ as

$$\begin{pmatrix} a_{11} & a_{12} \\ a_{21} & a_{22} \end{pmatrix} \to \begin{pmatrix} 1 & 0 & 0 & 0 & 0 & 0 \\ 0 & 1 & 0 & 0 & 0 & 0 \\ 0 & 0 & a_{11} & a_{12} & 0 & 0 \\ 0 & 0 & a_{21} & a_{22} & 0 & 0 \\ 0 & 0 & 0 & 0 & 1 & 0 \\ 0 & 0 & 0 & 0 & 0 & 1 \end{pmatrix}.$$

Then the action of H on $(M \setminus SO(4,2)^o)/\Gamma$ induces a Weyl chamber flow. To show an example of twisted Weyl chamber flow, observe that $\Gamma \subset SL(6, \mathbb{Z}[\sqrt{2}]$ has an embedding in $SL(12, \mathbb{Z})$ if we identify $\mathbb{Z}^{12}$ with the image in $\mathbb{R}^{12}$ of $\mathbb{Z}[\sqrt{2}]^6$ via the embedding $x \to (x, \sigma(x))$. Then the action of H on the manifold $N = (M \setminus SO(4,2)^o \times \mathbb{T}^{12})/\Gamma$ induces a twisted Weyl chamber flow.

2.4 Affine actions beyond tori and nilmanifolds

2.4.1 Non-invertible examples on tori and nilmanifolds and relations to solenoids

Let α be a $\mathbb{Z}_+^k$-action by endomorphisms of a torus $\mathbb{T}^m$ defined by matrices $A_1, \ldots, A_k$. The natural extension of α can be identified with a $\mathbb{Z}^k$-action $\alpha^* : \mathbb{Z}^k \to \operatorname{Aut}(S)$ by automorphisms of a *solenoid* S. The solenoid is a compact abelian group modeled locally on the product of an Euclidean space (the so called Archimedean directions) and several additive groups of p-adic integers (the so called non-Archimedean directions). The solenoid can be realized as a subset of $(\mathbb{T}^m)^{\mathbb{Z}^k}$ as follows. Let σ_i be the ith shift on $\mathbb{Z}^k$, that is, $\sigma_i(a_1, \ldots, a_i, \ldots, a_k) = (a_1, \ldots, a_i + 1, \ldots, a_k)$, and define

$$S = \{\omega \in (\mathbb{T}^m)^{\mathbb{Z}^k} | \omega_{\sigma_i j} = A_i \omega_j\}.$$

The solenoid is a compact subgroup of $(\mathbb{T}^m)^{\mathbb{Z}^k}$ considered with the product topology. Its dual is a subgroup of $\mathbb{Q}^m$ and is included in $(\mathbb{Z}(p_1, \ldots, p_l))^m$, where $p_1, \ldots, p_l$ are the prime integers that appear in the prime decomposition of the determinants of at least one of the $A_1, \ldots, A_k$ and $\mathbb{Z}(p_1, \ldots, p_l)$ is the set of rational numbers having denominators with prime divisors only $p_1, \ldots, p_l$. The group $\mathbb{Z}^k$ acts on S by coordinates shifts. The solenoid is a fibration over $\mathbb{T}^m$ with the projection given by $S \ni \omega \to \omega(0, \ldots, 0) \in \mathbb{T}^m$. The fibers are Cantor sets. The projection intertwines the restriction of α^* to $\mathbb{Z}_+^k$ and the action α. On the solenoid there exists a Hölder structure that can be introduced using any metric on the product space of the form

$$d_\lambda(\omega, \omega') = \sum_{j \in \mathbb{Z}^k} \frac{d(\omega_j, \omega'_j)}{\lambda^{\|j\|}},$$

where $\lambda > 1$ and d is a metric on the torus. The Hölder structure is independent of the constant λ. The structure can be used to define exponential convergence along the fibers and hence Lyapunov exponents. Note that the Lyapunov exponents split into Archimedean and non-Archimedean. The Weyl chamber analysis extends to this case although the space of the action is no longer a manifold.

One of the simplest and most famous example is Furstenberg's $\times 2$, $\times 3$ action on a circle [40].

Example 2.4.1 The action $E_{2,3}$ of $\mathbb{Z}_+^2$ on the circle is generated by the endomorphisms

$$E_2 : S^1 \to S^1, x \mapsto 2x \pmod 1,$$

and

$$E_2 : S^1 \to S^1, x \mapsto 3x \pmod 1.$$

Closed orbits of this actions are those of rational numbers whose denominators are relatively prime with 2 and 3. Thus, in particular the orbit of $\frac{1}{2^n-1}$ is closed if $2^n \equiv 2 \pmod 3$. Its E_2 orbit has period $n-1$ and consists of the points $\frac{2^k}{2^n-1}$, $k = 1, \dots, n-1$, and hence for large n is concentrated mostly around 0, while the E_3 orbit typically is almost uniformly distributed.

The natural extension $S_{2,3}$ of $E_{2,3}$ acts on the dual group of the discrete group $\mathbb{Z}[\frac{1}{2}, \frac{1}{3}]$. Topologically it is a connected but not locally connected one-dimensional compact space locally modeled on the direct product of $\mathbb{R}$ and the Cantor set. As a group it is an extension of S^1 with the product of dyadic integers and 3-adic integers $\mathbb{Z}_2 \times \mathbb{Z}_3$ in the fiber.

One can identify the discrete time $\mathbb{Z}^2$ with the integer lattice in the plane $\mathbb{R}^2$ with coordinates s, t. There are three Lyapunov exponents for $S_{2,3}$: one Archimedean:

$$t \log 2 + s \log 3,$$

and two non-Archimedean:

$$-t \log 2 \text{ and } -s \log 3.$$

This can be seen from the observation that multiplication by two acts as an isometry on $\mathbb{Z}_3$ and as a contraction with constant coefficient of contraction 1/2 on $\mathbb{Z}_2$, and, correspondingly, the multiplication by three acts as an isometry on $\mathbb{Z}_2$ and as a contraction with coefficient 1/3 on $\mathbb{Z}_3$. Thus, in this example there are three Lyapunov lines in general position:

$$\begin{aligned} t \log 2 + s \log 3 &= 0, \\ t &= 0, \\ s &= 0, \end{aligned} \tag{2.4.1}$$

and six Weyl chambers. Combinatorially, the picture looks exactly as for any Cartan action of $\mathbb{Z}^2$ on $\mathbb{T}^3$. The positive quadrant constitutes a Weyl chamber, namely the one where the Archimedean exponent is positive and the other two non-Archimedean exponents are negative.

Similar examples can by constructed on nilmanifolds. The simplest one is three dimensional.

Example 2.4.2 Let H be the Heisenberg group of 3×3 upper diagonal matrices, that is,

$$N = \begin{pmatrix} 1 & x & y \\ 0 & 1 & z \\ 0 & 0 & 1 \end{pmatrix}, x, y, z \in \mathbb{R}.$$

Let $\Lambda \subset H$ be the subgroup of integer matrices, $\rho_2 : H \to H$ be the automorphism

$$\begin{pmatrix} 1 & x & y \\ 0 & 1 & z \\ 0 & 0 & 1 \end{pmatrix} \to \begin{pmatrix} 1 & 2x & 4y \\ 0 & 1 & 2z \\ 0 & 0 & 1 \end{pmatrix}.$$

Then $\rho_2(\Lambda) \subset \Lambda$ and ρ_2 projects to a non-invertible map on the compact nilmanifold $X = H/\Lambda$ [67, Section 17.3]. One can similarly define the automorphism $\rho_3 : H \to H$ by replacing the multiplications by 2 and 4 with multiplications by 3 and 9. The projections of ρ_2 and ρ_3 to X define an expanding action of $\mathbb{Z}^2_+$.

In the previous example there are two Archimedean Lyapunov exponents. Using the coordinates from the previous example they can be written as

$$\chi_- = t \log 2 + s \log 3$$

and

$$\chi_+ = t \log 4 + s \log 9.$$

The exponent χ_- has multiplicity 2 and χ_+ is simple. The Lyapunov distribution of χ_- is non-integrable. The relation between the exponents $\chi_+ = 2\chi_-$ is a simple example of resonance. Note that the projection of the abelian action to the center gives the action from Example 2.4.1.

2.4.2 Automorphisms of other compact abelian groups

In this section we discuss $\mathbb{Z}^d$-actions by automorphisms of compact abelian groups. These actions are natural generalizations of the abelian actions by automorphisms of tori and solenoids discussed above. The key to their study is the interplay with commutative algebra and algebraic geometry. This is a well-developed area of dynamics. Our goal is to briefly describe some basic results and techniques, and to present several examples of such actions that do not fall in the category of automorphisms of a torus. This will allow the reader to make a comparison with the theory and examples presented earlier in this chapter. While the phase spaces for those actions are not any more finite-dimensional manifolds or even finite-dimensional objects in the sense that solenoids are, some of them exhibit similar rigidity properties. For a more complete discussion of the topics presented in the section we refer the reader to the monograph [157].

Definition 2.4.3 Let X be a compact abelian group and $d \geq 1$. An *algebraic* $\mathbb{Z}^d$*-action* on X, $\mathbf{n} \to \alpha^{\mathbf{n}}$, is a $\mathbb{Z}^d$ action by continuous automorphisms of X.

Any algebraic action α on the compact abelian group X has a natural invariant measure, namely the normalized Haar measure λ_X. Thus one can introduce in the standard way the notions of ergodicity, mixing, or Bernoulli with respect to this invariant measure.

We start to describe the relationship between abelian compact group actions and commutative algebra. For proofs of basic results in commutative algebra we refer to [90].

Let $d \geq 1$ integer, and let $R_d = \mathbb{Z}[u_1^{\pm}, \ldots, u_d^{\pm}]$ be the ring of Laurent polynomials with integral coefficients in the commuting variables $u_1, \ldots, u_d$. An element $f \in R_d$ can be described as

$$f = \sum_{\mathbf{n} \in \mathbb{Z}^d} c_f(\mathbf{n}) u^{\mathbf{n}}, \tag{2.4.2}$$

where $u^{\mathbf{n}} = u_1^{n_1} \ldots u_d^{n_d}$, $c_f(\mathbf{n}) \in \mathbb{Z}^d$ for every $\mathbf{n} = (n_1, \ldots, n_d) \in \mathbb{Z}^d$, and $c_f(\mathbf{n}) = 0$ for all but finitely many $\mathbf{n}$.

The dual group $M = \hat{X}$, consisting of the characters of X, and viewed as an additive group, is a module over the ring R_d with the scalar multiplication given by

$$f \cdot a = \sum_{\mathbf{n} \in \mathbb{Z}^d} c_f(\mathbf{n}) \widehat{\alpha^{\mathbf{n}}}(a),$$

for $f \in R_d, a \in M$, where $\widehat{\alpha^{\mathbf{n}}}$ is the automorphism of $\hat{X}$ dual to $\alpha^{\mathbf{n}}$. In particular, $u^{\mathbf{n}} \cdot a = \widehat{\alpha^{\mathbf{n}}}(a)$. The module M is called the *dual module* of the action α.

Conversely, any R_d-module M determines an algebraic $\mathbb{Z}^d$-action α_M on the compact abelian group $X_M = \widehat{M}$, with $\widehat{\alpha}_M^{\mathbf{n}}$ being the dual to multiplication by $u^{\mathbf{n}}$ on M for every $\mathbf{n} \in M$.

The simplest examples of R_d-modules that can be used to construct abelian actions are the cyclic ones, that is, those of the form $M = R_d/I$, where $I \subset R_d$ is an ideal. We recall that a module is called *Noetherian* if every strictly increasing sequence of submodules $M \subsetneq M_1 \subsetneq M_2 \subsetneq \cdots$ is finite. Since the ring R_d, as a module over itself, is Noetherian, a module over R_d is Noetherian if and only if it is finitely generated. In particular, M is Noetherian if and only if there exist elements $a_1, ..., a_n$ such that $M = R_d a_1 + \cdots + R_d a_n$, and every cyclic R_d-module is Noetherian.

We describe the abelian action induced by a Noetherian R_d-module.

Example 2.4.4 Let $M = R_d$. The module R_d is isomorphic as a group to the direct sum $\oplus_{\mathbb{Z}^d}\mathbb{Z}$, and its dual group $\widehat{R_d}$ is isomorphic to the cartesian product $\mathbb{T}^{\mathbb{Z}^d}$ of copies of the torus $\mathbb{T}$. If $x = (x_n)_{n\in\mathbb{Z}^d}$ is a generic element in $\oplus_{\mathbb{Z}^d}\mathbb{Z}$, one can identify $\widehat{R_d}$ and $\mathbb{T}^{\mathbb{Z}^d}$ via the duality bracket

$$< x, f >= e^{2\pi i \sum_{\mathbf{n}\in\mathbb{Z}^d} c_f(\mathbf{n})x_{\mathbf{n}}},$$

where f is given by (2.4.2). Under this identification the $\mathbb{Z}^d$-action α_{R_d} on $\mathbb{T}^{\mathbb{Z}^d}$ becomes the shift action given by

$$[(\alpha_{R_d})^{\mathbf{n}}(x)]_m = x_{n+m} \text{ for all } m, n \in \mathbb{Z}.$$

Example 2.4.5 More generally, let $I \subset R_d$ be an ideal and $M = R_d/I$. Note that any ideal is a submodule, hence an $\widehat{\alpha}_{R_d}$-invariant subgroup. The dual group $\widehat{M}$ is the α_{R_d}-invariant subgroup of $\widehat{R_d} = \mathbb{T}^{\mathbb{Z}^d}$

$$\begin{aligned} \widehat{M} &= \{x \in \mathbb{T}^{\mathbb{Z}^d} :< x, f >= 1 \text{ for all } f \in I\} \\ &= \{x \in \mathbb{T}^{\mathbb{Z}^d} : \sum_{\mathbf{n}\in\mathbb{Z}^d} c_f(\mathbf{n})x_{\mathbf{n}} \text{ (mod 1) for all } f \in I, \mathbf{n} \in \mathbb{Z}^d\}, \end{aligned} \qquad (2.4.3)$$

and the action $\alpha_{R_d/I}$ is the restriction of the action α_{R_d} to the shift invariant subgroup $\widehat{M}$.

Conversely, one can start with $X \subset \mathbb{T}^{\mathbb{Z}^d}$ closed subgroup, and let

$$X^{\text{ann}} = \{f \in R_d :< x, f >= 1 \text{ for every } x \in X\}$$

be the annihilator of X in $\widehat{R_d}$. Then X is shift invariant if and only if X^{ann} is an ideal in R_d.

Example 2.4.6 Let $d = 1$ and let I be the ideal generated by p, some prime number p. Then the dual action induced on R_1/I is the two-sided full shift on p symbols.

Example 2.4.7 Even more generally, let M be an R_d-Noetherian module, and let $\{a_1, ..., a_n\}$ be a set of generators for M. The surjective homomorphism

$(f_1, \ldots, f_k) \to f_1 a_1 + \cdots + f_k a_k$ from R_d^k to M induces a dual injective homomorphism $\phi : \widehat{M} \to \widehat{R_d^k} \cong (\mathbb{T}^k)^{\mathbb{Z}^d}$ such that $\phi \alpha_m^{\mathbf{n}} = \alpha_{R_d^k}^{\mathbf{n}} \phi$ for all $\mathbf{n} \in \mathbb{Z}^d$. In particular, ϕ embeds $\widehat{M}$ as a closed shift invariant subgroup of $(\mathbb{T}^k)^{\mathbb{Z}^d}$.

We observed that an algebraic $\mathbb{Z}^d$-action α is completely determined by the dual module M. Consequently, one can try to express the dynamical properties of α in terms of algebraical properties of the module M.

Definition 2.4.8 A prime ideal $\mathfrak{p} \subset R_d$ is said to be associated with the R_d-module M if

$$\mathfrak{p} = \{f \in R_d : f \cdot a = 0_M\},$$

for some $a \in M$. The set of prime ideals associated with a Noetherian R_d-module M is finite.

Definition 2.4.9 If $I \subset R_d$ is an ideal we denote by $V_{\mathbb{C}}(I)$ the variety of I, that is,

$$V_{\mathbb{C}}(I) = \{c = (c_1, \ldots, c_d) \in (\mathbb{C}^*)^d : f(c) = 0 \text{ for all } f \in I\},$$

where $\mathbb{C}^* = \mathbb{C} - \{0\}$.

The following basic results are part of Theorem 6.5 in [157].

Theorem 2.4.10 *Let $\mathfrak{p} \subset R_d$ be a prime ideal and $\alpha = \alpha_{R_d/\mathfrak{p}}$ be the algebraic $\mathbb{Z}^d$-action on $X := \widehat{R_d/\mathfrak{p}}$.*

(i) For every $\mathbf{n} \in \mathbb{Z}^{\mathbf{d}}$ the following are equivalent:
  (a) $\alpha^{\mathbf{n}}$ is ergodic;
  (b) $\mathfrak{p} \cap \{u^{l\mathbf{n}} - 1 : l \geq 1\} = \emptyset$.

(ii) The following conditions are equivalent:
  (a) α is ergodic;
  (b) $\alpha^{\mathbf{n}}$ is ergodic for some $\mathbf{n} \in \mathbb{Z}^{\mathbf{d}}$;
  (c) $\{u^{\mathbf{n}} - 1 : \mathbf{n} \in \Gamma\} \not\subseteq \mathfrak{p}$ for every subgroup of finite index $\Gamma \subset \mathbb{Z}^d$.

(iii) The following conditions are equivalent:
  (a) α is mixing (either topologically or w.r.t. the Haar measure λ_X);
  (b) for every non-zero $\mathbf{n} \in \mathbb{Z}^n$, $\alpha^{\mathbf{n}}$ is ergodic;
  (c) for every non-zero $\mathbf{n} \in \mathbb{Z}^n$, $\alpha^{\mathbf{n}}$ is mixing;
  (d) $\mathfrak{p} \cap \{u^{\mathbf{n}} - 1 : \mathbf{n} \in \mathbb{Z}^d\} = \{0\}$.

(iv) The following are equivalent:
  (a) α is expansive;
  (b) $V_{\mathbb{C}}(\mathfrak{p}) \cap \mathbb{S}^d = \emptyset$, where $\mathbb{S} \subset \mathbb{T}$ is the unit circle.

We show now more examples of abelian actions and determine their dynamics using the last theorem. We first consider the case of a toral automorphism and $d = 1$, which is instructive in itself.

Example 2.4.11 Let α be the automorphism of $\mathbb{T}^2$ determined by the matrix $A = \begin{pmatrix} 0 & 1 \\ 1 & 1 \end{pmatrix} \in \mathrm{GL}(n, \mathbb{Z})$ with characteristic irreducible polynomial $f(u) = u^2 - u - 1$. Note that A is equal to the companion matrix of f. It follows from Example 2.4.5 that

$$\widehat{R_1/(f)} = \{x = (x_n) \in \mathbb{T}^{\mathbb{Z}} : x_n + x_{n+1} - x_{n+2} = 0 \text{ (mod 1) for all } n \in \mathbb{Z}\},$$

where $(f) = fR_1 \subset R_1$ is the principal prime ideal generated by f and the action $\alpha_{R_1/(f)}$ is the shift on $\widehat{R_1/(f)} \subset \mathbb{T}^{\mathbb{Z}}$. The projection on the coordinates $0, 1$, $\pi_{0,1} : \widehat{R_1/(f)}(\subset \mathbb{T}^{\mathbb{Z}}) \to \mathbb{T}^2$ gives an algebraic conjugacy between the actions α and $\alpha_{R_1/(f)}$. Theorem 2.4.10 (iiid) immediately implies that α is mixing and expansive.

Example 2.4.12 More generally, let α be an automorphism of $\mathbb{T}^n$ determined by a matrix $A \in GL(n, \mathbb{R})$, and let f be the characteristic polynomial of A. The associated prime ideals of the module $\widehat{\mathbb{T}^n}$ are in one-to-one correspondence with the principal prime ideals arising from the irreducible divisors of f. Theorem 2.4.10 implies now that A is ergodic (or mixing) if and only if A does not have any eigenvalue that is a root of unity. Moreover, α is expansive if f has no eigenvalue of absolute value 1.

We recall that if $f = c_0 + \cdots + c_{n-1}u^{n-1} + u^n$ is the characteristic polynomial of A, then its companion matrix is

$$\begin{pmatrix} 0 & 1 & 0 & \cdots & 0 & 0 \\ 0 & 0 & 1 & \cdots & 0 & 0 \\ \cdots & \cdots & \cdots & \cdots & \cdots & \cdots \\ 0 & 0 & 0 & \cdots & 0 & 1 \\ -c_0 & -c_1 & -c_3 & \cdots & -c_{n-2} & -c_{n-1} \end{pmatrix}.$$

If A is equal to its companion matrix, then the projection

$$\phi : \widehat{R_1/(f)}(\subset \mathbb{T}^{\mathbb{Z}}) \to \mathbb{T}^n$$

on the coordinates $0, 1, \ldots, n-1$ is an algebraic conjugacy between the actions $\alpha_{R_1/(f)}$ and α.

In general, the matrix A is conjugate to the companion matrix of f over $\mathbb{Q}$, but not over $\mathbb{Z}$. The dual module $M = \widehat{\mathbb{T}^n}$ of α has a submodule of finite index $N \subset M$ that is isomorphic to $R_1/(f)$. From this it follows that there are continuous surjective finite-to-one group homomorphisms $\phi_1 : \mathbb{T}^n \to \widehat{R_1/(f)}$ and $\phi_2 : \widehat{R_1/(f)} \to \mathbb{T}^n$ such that $\phi_1\alpha = \alpha_{R_1/(F)}\phi_1$ and $\alpha\phi_2 = \phi_2\alpha_{R_1/(f)}$. See Section 9 in [157] for details.

Example 2.4.13 Let I be the ideal generated in R_2 by $u_1 - 2$ and $u_2 - 3$. Then

$$\widehat{R_2/I} = \{x \in \mathbb{T}^{\mathbb{Z}^2} | x_{m+1,n} = 2x_{m,n}, x_{m,n+1} = 3x_{m,n} \text{ for all } m, n \in \mathbb{Z}\}.$$

This system is the invertible extension of the $\times 2$, $\times 3$ abelian action in Example 2.4.1.

We show now an example of higher rank algebraic $\mathbb{Z}^d$-action that is not given by homomorphisms of a torus.

Example 2.4.14 Let

$$X = \{x \in (\mathbb{Z}/2\mathbb{Z})^{\mathbb{Z}^2} : x_{m,n} + x_{m+1,n} + x_{m,n+1} = 0 \text{ for all } m, n \in \mathbb{Z}\}.$$

Equivalently, X is the dual group of R_2/I, where I is the ideal generated by 2 and $1 + u_1 + u_2$.

The set X is a compact totally disconnected group which is invariant under both horizontal and vertical shifts on $(\mathbb{Z}/2\mathbb{Z})^{\mathbb{Z}^2}$. We define α to be the $\mathbb{Z}^2$-action induced on X by these shifts.

It was shown by Ledrappier [93] that α is mixing of order two but not mixing of order three. This is in contrast to the case of a $\mathbb{Z}^d$-action by automorphisms of a torus. Indeed, it was shown by Schmidt and Ward [159] that every mixing $\mathbb{Z}^d$-action by automorphisms of a compact, connected, abelian group is mixing of all orders.

3

Preparatory results from analysis

3.1 Introduction

A classical and fairly elementary result from Fourier analysis asserts that a continuous function $f : \mathbb{R}^2 \to \mathbb{R}$ is actually C^∞ assuming only that the partial derivatives $\partial^n f/\partial x^n$ and $\partial^n f/\partial y^n$ exist for all positive integers n. To prove this result, one can use, for example, techniques from Chapter 1 of the classical book of Stein and Weiss [167]. No mixed partial derivatives are necessary here. For a smooth manifold M, this result can be easily extended to functions $f : M \to \mathbb{R}$ that are smooth along pairs of transverse smooth foliations on M, or even to functions that are smooth along a web of smooth foliations with tangent distributions spanning the tangent bundle TM. Moreover, general results can be obtained for functions that are smooth along pairs of transverse Hölder foliations with smooth leaves. The later type of results are often used in rigidity in order to show higher regularity for solutions of cohomological equations over hyperbolic and partially hyperbolic systems. In this setup, the corresponding transverse foliations are the stable and unstable foliations associated to the system. A more complete discussion of this relationship can be found in Chapters 4 and 5.

The main goal of this chapter is to present, as far as possible, complete proofs and references for several results along this theme that appear in the literature. We start with the well-known result of Journé [58], about $C^{n,\alpha}$ regularity of a continuous function that is $C^{n,\alpha}$ along two transverse continuous foliations with $C^{n,\alpha}$ leaves. This result is proved in Section 3.3. An early preliminary technical result needed in the proof is the embedding result of Campanato [14], which is presented in Section 3.2. Journé's result has the advantage of guaranteeing optimal regularity in the finite regularity setup.

If one is interested only in smooth (C^∞) regularity, a convenient alternative to Journé's approach is the smooth regularity result of Hurder and Katok [56], which we show in Section 3.5. Besides having a simpler proof, this result can be applied to the more general situation in which the function is regular along a web of foliations. An alternative approach, using deeper regularity results for elliptic differential operators, is due to de la Llave *et al.* and can be found in [104]. A real analytic regularity result by de la Llave [100] is proved in Section 3.6. The last result also can be applied to a web of foliations.

The applicability of the results presented so far is mainly to problems in which we can control the regularity in enough directions to span the whole tangent space at a point in the manifold. A different situation appears when one works with partially hyperbolic actions induced by left multiplication by elements in Lie groups on homogenous manifolds. In this case the stable and unstable directions are not jointly integrable and their sum has a dimension strictly less than the ambient tangent space. A smooth regularity result for this case was proved by Katok and Spatzier [79]. In order to apply this result the foliations need to be smooth not only along the leaves, but also transversally. We note also that in this result the function can be replaced by a distribution. This requirement is necessary in certain cocycle rigidity results. We present a proof in Section 3.7.

The material in this chapter will be used later in Chapters 4 and 5 in order to show the existence and to prove higher regularity for the transfer map between two cohomologous cocycles.

3.2 Preparatory norm estimates

In this section we follow closely the paper of Campanato [14]. Certain minor simplifications are available to us, since we do not need Campanato's result in full generality. In particular, we replace the space of measurable functions $\mathcal{L}_k^{\lambda,q}$ from [14] by a space of continuous functions. Also, in our presentation it is enough to choose the constant q from [14] equal to 1. The main result of this section, which will be used later to prove Journé's result, is Theorem 3.2.10.

Let Ω be a convex bounded set in $\mathbb{R}^n$, with diameter $d(\Omega)$ and closure $\bar{\Omega}$. If $x_0 \in \mathbb{R}^n$ and $\rho > 0$, let $B(x_0, \rho)$ be the closed ball of radius ρ centered in x_0, and let $\Omega(x_0, \rho) = B(x_0, \rho) \cap \Omega$.

Let μ be the Lebesgue measure on $\mathbb{R}^n$. Throughout this section we assume that there exists a constant $A > 0$ such that for any $x_0 \in \Omega$ and any $\rho \in [0, d(\Omega)]$

$$\mu(\Omega(x_0, \rho)) \geq A\rho^n. \tag{3.2.1}$$

For $k \geq 1$ integer, fixed for the rest of the section, denote by $\mathcal{P}_k$ the set of polynomials in n real variables $x_1, \ldots, x_n$ with real coefficients and of degree less or equal to k. If $p = (p_1, \ldots, p_n)$ is an n-tuple of positive integers, denote

$$p! = p_1! \cdots p_n!, \quad |p| = p_1 + \cdots + p_n, \quad x^p = x_1^{p_1} \cdots x_n^{p_n}.$$

If $u(x)$, $x \in \mathbb{R}^n$, is a differentiable real valued function, denote

$$D^p u(x) = \frac{\partial^{|p|} u}{\partial x_1^{p_1} \cdots \partial x_n^{p_n}}.$$

Let $\lambda > 0$. A function u, which is continuous on $\bar{\Omega}$, is said to belong to $\mathcal{L}_k^\lambda(\Omega)$ if

$$|||u|||_{k,\lambda} := \sup_{x_0 \in \bar{\Omega}, 0 < \rho \leq d(\Omega)} \rho^{-\lambda} \inf_{P \in \mathcal{P}_k} \int_{\Omega(x_0,\rho)} |u(x) - P(x)| d\mu < \infty.$$

The following lemma is attributed to Di Giorgi.

Lemma 3.2.1 *If $P(x) \in \mathcal{P}_k$, $E \subset B(x_0, \rho)$ measurable, and $\mu(E) \geq A\rho^n$, then there exists a constant $C_1(k, n, A)$ such that for any $p = (p_1, \ldots, p_n)$*

$$|D^p P(x)_{|x=x_0}| \leq \frac{C_1}{\rho^{n+|p|}} \int_E |P(x)| d\mu. \tag{3.2.2}$$

Proof Let $\mathcal{P}_k^0$ be the set of polynomials $P(x) = \sum_{|p| \leq k} a_p x^p$ such that

$$\sum_{|p| \leq k} |a_p|^2 = 1. \tag{3.2.3}$$

Let $\mathcal{F}$ be the set of measurable real valued functions on $\mathbb{R}^n$ supported on $B(0, 1)$ that satisfy

$$0 \leq f(x) \leq 1, \quad \int_{\mathbb{R}^n} f(x) d\mu \geq A. \tag{3.2.4}$$

Define

$$\gamma(A) = \inf_{P \in \mathcal{P}_k, f \in \mathcal{F}} \int_{B(0,1)} |P(x)| f(x) d\mu. \tag{3.2.5}$$

We show that $\gamma(A)$ is reached for a choice of $P(x)$ and $f(x)$. From (3.2.5), for any integer n there exist $P_n(x)$ and $f_n(x)$ such that

$$\gamma(A) \leq \int_{B(0,1)} |P_n(x)| f_n(x) d\mu < \gamma(A) + \frac{1}{n}. \tag{3.2.6}$$

By a compactness argument, from (3.2.3) there exists a subsequence $\{P_m(x)\}_m$ that converges uniformly on $\bar{\Omega}$ to a polynomial $p_0 \in \mathcal{P}_k^0$. Similarly, from (3.2.4) there exists a subsequence $\{f_p(x)\}_p$ of $\{f_m(x)\}_m$ that converges weakly in $L^2(B(0,1))$ to a function $f_0(x) \in \mathcal{F}$. Taking limit in (3.2.6) as $p \to \infty$ implies

$$\gamma(A) = \int_{B(0,1)} |P_0(x)| f_0(x) d\mu.$$

In particular, $\gamma(A) > 0$, and if $E \subset B(0,1)$ is measurable, with $\mu(E) \geq A$, then for any $P(x) \in \mathcal{P}_k^0$, choosing $f(x)$ to be the characteristic function of E in (3.2.5), gives

$$\int_E |P(x)| d\mu \geq \gamma(A). \tag{3.2.7}$$

More generally, if $P \in \mathcal{P}_k$, then (3.2.7) becomes

$$\left(\sum_{|p| \leq k} |a_p|^2 \right)^{1/2} \leq \frac{1}{\gamma(A)} \int_E |P(x)| d\mu,$$

which implies, for all $|p| \leq k$

$$|a_p| \leq \frac{1}{\gamma(A)} \int_E |P(x)| d\mu. \tag{3.2.8}$$

Assume now that $P(x) \in \mathcal{P}_k$ and $E \subset B(x_0, \rho)$, E measurable, $\mu(E) \geq A\rho^n$. The change of variable $y = T(x) = (x - x_0)/\rho$ gives

$$\int_E |P(x)| d\mu = \rho^n \int_{T(E)} |P(x_0 + \rho y)| d\mu \tag{3.2.9}$$

and $T(E) \subset B(0,1)$, $\mu(T(E)) \geq A$. Note also that

$$P(x_0 + \rho y) = \sum_{|p| \leq k} \frac{\rho^{|p|} D^p P(x)_{|x=x_0}}{p\,!} y^p. \tag{3.2.10}$$

Finally it follows from (3.2.8), (3.2.9), and (3.2.10) that (3.2.2) is true. □

For the rest of the section, if $x_0 \in \mathbb{R}^n$ is specified, we assume that a polynomial $P(x) \in \mathcal{P}_k$ is written in the form

$$P(x) = \sum_{|p| \leq k} \frac{a_p}{p\,!} (x - x_0)^p.$$

Lemma 3.2.2 *Let $u \in \mathcal{L}_k^\lambda(\Omega)$. Then for any $x_0 \in \bar{\Omega}$ and any $\rho \in [0, d(\Omega)]$ there exists a unique polynomial $P_k(x, x_0, \rho, u)$ such that*

$$\inf_{P \in \mathcal{P}_k} \int_{\Omega(x_0,\rho)} |u(x) - P(x)| d\mu = \int_{\Omega(x_0,\rho)} |u(x) - P_k(x, x_0, \rho, u)| d\mu.$$

Proof Let m be the number of coefficients of a polynomial $P \in \mathcal{P}_k$. The function $F : \mathbb{R}^m \to \mathbb{R}$, defined by

$$F(a_1, \dots, a_m) := \|u(x) - P(x)\|_{L^1(\Omega(x_0,\rho))},$$

where $P(x)$ has coefficients $(a_1, \dots, a_m)$, is a positive continuous function. F reaches its infimum on a bounded compact set. Indeed, if

$$\|(a_1, \dots, a_n)\| = \left(\sum_{|p|\le k} |a_p|^2\right)^{1/2} \to \infty,$$

then Lemma 3.2.1 implies that $\|P(x)\|_{L^1(\Omega(x_0,\rho))} \to \infty$, and consequently

$$|\|u\|_{L^1(\Omega(x_0,\rho))} - \|P(x)\|_{L^1(\Omega(x_0,\rho))}| \le F(a_1, \dots, a_n) \to \infty.$$

The uniqueness of P_k now follows from the uniform convexity of the space L^1. □

In what follows, if the function u is specified, we denote $P_k(x, x_0, \rho, u)$ by $P_k(x, x_0, \rho)$ and denote $a_p(x_0, \rho) := D^p P_k(x, x_0, \rho)_{|x=x_0}$.

Lemma 3.2.3 *Let $u \in \mathcal{L}^k_\lambda(\Omega)$. There exists a constant $C_2(\lambda)$ such that for any $x_0 \in \bar{\Omega}$, $0 < \rho \le d(\Omega)$, and r positive integer,*

$$\int_{\Omega(x_0,2^{-r-1}\rho)} |P_k(x, x_0, 2^{-r}\rho) - P_k(x, x_0, 2^{-r-1}\rho)|d\mu \le C_2|||u|||_{k,\lambda}2^{-r\lambda}\rho^\lambda.$$

Proof

$$\begin{aligned}
\int_{\Omega(x_0,2^{-r-1}\rho)} &|P_k(x, x_0, 2^{-r}\rho) - P_k(x, x_0, 2^{-r-1}\rho)|d\mu \\
&\le \int_{\Omega(x_0,2^{-r}\rho)} |P_k(x, x_0, 2^{-r}\rho) - u(x)|d\mu \\
&\quad + \int_{\Omega(x_0,2^{-r-1}\rho)} |P_k(x, x_0, 2^{-r-1}\rho) - u(x)|d\mu \\
&\le |||u|||_{k,\lambda}2^{-r\lambda}\rho^\lambda + |||u|||_{k,\lambda}2^{-(r+1)\lambda}\rho^\lambda \\
&= (1 + 2^{-\lambda})|||u|||_{k,\lambda}2^{-r\lambda}\rho^\lambda.
\end{aligned}$$
□

Lemma 3.2.4 *Let $u \in \mathcal{L}^k_\lambda(\Omega)$, $x_0, y_0 \in \bar{\Omega}$ and $p = (p_1, \dots, p_n)$, $|p| = k$. Then*

$$\begin{aligned}
&|a_p(x_0, 2|x_0 - y_0|) - a_p(y_0, 2|x_0 - y_0|)| \\
&\quad \le C_1 2^{1+\lambda}|||u|||_{k,\lambda}|x_0 - y_0|^{\lambda-n-k}.
\end{aligned} \tag{3.2.11}$$

Proof Let $\rho = |x_0 - y_0|$ and $B_\rho = \Omega(x_0, 2\rho) \cap \Omega(y_0, 2\rho)$. Then, for any $x \in B_\rho$,

$$\begin{aligned}&\int_{\Omega(x_0,\rho)} |P_k(x, x_0, 2\rho) - P_k(x, y_0, 2\rho)| d\mu \\ &\leq \int_{\Omega(x_0,2\rho)} |P_k(x, x_0, 2\rho) - u(x)| d\mu + \int_{\Omega(y_0,2\rho)} |P_k(x, y_0, 2\rho) - u(x)| d\mu \\ &\leq 2^{\lambda+1} |||u|||_{k,\lambda} \rho^\lambda. \end{aligned} \tag{3.2.12}$$

From Lemma 3.2.1 applied to the polynomial $P_k(x, x_0, 2\rho) - P_k(x, y_0, 2\rho)$, follows that

$$\begin{aligned}&|a_p(x_0, 2\rho) - a_p(y_0, 2\rho)| \\ &\quad \leq \frac{C_1}{\rho^{n+k}} \int_{\Omega(x_0,\rho)} |P_k(x, x_0, 2\rho) - P_k(x, y_0, 2\rho)| d\mu. \end{aligned} \tag{3.2.13}$$

Now (3.2.12) and (3.2.13) imply (3.2.11). □

Lemma 3.2.5 *Let $u \in \mathcal{L}^k_\lambda(\Omega)$. There exists a constant $C_3(k, \lambda, n, A)$ such that for any $x_0 \in \Omega$, $0 < \rho \leq d(\Omega)$ and $p = (p_1, \ldots, p_n)$, $|p| \leq k$, one has*

$$|a_p(x_0, \rho) - a_p(x_0, 2^{-i}\rho)| \leq C_3 |||u|||_{k,\lambda} \sum_{j=0}^{i-1} 2^{j(n+|p|-\lambda)} \rho^{\lambda-n-|p|}. \tag{3.2.14}$$

Proof

$$\begin{aligned}|a_p(x_0, \rho) - a_p(x_0, 2^{-i}\rho)| &\leq \sum_{j=0}^{i-1} |a_p(x_0, 2^{-j}\rho) - a_p(_0, 2^{-j-1}\rho)| \\ &= \sum_{j=0}^{i-1} |D^p_{|x=x_0} [P_k(x, x_0, 2^{-j}\rho) - P_k(x, x_0, 2^{-j-1}\rho)]|. \end{aligned}$$

Lemma 3.2.1 applied to the function

$$P_k(x, x_0, 2^{-j}\rho) - P_k(x, x_0, 2^{-j-1}\rho),$$

together with Lemma 3.2.3 imply

$$\begin{aligned}&|a_p(x_0, \rho) - a_p(x_0, 2^{-i}\rho)| \leq C_1 \rho^{-n-|p|} \\ &\cdot \sum_{j=1}^{i-1} \left[\int_{\Omega(x_0, 2^{-j-1}\rho)} |P_k(x, x_0, 2^{-j}\rho) - P_k(x, x_0, 2^{-j-1}\rho)| d\mu \right] 2^{(j+1)(n+|p|)} \\ &\quad \leq C_3 |||u|||_{k,\lambda} \sum_{j=0}^{i-1} 2^{j(n+|p|-\lambda)} \rho^{\lambda-n-|p|}. \end{aligned}$$

□

Lemma 3.2.6 *Let $u \in \mathcal{L}^k_\lambda(\Omega)$ and $r \le k$ an integer such that $n + r < \lambda \le n + r + 1$. There exists a family of functions $\{v_p(x_0)\}_p$, $|p| \le r$, defined on $\bar{\Omega}$ and such that for any $0 < \rho \le d(\bar{\Omega})$ and any $|p| \le r$*

$$|a_p(x_0, \rho) - v_p(x_0)| \le C_4(\lambda, k, n, A)|||u|||_{k,\lambda}\rho^{\lambda-n-|p|}. \tag{3.2.15}$$

Consequently, $\lim_{\rho\to 0} a_p(x_0, \rho) = v_p(x_0)$ uniformly with respect to x_0.

Proof For fixed p, ρ, x_0, the sequence $\{a_p(x_0, \rho 2^{-i})\}_i$ converges as $i \to \infty$. Indeed, if $j > i$ are two positive integers, Lemma 3.2.5 implies

$$|a_p(x_0, 2^{-j}\rho) - a_p(x_0, 2^{-i}\rho)| \le C_3|||u|||_{k,\lambda} \sum_{h=i}^{j-1} 2^{h(n+|p|-\lambda)}\rho^{\lambda-n-|p|}, \tag{3.2.16}$$

and since $|p| \le r$ and $\lambda > n + r$, the series $\sum_{h=0}^{\infty} 2^{h(n+|p|-\lambda)}$ is convergent.

We show that the limit is independent of $\rho \in (0, d(\Omega)]$. Let $0 < \rho_1 \le \rho_2 \le d(\Omega)$. Lemma 3.2.1 and the definition of the space $\mathcal{L}^\lambda_k(\Omega)$ imply that

$$\begin{aligned}
&|a_p(x_0, \rho_1 2^{-i}) - a_p(x_0, \rho_2 2^{-i})| \\
&\quad\le \frac{C_1 2^{i(n+|p|)}}{\rho_1^{n+|p|}} \int_{\Omega(x_0,\rho_1 2^{-i})} |P_k(x, x_0, \rho_1 2^{-i}) - P_k(x, x_0, \rho_2 2^{-i})|d\mu \\
&\quad\le \frac{C_1 2^{i(n+|p|)}}{\rho_1^{n+|p|}} \Big[\int_{\Omega(x_0,\rho_1 2^{-i})} |P_k(x, x_0, \rho_1 2^{-i}) - u(x)|d\mu \\
&\qquad + \int_{\Omega(x_0,\rho_2 2^{-i})} |P_k(x, x_0, \rho_2 2^{-i}) - u(x)|d\mu\Big] \\
&\quad\le C_1|||u|||_{k,\lambda} \frac{\rho_1^\lambda + \rho_2^\lambda}{\rho_1^{n+|p|}} 2^{-i(\lambda-n-|p|)},
\end{aligned}$$

and the last term converges to 0 as $i \to \infty$ because $\lambda - n - |p| > 0$. We define $v_p(x_0) := \lim_{i\to\infty} a_p(x_0, \rho 2^{-i})$ for $x_0 \in \bar{\Omega}$. To deduce the inequality (3.2.15) take limit as $i \to \infty$ in formula (3.2.14). □

Lemma 3.2.7 *Let $u \in \mathcal{L}^\lambda_k(\Omega)$ and $\lambda > n + k$. If $p = (p_1, \dots, p_n)$, $|p| = k$, then the function $v_p(x_0)$ satisfies*

$$|v_p(x) - v_p(y)| \le C_5(\lambda, k, n, A)|||u|||_{k,\lambda}|x - y|^{\lambda-n-k}. \tag{3.2.17}$$

Proof Let $x, y \in \bar{\Omega}$, $\rho := |x - y| \le d(\Omega)/2$. Then

$$\begin{aligned}
|v_p(x) - v_p(y)| &\le |v_p(x) - a_p(x, 2\rho)| \\
&\quad + |v_p(y) - a_p(y, 2\rho)| + |a_p(x, 2\rho) - a_p(y, 2\rho)|.
\end{aligned} \tag{3.2.18}$$

From (3.2.15) in Lemma 3.2.6 follows

$$\begin{aligned}\max\{|v_p(x) - a_p(x, 2\rho)|, |v_p(y) - a_p(y, 2\rho)|\} \\ \leq C_4 2^{\lambda-n-k} |||u|||_{k,\lambda} \rho^{\lambda-n-k}.\end{aligned} \tag{3.2.19}$$

From (3.2.11) in Lemma 3.2.4 follows

$$|a_p(x, 2\rho) - a_p(y, 2\rho)| \leq C_1 2^{1+\lambda} |||u|||_{k,\lambda} \rho^{\lambda-n-k}. \tag{3.2.20}$$

If $|x - y| \leq d(\Omega)/2$, the theorem follows from (3.2.18), (3.2.19), and (3.2.20). If $|x-y| > d(\Omega)/2$, construct a polygonal line from x to y, contained in Ω, that has all edges of length less than $d(\Omega)/2$. Then formula (3.2.17) holds for each edge, and consequently for x and y. □

To simplify future notation, we denote by (0) the n-tuple $(0, \ldots, 0)$ and by e_i the ith vector in the standard basis of $\mathbb{R}^n$.

Lemma 3.2.8 *Let* $u \in \mathcal{L}_k^\lambda(\Omega)$, $k \geq 1, \lambda > n + k$, *and* $p = (p_1, \ldots, p_n)$, $|p| \leq k - 1$. *Then*

$$\frac{\partial v_p(x)}{\partial x_i} = v_{(p+e_i)}(x), \quad i = 1, 2, \ldots, n.$$

Proof By Lemma 3.2.7 the functions $v_p(x)$ with $|p| = k$ are continuous on $\bar{\Omega}$. Lemma is proved by induction under the additional assumption that the functions $v_{(p+\delta e_i)}(x)$ are continuous on $\bar{\Omega}$ for $\delta = 1, 2, \ldots, k - |p|$.

If $x_0 \in \Omega$ choose ρ small enough so that $B(x_0, |\rho|) \subset \Omega$. One has

$$\begin{aligned}&\frac{a_p(x_0 + e_i\rho, 2|\rho|) - a_p(x_0, 2|\rho|)}{\rho} \\ &= \frac{\{D^p[P_k(x, x_0 + e_i\rho, 2|\rho|) - P_k(x, x_0, |2\rho|)]\}_{|x=x_0}}{\rho} \\ &- \sum_{\delta=1}^{k-|p|} \frac{(-1)^\delta}{\delta\,!} \rho^{\delta-1} a_{(p+\delta e_i)}(x_0 + e_i\rho, 2|\rho|).\end{aligned} \tag{3.2.21}$$

From Lemma 3.2.1 and formula (3.2.12) follows that

$$\begin{aligned}&\left|\frac{\{D^p[P_k(x, x_0 + e_i\rho, 2|\rho|) - P_k(x, x_0, |2\rho|)]\}_{|x=x_0}}{\rho}\right| \\ &\leq \frac{C_1}{\rho^{n+|p|}} \int_{\Omega(x_0,|\rho|)} |P_k(x, x_0 + e_i\rho, 2|\rho|) - P_k(x, x_0, |2\rho|)| d\mu \\ &\leq C_1 2^{\lambda+1} |||u|||_{k,\lambda} |\rho|^{\lambda-n-|p|} \to 0 \text{ as } \rho \to 0.\end{aligned}$$

For any $1 \le \delta \le k - |p|$, by (3.2.15) follows

$$\begin{aligned}
&|a_{(p+\delta e_i)}(x_0 + e_i\rho, 2|\rho|) - v_{(p+\delta e_i)}(x_0)| \\
&\quad \le |a_{(p+\delta e_i)}(x_0 + e_i\rho, 2|\rho|) - v_{(p+\delta e_i)}(x_0 + \rho e_i)| \\
&\qquad + |v_{(p+\delta e_i)}(x_0 + \rho e_i) - v_{(p+\delta e_i)}(x_0)| \\
&\quad \le C_4 2^{\lambda-n-|p|-\delta} |||u|||_{k,\lambda} |\rho|^{\lambda-n-|p|-\delta} \\
&\qquad + |v_{(p+\delta e_i)}(x_0 + \rho e_i) - v_{(p+\delta e_i)}(x_0)|.
\end{aligned} \tag{3.2.22}$$

Since the functions $v_{(p+\delta e_i)}(x_0)$ are continuous for $\delta = 1, 2, \ldots, k - |p|$, one has

$$\lim_{\rho\to 0} a_{(p+\delta e_i)}(x_0 + e_i\rho, 2|\rho|) = v_{(p+\delta e_i)}(x_0), \quad \delta = 1, 2, \ldots, k - |p|.$$

Hence by (3.2.21) one has, uniformly with respect to x_0, that

$$\lim_{\rho\to 0} \frac{a_p(x_0 + e_i\rho, 2|\rho|) - a_p(x_0, 2|\rho|)}{\rho} = v_{(p+\delta e_i)}(x_0).$$

It remains to check that

$$\lim_{\rho\to 0} \frac{v_p(x_0 + e_i\rho) - v_p(x_0)}{\rho} = \lim_{\rho\to 0} \frac{a_p(x_0 + e_i\rho, 2|\rho|) - a_p(x_0, 2|\rho|)}{\rho}.$$

Recall that $|p| \le k - 1$. One can write:

$$\begin{aligned}
\frac{v_p(x_0 + e_i\rho) - v_p(x_0)}{\rho} &= \frac{v_p(x_0 + e_i\rho) - a_p(x_0 + e_i\rho, 2|\rho|)}{\rho} \\
&+ \frac{a_p(x_0 + e_i\rho, 2|\rho|) - a_p(x_0, 2|\rho|)}{\rho} + \frac{a_p(x_0, 2|\rho|) - v_p(x_0)}{\rho}.
\end{aligned}$$

The estimate (3.2.15) implies now that the first and third terms of the sum converge to 0 as $\rho \to 0$. □

Lemma 3.2.9 *Let $u \in \mathcal{L}_k^\lambda(\Omega)$ and assume $\lambda > n + k$. Then the function $v_{(0)}(x) \in C^{n,\alpha}(\bar{\Omega})$, where $\alpha = \lambda - n - k$ and*

$$D^p v_{(0)}(x) = v_p(x), \ x \in \Omega, \ |p| \le k.$$

Proof The lemma is a corollary of Lemma 3.2.7 and Lemma 3.2.8. □

Theorem 3.2.10 *Let $u \in \mathcal{L}_k^\lambda(\Omega)$ and assume $n + k < \lambda \le n + k + 1$. Then $u \in C^{n,\alpha}(\Omega)$, where $\alpha = \lambda - n - k$ and*

$$\|u\|_{C^{k,\alpha}} \le C_6 |||u|||_{k,\lambda}. \tag{3.2.23}$$

Proof Let $x_0 \in \Omega$. Then

$$|a_{(0)}(x_0, \rho) - u(x_0)|$$
$$\leq |a_{(0)}(x_0, \rho) - P_k(x, x_0, \rho)| + |P_k(x, x_0, \rho) - u(x)| + |u(x) - u(x_0)|.$$

Now integration over Ω gives

$$\begin{aligned}
&|a_{(0)}(x_0, \rho) - u(x_0)| \\
&\quad \leq \frac{1}{\mu(\Omega(x_0, \rho))} \int_{\Omega(x_0,\rho)} |a_{(0)}(x_0, \rho) - P_k(x, x_0, \rho)| d\mu \\
&\quad\quad + \frac{1}{\mu(\Omega(x_0, \rho))} \int_{\Omega(x_0,\rho)} |P_k(x, x_0, \rho) - u(x)| \\
&\quad\quad + \frac{1}{\mu(\Omega(x_0, \rho))} \int_{\Omega(x_0,\rho)} |u(x) - u(x_0)| \\
&\quad = I_1 + I_2 + I_3.
\end{aligned}$$

Since u is continuous, I_3 converges to 0 as $\rho \to 0$. For I_2 use the definition of $\mathcal{L}_k^\lambda(\Omega)$ and the fact that $\mu(\Omega(x_0, \rho)) \geq A\rho^n$ to conclude that

$$I_2 \leq \frac{1}{A\rho^n} \int_{\Omega(x_0,\rho)} |P_k(x, x_0, \rho) - u(x)| d\mu \leq \frac{\rho^{k+n}}{A\rho^n} |||u|||_{k,\lambda} \to 0 \text{ as } \rho \to 0.$$

Since $a_{(0)}(x_0, \rho) = P_k(x_0, x_0, \rho)$, there is a constant $C_7(A, n, k) > 0$ such that

$$\begin{aligned}
I_1 &\leq \int_{\Omega(x_0,\rho)} |a_{(0)}(x_0, \rho) - P_k(x_0, x, \rho)| d\mu \\
&\leq C_7 \sum_{1 \leq |p| \leq k} |a_p(x_0, \rho)| \rho^{|p|} \to 0 \text{ as } \rho \to 0.
\end{aligned}$$

So $v_{(0)}(x_0) = \lim_{\rho \to 0} a_{(0)}(x_0, \rho) = u(x_0)$, where the first equality follows from Lemma 3.2.6.

Now the estimate from the theorem follows from (3.2.17). □

3.3 Journé's theorem

In this section we present a proof of Journé's theorem [58]. We follow closely the presentation from [124], which contains a more explicit proof of the result than the original paper, as well as applications to dynamical systems over fractal sets. Another paper that discusses aspects of the proof is [99].

For the simplicity of the presentation we assume that the manifold supporting the foliations is two dimensional, and that the foliations are one dimensional. Moreover, since the argument is local, we can assume that the manifold is an open set in $\mathbb{R}^2$.

Let $n \geq 1, \alpha \in (0, 1)$. For an open set $U \in \mathbb{R}^2$, the space $C^{n,\alpha}(U)$ consists of all functions that are differentiable of order n, have all derivatives bounded, and have the nth partial derivatives α-Hölder. A $C^{n,\alpha}$ lamination of U is a partition of U into $C^{n,\alpha}$ submanifolds of the same dimension, which varies continuously in the $C^{n,\alpha}$-topology.

The main result can be stated as follows. Note that the term "uniformly" with respect to a given property means that the property holds uniformly on a given open set U.

Theorem 3.3.1 (Journé's theorem) *Let $U \subset \mathbb{R}^2$ open set, and W^1, W^2 two transverse uniformly $C^{n,\alpha}$ laminations of U. Suppose that a continuous function $u : U \to \mathbb{R}$ is uniformly $C^{n,\alpha}$ when restricted to each local leaf $W^1_\epsilon(p)$, $W^2_\epsilon(p)$, $p \in U$. Then u is $C^{n,\alpha}$ on U.*

Remark 3.3.2 One should note that the method of proof presented below does not extend to $C^{n,\mathrm{Lip}}$ regularity. That is, the constant α cannot be 1. Several observations about the extension of Journé's theorem to Dini regularity can be found in [131].

Journé constructed approximating polynomials $Q(p; q)$ indexed by $p \in U$, which satisfies

$$|Q(p; q) - u(q)| \leq C|p - q|^{n+\alpha}, \text{ for all } q \in U,$$

and then applied Campanato's result Theorem 3.2.10 to conclude that u is $C^{n,\alpha}$.

All constants that are uniform on U are denoted by C. The local leaves of the foliations W^1 and W^2 are denoted by $W^1_\epsilon(p)$ and $W^2_\epsilon(p)$, and $[p, q] := W^1_\epsilon(p) \cap W^2_\epsilon(q)$. Note that due to transversality one can assume that for $p, q \in U$ close enough $[p, q]$ consists of a unique point.

Theorem 3.3.3 *Under the assumption of Theorem 3.3.1, there are constants $\epsilon', C' > 0$, and a polynomial $Q(\cdot; p)$ of degree n , for each $p \in U$, such that*

$$|Q(p; q) - u(q)| \leq C'|p - q|^{n+\alpha}, \text{ for } q \in U, |q - p| < \epsilon'.$$

Let $p \in U$. Since the argument is local, one can make a $C^{n,\alpha}$ change of variable that maps $W^1_\epsilon(p)$ and $W^2_\epsilon(p)$ into the coordinate axis through the origin.

Theorem 3.3.3 is a corollary of the following proposition that will be proved later.

Proposition 3.3.4 *Given $k > 0$ and a cone*

$$\mathcal{K}(k) = \{(x, y) \in \mathbb{R}^2 : |y| \leq k|x|\},$$

there exists a polynomial $Q = Q(\mathcal{K})$ of degree n and constants $C_1, \epsilon_1 >$, depending on k and uniform with respect to $p \in U$, such that

$$|u(z) - Q(z)| \leq C_1 |z|^{n+\alpha}, \text{ for } z \in U \cap \mathcal{K} \cap B(0, \epsilon_1).$$

Proof of Theorem 3.3.3 Using Proposition 3.3.4 one can also construct an approximating polynomial Q' on the cone $\mathcal{K}' = \{(x, y) : |x| \leq k'|y|\}$ centered on the $W^1_\epsilon(p)$. By choosing k' large enough, $V = U \cap \mathcal{K} \cap \mathcal{K}' \cap B(0, \epsilon_1)$ becomes a nonempty open set. This implies that the nth degree polynomials Q and Q' have to coincide, because they have a contact higher than n.

Thus at the point p, in the linearizing coordinates, Q is a local approximation of u of order $n + \alpha$. Back to the original coordinates, the polynomial Q becomes a $C^{n,\alpha}$ function $\mathcal{Q}$, but the local approximating property still holds. Denote by $Q(\cdot; p)$ the nth order Taylor polynomial of $\mathcal{Q}$ at p. Then $R(q) = \mathcal{Q}(q) - Q(q; p)$ satisfies

$$\begin{aligned} |R(q)| &= \left| \frac{1}{(n+1)!} \int_0^1 (1-t)^{n-1} \frac{d^n}{dt^n} [R(p + t(q-p))] dt \right| \\ &\leq C \left\| \frac{d^n}{dt^n} \right\| |q - p|^{n+\alpha}. \end{aligned}$$

□

In the reminder of this section we prove Proposition 3.3.4. By the reduction done so far, the point p is the origin in $\mathbb{R}^2$ and the coordinate axis are leaves of the two laminations.

The following lemma is an immediate corollary of the Lagrange interpolation formula.

Lemma 3.3.5 *Let $n \geq 1$ and $B > 0$. Then there are constants $\epsilon = \epsilon(B) > 0, C = C(B) > 0$ such that for any collections of points $\{z_{k,l} : 0 \leq k, l \leq n\} \subset \mathbb{R}^2$, $\{x_k : 0 \leq k \leq n\} \subset \mathbb{R}$, and $\{y_l : 0 \leq l \leq n\} \subset \mathbb{R}$ that satisfy*

$$R/\eta < B \text{ and } |z_{k,l} - (x_k, y_l)| \leq \epsilon\eta,$$

where

$$R = \sup_{k,l} |z_{k,l}|, \quad \eta = \inf_{(k,l) \neq (k',l')} |z_{k,l} - z_{k',l'}|,$$

and for any set of real numbers $\{b_{k,l} : 0 \leq k, l \leq n\}$, there exists a unique polynomial $p(x, y) = \sum_{p,q \leq n} c_{pq} x^p y^q$ with $p(z_{k,l}) = b_{k,l}$. Moreover, one has the estimate

$$\sum_{p,q} |c_{pq}| R^{p+q} \le C \sup_{k,l} |b_{k,l}|. \tag{3.3.1}$$

Since $u(\cdot, 0), u(0, \cdot)$ are $C^{n,\alpha}$, and Proposition 3.3.4 is true for them by approximating with the Taylor polynomial, one can replace $u(x, y)$ by $u(x, y) - u(x, 0) - u(0, y) + u(0, 0)$, and assume that u vanishes along the coordinate axis.

Then one selects a convenient grid on each axis: $(r'_k, 0)$ along the x-axis, and $(0, s'_k)$ along the y-axis, such that both sequences converge to $(0, 0)$, both $\{|r'_k|\}, \{|s'_k|\}$ are geometric progressions with same fixed ratio $0 < \omega < 1$, and $|r'_1|, |s'_1|$ are both of order ϵ_1, uniformly with respect to $p \in U$. The quantity ϵ_1 will be specified later.

For $k, l \ge 1$, use the local product structure to define the points $z_{k,l} := [(0, s_l), (r_k, 0)]$. Note that the C^1-continuity of the stable and unstable leaves and uniform transversality implies that

$$|[(x, 0), (0, y)] - (x, y)| = O(|(x, y)|).$$

The polynomial Q will be found as a limit of polynomials obtained by interpolating over parts of the infinite grid above, and truncating the terms of higher order.

Consider now the rectangular grid $S_{k,l} = \{(0, 0)\} \cup \{(r_{k'}, 0) : k \le k' < k + n\} \cup \{(0, s_{l'}) : l \le l' < l + n\} \cup \{z_{k',l'} : k \le k + n, l \le l' \le l + n\}$.

For each k, l introduce

$$\begin{aligned} \eta_{k,l} &:= \inf\{|z - z'| : z, z' \in S_{k,l}, z \ne z'\}, \\ R_{k,l} &:= \sup\{|z| : z \in S_{k,l}\}, \\ T_k &= \max\{|r_k|, |s_k|\}. \end{aligned}$$

Then there are uniform constants C_0, k_1 such that for $k, l \ge k_1$,

$$\begin{aligned} \frac{R_{k,l}}{\eta_{k,l}} &< C_0, \text{ for } |k - l| \le 1, \\ |z_{k,l} - (r_k, s_l)| &\le \frac{\epsilon(C_0)}{C_0} |(r_k, s_l)| \text{ for } |k - l| \le n, \\ C_0^{-1} \omega^{2k} &\le T_k \le C_0 \omega^{2k}, \end{aligned}$$

where $\epsilon(C_0)$ is the constant from Lemma 3.3.5.

One can apply now Lemma 3.3.5 to conclude that for $k > k_1$ there exists a unique degree n polynomial $P(x, y) = \sum_{p,q \le n} c_{pq} x^p y^q$ which interpolates u on the grid $S := S_{k,k}$. Moreover,

$$\sum_{p,q} R_S^{p+q} \le C \sup\{|u(z)| : z \in S\}, \tag{3.3.2}$$

where $C = C(C_0)$ appears in Lemma 3.3.5, and $R_S = R_{k,k}$. Similar observations are valid for the grid $S_{k,k+1}$.

Consecutive interpolations Assume now $k \geq k_1$. Let P, P' be the interpolating polynomials over $S_{k,k}$, $S_{k,k+1}$ respectively, with coefficients $c_{pq}, c'_{p,q}$ respectively. The next goal is to show that

$$|c'_{pq} - c_{pq}| < O(T_k^{n+\alpha-p-q}). \tag{3.3.3}$$

By Lemma 3.3.5 applied to $P - P'$, (3.3.3) follows if one obtains a good upper bound for $|P - P'|$ on the grid $S_{k,k+1}$. It is enough to show that for $(x, y) \in S_{k,k+1}$:

$$|P'(x, y) - P(x, y)| = O(T_k^{n+\alpha}).$$

Instead, one shows the intermediary result

$$\sum_{p,q} |c'_{p,q} - c_{p,q}| T_k^{p+q} \leq C(T^{n+\alpha} + \delta \sum_{p+q>n} |c_{pq}| T_k^{p+q} + \sum_{p+q\leq n} |c_{pq}| T_k^{n+\alpha}). \tag{3.3.4}$$

Again, due to Lemma 3.3.5 applied to $P - P'$, it is enough to obtain a good estimate as above for $|P' - P|$ on the grid $S_{k,k+1}$. Moreover, since P, P' agree on $S_{k,k+1}$ except for the points $z_{k',k+n}$, $k \leq k' < k+n$, and since $P'(z_{k',k+n}) = u(z_{k',k+n})$, it is enough to find estimates for $|u(z_{k',k+n}) - P(z_{k',k+n})|$ for $k \leq k' \leq k + n$. We parameterize the leaf $W^2_\epsilon(r_{k'}, 0)$ by the y coordinate, and denote a point on it by $z_{k'}(y) = (x_{k'}(y), y)$. Then choose a constant $C_2 > 0$, independent of k, such that the interval I_k of length $2C_2T_k$ along $W^2_\epsilon(r_{k'}, 0)$, centered in the origin, contains all the points from the grid $S_{k,k}$ belonging to $W^2_\epsilon(r_{k'}, 0) \cap \mathcal{K}$. This is the place where we use the cone condition. Note that the functions $x_{k'}(y)$ are $C^{n,\alpha}$ on I_k.

Fix k' as above. If f is a function, to simplify the notation, we denote $\tilde{f}(y) = f(z_{k'}(y))$, and assume the domain of $\tilde{f}$ to be I_k.

Lemma 3.3.6 *(i) There is $C > 0$ such that if $k \geq k_1$, $k \leq k' \leq k + n$, and $y \in [-C_2T_k, C_2T_k]$, then*

$$|(\tilde{u} - \tilde{P})(y)| \leq C T_k^{n+\alpha} \| \frac{d^n}{dy^n}(\tilde{u} - \tilde{P}) \|_\alpha. \tag{3.3.5}$$

(ii) If $p, q \leq n$ and $p + q > n$ then

$$\| \frac{d^n}{dy^n} x^p_{k'}(y) y^p \|_\alpha \leq C T^{p+q-n-\alpha} \| x_{k'} \|_{C^{n,\alpha}(I_k)}.$$

(iii) If $p+q \leq n$

$$\|\frac{d^n}{dy^n} x_{k'}^p(y) y^p\|_\alpha \leq C.$$

Consequently:

$$\begin{aligned}&\|\frac{d^n}{dy^n}\tilde{P}\|_\alpha \\ &\leq C\|x_{k'}\|_{C^{n,\alpha}(I_k)} \sum_{p+q>n} |c_{pq}| T_k^{p+q-n-\alpha} + C \sum_{p+q\leq n} |c_{pq}|.\end{aligned} \tag{3.3.6}$$

Proof (i) Note that $u - P$ is $C^{n,\alpha}$ along $W^2_\epsilon(r_{k'}, 0)$ and has $(n+1)$ distinct zeroes. Thus each derivative $(\tilde{u} - \tilde{P})^{(j)}$, $j = 0, \ldots, n$, has at least one zero, say t_i, in the interval I_k. One has

$$\begin{aligned}\left|\frac{d^n}{dy^n}(\tilde{u} - \tilde{P})(t)\right| &= \left|\frac{d^n}{dy^n}(\tilde{u} - \tilde{P})(t) - \frac{d^n}{dy^n}(\tilde{u} - \tilde{P})(t_n)\right| \\ &\leq \left\|\frac{d^n}{dy^n}(\tilde{u} - \tilde{P})\right\|_\alpha (C_2 T_k)^\alpha,\end{aligned}$$

and, similarly,

$$\begin{aligned}&\left|\frac{d^j}{dy^j}(\tilde{u} - \tilde{P})(t)\right| \\ &= \left|\int_{t_i}^t \frac{d^{j+1}}{dy^{j+1}}(\tilde{u} - \tilde{P})(\tau) d\tau\right| \leq \left\|\frac{d^{j+1}}{dy^{j+1}}(\tilde{u} - \tilde{P})\right\|_{C^0(I_k)} C_2 T_k,\end{aligned}$$

and (i) follows.

(ii) Notice that, by the Leibnitz product formula, $\frac{d^n}{dy^n} x_{k'}^p(y) y^p$ is a sum of terms $\Pi x_{k'}^{p'} y^{q'}$, where Π is a product of derivatives of $x_{k'}$ terms, and $p' + q' \geq p + q - n$. Its Hölder norm can be bounded by

$$\begin{aligned}&\|\Pi x_{k'}^{p'} y^{q'}\|_\alpha \leq \|\Pi\|_\alpha \|x_{k'}^{p'}\|_{C^0} \|y^{q'}\|_{C^0} \\ &\quad + \|\Pi\|_{C^0} \|x_{k'}^{p'}\|_\alpha \|y^{q'}\|_{C^0} + \|\Pi\|_{C^0} \|x_{k'}^{p'}\|_{C^0} \|y^{q'}\|_\alpha.\end{aligned} \tag{3.3.7}$$

We use now the uniform estimates of the $C^{n,\alpha}$-norm of $x_{k'}$, which are given by W^2 being a uniform lamination. Note that $|x_{k'}(0)| \leq T_k$, and the derivative of $x_{k'}$ is bounded, so $\|x_{k'}\|_{C^0} \leq C T_k$. Also, note that $\|f\|_\alpha \leq C\|f'\|_{C^0} T_k^{1-\alpha}$ for any differentiable function $f : I_k \to \mathbb{R}$. Thus:

$$\begin{aligned} \|x_{k'}^{p'}\|_\alpha &\le C\|x_{k'}\|_{C^0} T_k^{p'-\alpha}, \\ \|x_{k'}^{p'}\|_{C^0} &\le C T_k^{p'}, \\ \|y^{q'}\|_\alpha &\le C T_k^{q'-\alpha}, \\ \|y^{q'}\|_{C^0} &\le C T_k^{q'}, \\ \|\Pi\|_\alpha &\le C \left\| \frac{d^n}{dy^n} x_{k'} \right\|_\alpha + C\|x_{k'}\|_{C^n} T_k^{1-\alpha}. \end{aligned} \tag{3.3.8}$$

The estimates (3.3.8) imply that the first term in the right-hand side of formula (3.3.7) is bounded by $CT_k^{p+q-n-\alpha}\|x_{k'}\|_{C^{n,\alpha}}$. If Π is not a constant then $\|\Pi\|_{C^0} \le C\|x_{k'}\|_{C^n}$, and one obtains the claimed bounds for the other two terms in the right-hand side of (3.3.7). If Π is a constant, then $q = n, q' = 0, p = p'$ hence the first term vanishes, the second term has the claimed bound, and the Hölder norm in the third term of (3.3.7) is zero.

(iii) This is proven similarly to (ii). □

Observe now that the local leaves $W^2(x)$ are continuous in $C^{n,\alpha}$-topology and the leaf through the origin coincides with the vertical axis. So for any $\delta > 0$, by choosing the constant $\epsilon_1 > 0$ small enough, one has

$$\|x_{k'}\|_{C^{n,\alpha}} < \delta, \text{ for } |r_{k'}| < \epsilon_1, y \in I_k. \tag{3.3.9}$$

The constant k_1 can be increased so that the last estimate is true for all $k \ge k_1$.

Since $u \in C^{n,\alpha}(W_\epsilon^2(r_{k'}, 0))$ uniformly, one has also

$$\left\| \frac{d^n}{dy^n} \tilde{u}(y) \right\| \le C. \tag{3.3.10}$$

Finally, from (3.3.10), (3.3.6) and (3.3.5) follows

$$|(u - P)(z_{k'}(y))| \le CT_k^{n+\alpha} + C\delta \sum_{p+q>n} |c_{pq}|T_k^{p+q} + C \sum_{p+q\le n} |c_{pq}|T_k^{n+\alpha}, \tag{3.3.11}$$

and the intermediary result (3.3.4) holds.

Convergence of interpolating polynomials For any k we denote by P_{2k} the interpolating polynomial corresponding to the grid $S_{k,k}$ and by P_{2k+1} the interpolating polynomial corresponding to the grid $S_{k,k+1}$. Then formula (3.3.4) relates the coefficients of P_{2k+1} and P_{2k}. Using the smoothness of u along the other foliation, a similar formula holds relating the coefficients of P_{2k+2} and P_{2k+1}.

Recall that T_k is comparable to ω^k. By abuse of notation, we denote $T_{[j/2]}$ by $T_{j/2}$. Let c^m_{pq} be the coefficient of $x^p y^q$ in P_m.

We show now that, by reducing ϵ_1, and by taking K, m_0 large enough, one has

$$|c^m_{pq}| \leq K \sum_{j=m_0}^{m-1} (T_{j/2})^{n+\alpha-p-q} \text{ for } m \geq m_0, \tag{3.3.12}$$

and

$$|c^{m+1}_{pq} - c^m_{pq}| \leq K T^{n+\alpha-p-q}_{m/2} \text{ for } m \geq m_0. \tag{3.3.13}$$

The proof proceeds by induction on m. It is clear that for fixed m_0 both (3.3.12) and (3.3.13) are true for K large enough. We show now that for K, m_0 large enough, and δ small enough, where these quantities depend only on the constant C from (3.3.4), one has that (3.3.12) and (3.3.13) true for $m \geq m_0$ implies that they are true for $m+1$. This will finish the proof by induction.

From equation (3.3.4), and from the bound for $|c^m_{pq}|$ given by (3.3.12), relation (3.3.12) will hold for $m+1$ provided that

$$\begin{aligned} C\Big(T^{n+\alpha}_{m/2} &+ K\delta \sum_{p+q>n} \sum_{j=m_0}^{m-1} (T_{j/2})^{n+\alpha-p-q} T^{p+q}_{m/2} \\ &+ K \sum_{p+q\leq n} \sum_{j=m_0}^{m-1} (T_{j/2})^{n+\alpha-p-q} T^{n+\alpha}_{m/2}\Big) \leq K T^{n+\alpha}_{m/2}. \end{aligned} \tag{3.3.14}$$

Note also that by (3.3.4), formula (3.3.14) implies (3.3.13) as well. If K is large enough, the first term in (3.3.14) is less than $(1/3)K T^{n+\alpha}_{m/2}$.

Consider now the second term divided by the right-hand side of (3.3.14):

$$C\delta \sum_{p+q>n} \sum_{j=m_0}^{m-1} \left(\frac{T_{m/2}}{T_{j/2}}\right)^{p+q-n-\alpha} \leq C\delta \sum_{p+q>n} \sum_{p=m-m_0}^{m-1} (\omega^u)^{p+q-n-\alpha}.$$

By taking δ small enough, the last quantity becomes less than $1/3$. Observe also that the third term can be bounded by the geometric series $CKT^{n+\alpha}_{m/2} \sum_{p+q\leq n} T^{n+\alpha-p-q}_{m_0/2}$. By taking m_0 sufficiently large, this term becomes less than $(1/3)KT^{n+\alpha}_{m/2}$ as well.

End of proof of Proposition 3.3.4 Let K, m_0 as above, uniform on U. By reducing ϵ_1 further, one can assume that $k_1 \geq m_0$. Recall that $P_m(x, y) = \sum_{p,q} c^m_{pq} x^p y^q$.

Denote by l_{pq} the coefficients of the polynomial Q that appears in Lemma 3.3.4. If $p+q > n$ define $l_{pq} = 0$, and if $p+q \leq n$ define $l_{pq} = \lim_{m\to\infty} c^m_{pq}$. The limit exists by (3.3.13).

We claim now that for any $M > 0$ there exists $\eta(M) > 0$ such that for all $m \geq m_0$

$$|Q - P_m| \leq \eta(M) T_{m/2}^{n+\alpha} \quad \text{if} \quad |x|, |y| \leq M T_{m/2}. \tag{3.3.15}$$

By (3.3.12), if $p + q > n$, then

$$|c_{pq}^m T_{m/2}^{p+q}| = O(T_{m/2}^{n+\alpha}). \tag{3.3.16}$$

By iterating (3.3.13), if $p + q \leq n$, it follows that

$$|c_{pq}^m - c_{pq}^{m+k}| \leq K \sum_{j=m}^{m+k-1} T_{j/2}^{n+\alpha-p-q},$$

and taking the limit as $k \to \infty$ gives

$$|(c_{pq}^m - l_{pq}) T_{m/2}^{p+q}| \leq K T_{m/2}^{n+\alpha} \sum_{j=0}^{\infty} \left(\frac{T_{(m+j)/2}}{T_{m/2}} \right)^{n+\alpha-p-q}$$

$$\leq CK T_{m/2}^{n+\alpha} \sum_{j=0}^{\infty} (\omega^j)^{n+\alpha+p+q} = O(T_{m/2}^{n+\alpha}),$$

and (3.3.15) holds.

Now from (3.3.12) one obtains the following upper bounds for $p + q \leq n$:

$$|c_{pq}^m| \leq K, \text{ if } m \geq m_0. \tag{3.3.17}$$

Using (3.3.16) and (3.3.17), formula (3.3.11) implies

$$|(u - P)(z_{k'}(y))| \leq C T_{m/2}^{n+\alpha}, \tag{3.3.18}$$

if $m \geq m_0$, $m/2 \leq k' \leq m/2 + n$ and $y \in [-C_2 T_m/2, C_2 T_m/2]$.

Finally, by the triangle inequality, formulas (3.3.15) and (3.3.18) imply

$$|Q(w) - u(w)| \leq C|w|^{n+\alpha}, \tag{3.3.19}$$

for $w \in \mathcal{K} \cap (\cup_{k \geq m_0} W_\epsilon^2(r_k, 0)) \cap B(0, \epsilon_1)$.

To finish the proof of Lemma 3.3.4 we need to show (3.3.19) for any $w \in \mathcal{K} \cap B(0, \epsilon_1)$. We do this by including in the grid arbitrary elements $w \in \mathcal{K} \cap B(0, \epsilon_1)$ that are close enough to the origin, without disturbing the exponential decay of the geometric series $\{r_m\}_m$, $\{s_m\}_m$.

Let $w \in K \cap B(0, \epsilon_1)$, $(x_w, 0) = [p, w]$, $(0, y_w) = [w, p]$. Let $\{r_m\}_m$, $\{s_m\}_m$ be the initial geometric series. Assume that $r_{m+1} < x_w < r_m$. Then replace $\{r_m\}_m$ by a new sequence $\{r'_m\}_m$ given by $r'_n = r_n$ if $n \geq m + 2$, $r'_{m+1} = X_w$ and $r'_n = r_{n-2}$ is $n \leq m$. Proceed similarly for the sequence $\{s_m\}_m$. Finally, observe that the new sequences differ from the initial ones only at a finite number of positions. So the limit polynomial Q remains the same. □

3.4 The Jacobian along the stable leaves of a partially hyperbolic diffeomorphism

In this section we introduce the Jacobian along the stable leaves of a partially hyperbolic diffeomorphism, which is defined below in Theorem 3.4.4, and show some of its basic properties. Most of the results are classical and originate in the book by Anosov [2], where they are a fundamental tool in the proof of the ergodicity of geodesic flows on manifolds of strictly negative sectional curvature. See also [143] for an extension to partially hyperbolic diffeomorphisms.

The technical result that we need later in this chapter is the high regularity for the Jacobian. This was proved in [104] for smooth diffeomorphisms and in [100] for real analytic ones. We follow the presentation in [127].

In what follows, if not stated otherwise, K is one of $\{2, ..., \infty, \omega\}$.

Definition 3.4.1 Let $U \subset \mathbb{R}^n$ be an open set. A function $f : U \to \mathbb{R}^\ell$ is C^K if it has continuous derivatives of order K. The space $C^K(U)$ of all C^K functions on U, endowed with the norm defined as the supremum of the derivatives of order up to K, has a structure of Banach space.

Definition 3.4.2 Let $U \subset \mathbb{R}^n$ be an open set. The (vector) space $(U, \mathbb{R}^\ell)$ of analytic mappings from U to $\mathbb{R}^\ell$ is topologized by the following system of neighborhoods of the origin: f is in $V_{r,L}$ if it admits a complex-analytic extension $\widetilde{f} : U_r \to \mathbb{C}^\ell$ such that $\sup_{z \in U_r} \|\widetilde{f}(z)\| < L$, where

$$U_r := \{z = (z_1, z_2, \ldots, z_n) \in \mathbb{C}^n | |\text{Im } z_i| \leq r, \\ (\text{Re } z_1, \text{Re } z_2, \ldots, \text{Re } z_n) \in U\}.$$

Assume that M is a C^K compact manifold, with a C^K Riemannian metric. Consider a partially hyperbolic diffeomorphism $f \in \text{Diff}^K(M)$. Denote the splitting of TM by $TM = E^u \oplus E^{cs}$, where $E^{cs} = E^c \oplus E^s$, and denote the strong unstable foliation by W^u. By Theorem 1.8.14, W^u is a continuous foliation with C^K leaves.

Definition 3.4.3 Given a C^K-foliation, by a C^K foliated chart we mean a map $\chi : U \times V \to M$ which is a homeomorphism onto an open subset of M and such that, for each $v \in V$, $\chi_v := \chi(\cdot, v) : U \to M$ describes locally a leaf of the foliation, χ_v is C^K, and depends continuously on $v \in V$ in the C^K topology.

Let μ be the measure induced by the Riemannian metric on M.

Theorem 3.4.4 *Given any point $x_0 \in M$, there is a C^K foliated chart $\chi : U \times V \to M$ of W^u, centered around x_0, such that the pull-back $\chi^*\mu$ of the measure μ is absolutely continuous with respect to the volume on transversals by μ. Moreover, $\chi^*\mu = \rho(u, v)dudv$, with the Jacobian ρ continuous on $U \times V$, C^K in the U-variable, and such that $v \in V \to \rho(\cdot, v) \in C^K(U)$ is continuous.*

Proof Let D' and D'' be two smooth transversals to W^u, and denote by $h : D' \to D''$ the holonomy map along W^u. The map h is absolutely continuous with respect to the measure induced on the transversals by the Riemannian metric. The Radon–Nikodym derivative is given by

$$J_{D''}^{D'}(y) = \lim_{n\to\infty} \frac{\det(f^{-n}|T_y D')}{\det(f^{-n}|T_{h(y)} D'')}, \quad y \in D', \tag{3.4.1}$$

the limit being uniform (see [143, Theorem 2.1]).

Denote by $\det(f|_V)$, where $V \subset T_x M$ is a vector space, the absolute value of the determinant of $Df_x|_V$ with respect to a pair of orthonormal bases of $V \subset T_x M$, respectively $D_x f(V) \subset T_{f(x)} M$. We want to show that the Jacobian $J_{D''}^{D'}$ is C^K along the leaves of W^u.

First we have to write (3.4.1) in a different form. Choose a continuous bundle F which is complementary to E^u. Denote by $\tilde{\pi} : TM \to F$ the projection along E^u. Denote by $\tilde{f} := \tilde{\pi} \circ Df|_F$ the compression of Df to F. Since $\tilde{f}$ preserves E^u, the matrix of Df with respect to the decomposition $TM = E^u \oplus F$ has the form

$$Df = \begin{pmatrix} * & * \\ 0 & \tilde{f} \end{pmatrix}. \tag{3.4.2}$$

Note that $\tilde{f}$ is an invertible bundle map on F.

We first show that

$$J_{D''}^{D'}(y) = \lim_{n\to\infty} \frac{\det(\tilde{f}^{-n}|F_y)}{\det(\tilde{f}^{-n}|F_{h(y)})} \cdot \frac{\det(\tilde{\pi}|T_y D')}{\det(\tilde{\pi}|T_{h(y)} D'')}, \quad y \in D', \tag{3.4.3}$$

the limit being uniform.

Denote by $\pi : TM \to E^{cs}$ the projection along E^u. Then

$$J_{D''}^{D'}(y) = \lim_{n\to\infty} \frac{\det(f^{-n}|E_y^{cs})}{\det(f^{-n}|E_{h(y)}^{cs})} \cdot \frac{\det(\pi|T_y D')}{\det(\pi|T_{h(y)} D'')}, \quad y \in D',$$

the limit being uniform. Note that this is equivalent to (3.4.3): π commutes with Df, hence

$$Df^{-n}|T_y D' = (\pi|T_{f^{-n}(y)} f^{-n}(D'))^{-1} \circ Df^{-n}|E_y^{cs} \circ \pi|T_y D',$$

which implies that

$$\det(f^{-n}|T_yD') = \frac{\det(f^{-n}|E^{cs}_y)\det(\pi|T_yD')}{\det(\pi|T_{f^{-n}(y)}f^{-n}(D'))},$$

and the denominator converges uniformly to 1 as $n \to \infty$. See [143].

Since $\tilde{f}^k \circ \tilde{\pi}|E^{cs} = \tilde{\pi} \circ Df^k|E^{cs}$, one obtains that

$$\det(f^{-n}|E^{cs}_z) = \det(\tilde{f}^{-n}|F_z)\det(\tilde{\pi}_z|E^{cs}_z)\det(\tilde{\pi}_{f^{-n}(z)}|E^{cs}_{f^{-n}(z)})^{-1},$$

and since $\tilde{\pi} \circ \pi = \tilde{\pi}$, one has

$$\det(\pi_z|T_zD) = \det(\tilde{\pi}_z|T_zD)\det(\tilde{\pi}_z|E^{cs}_z)^{-1}.$$

Therefore, for $y \in D'$

$$\begin{aligned}&\frac{\det(f^{-n}|E^{cs}_y)\det(\pi|T_yD')}{\det(f^{-n}|E^{cs}_{h(y)})\det(\pi|T_{h(y)}D'')}\\ &\quad= \frac{\det(\tilde{f}^{-n}|F_y)\det(\tilde{\pi}|T_yD')}{\det(\tilde{f}^{-n}|F_{h(y)})\det(\tilde{\pi}|T_{h(y)}D'')} \cdot \frac{\det(\tilde{\pi}|E^{cs}_{f^{-n}(h(y))})}{\det(\tilde{\pi}|E^{cs}_{f^{-n}(y)})},\end{aligned}$$

and (3.4.3) follows because $h(y) \in W^u(y)$ and $d_M(f^{-n}(h(y)), f^{-n}(y))$ converges uniformly to zero as $n \to \infty$.

We introduce the coordinate system χ as follows. Consider a C^K system of coordinates around x_0, $\rho : \bar{U} \times \bar{V} \subset \mathbb{R}^{\dim E^u} \times \mathbb{R}^{\dim E^{cs}} \to M$, such that $x_0 = \rho(0,0)$ and $\rho_u := \rho(u,.) : \bar{V} \to M$ are transversals to W^u for each $u \in \bar{U}$. Denote the image of $\rho(u,\cdot)$ by $D_u \subset M$. For a point x close to x_0 define its coordinates $\chi^{-1}(x) = (u,v) \in \bar{U} \times \bar{V}$ by the condition that $x = D_u \cap W_{0,\mathrm{loc}}(\rho(0,v))$. Then $\chi(u,v) = h_u(\rho(0,v))$, where $h_u : D_0 \to D_u$ is the holonomy map along W^u. The domain of χ will be an open subset $U \times V \subset \bar{U} \times \bar{V}$ containing the origin. The map χ is C^K. It remains to show that its Jacobian is C^K.

Note that $\rho^{-1} \circ \chi(u,v) = (u, \tilde{h}_u(v))$, where $\tilde{h}_u = \rho_u^{-1} \circ h_u \circ \rho_0$. Hence, up to composition by some C^K functions and changes of variable, the Jacobian of χ is given by $J(u,y) := J^{D_0}_{D_u}(y)$, with $u \in U$ and $y \in D_0$. We are done once we prove that $J(u,y)$ is C^K in $u \in U$, uniformly with respect to $y \in D_0$.

Choose as F a C^K bundle on M which is complementary to E^u, and denote $\mathcal{D}(z) := \det(\tilde{f}|F_z)$. In view of (3.4.2), one can rewrite (3.4.3) as

$$J(u,y) = \lim_{n\to\infty} \frac{\det(\tilde{f}^{-n}|F_y)}{\prod_{k=0}^{n-1} \mathcal{D} \circ f^{-k}(h_u(y))} \cdot \frac{\det(\tilde{\pi}|T_yD_0)}{\det(\tilde{\pi}|T_{h_u(y)}D_u)}.$$

Note that $\mathcal{D}$ is C^K along the leaves of W^u. Indeed, one can choose locally trivializing frames of TM consisting of the union of a frame for E^u and a C^K

frame for F. The vector fields in the former can be chosen to be C^K along the leaves of W^u. Then all the entries of the matrix of Df with respect to these frame are C^K along the leaves of W^u, hence the same holds for $\tilde{f}$ as well. We also obtain that $\tilde{\pi} : TM \to F$ is C^K along the leaves of W^u.

Given a product of positive real valued functions $P = a_{n-1} \cdots a_1 a_0$, its derivative is given by

$$DP = P[D(\log a_{n-1}) + \cdots + D(\log a_1) + D(\log a_0)].$$

Denote by D_u the differential in the U-variable (i.e., along W^u), and let

$$J_n(u, y) := \frac{\det(\tilde{f}^{-n}|F_y)}{\prod_{k=0}^{n-1} \mathcal{D} \circ f^{-k}(h_u(y))} \cdot \frac{\det(\tilde{\pi}|T_y D_0)}{\det(\tilde{\pi}|T_{h_u(y)} D_u)}.$$

Then

$$\begin{aligned} &D_u J_n(u, y) \\ &\quad = -J_n(u, y)\Big(D_u \log(\det(\tilde{\pi}|T_{h_u(y)} D_u)) + \sum_{k=0}^{n-1} D_u \log(\mathcal{D} \circ f^{-k}(h_u(y)))\Big) \\ &\quad = -J_n(u, v)\Big(D_u \log(\det(\tilde{\pi}|T_{h_u(y)} D_u)) \\ &\qquad + \sum_{k=0}^{n-1} D_u \log \mathcal{D}|_{f^{-k}(h_u(y))} D_u f^{-k}|_{h_u(y)} D_u h_u|_y\Big). \end{aligned} \tag{3.4.4}$$

Due to the factor $D_u f^{-k}$ and the uniform convergence of J_n one obtains that $D_u J_n$ converges uniformly, hence J is differentiable along the leaf, and

$$\begin{aligned} &D_u \log(J(u, y)) \\ &\quad = -\Big(D_u \log(\det(\tilde{\pi}|T_{h_u(y)} D_u)) + \sum_{k=0}^{\infty} D_u \log(\mathcal{D} \circ f^{-k}(h_u(y)))\Big). \end{aligned} \tag{3.4.5}$$

We have proved so far that the Jacobian of W^s is C^1.

To obtain that it is C^K, K finite or infinite, differentiate further in formula (3.4.5). The first derivatives of f^{-n} give enough contraction to compensate for the growth due to the increase in the number of terms in the derivative that appear due to the product rule. Complete details for a similar computation are shown in Section 4.2.2.

To obtain that the Jacobian is real analytic, notice that the right-hand side in (3.4.5) is real analytic, uniformly with respect to y. Indeed, each factor involved in the sum in formula (3.4.5) is analytic in u, uniformly with respect to y, and the sum remains convergent when evaluated on the complexification of the unstable leaves. □

3.5 Smooth regularity by Fourier method

In this section we present a regularity result from [56]. Similar to Journé's theorem (Theorem 3.3.1), it allows to show overall regularity of a real valued function once we know its regularity along the leaves of several foliations. Even though it does not allow for the optimal regularity in the finite differentiability setup, it has several advantages: it is simpler than Journé's theorem and the pair of foliations can be replaced by a web of foliations for which the tangent distributions generate the whole tangent space. The proof elaborates upon an unpublished idea of C. Toll. It uses a direct approach based on estimates for the decay rate of the Fourier transform via a cone method.

We start with a definition that summarizes the properties needed for the web of foliations. Briefly, the foliations need to be continuous, to have C^k-leaves and have the Jacobians along the leaves uniformly C^k.

Definition 3.5.1 Let $W_1, W_2, \dots, W_r$ be a family of continuous foliations on a smooth n-dimensional manifold M with leaves of dimensions n_i, $n_1 + n_2 + \cdots + n_r = n$. The family is called a *C^k-regular web of foliations* if the following are true:

(i) For each $1 \le i \le r$ and $p \in M$ the leaf $W_i(p)$ is a C^k immersed submanifold of M, which depends continuously in C^k topology on the point p.

(ii) The tangential distributions TW_i are pairwise transversal and $TM = TW_1 \oplus \cdots \oplus TW_r$.

(iii) For each $1 \le i \le r$ there exists an $\epsilon > 0$ such that for each $p \in M$ there exists a coordinate system $\Phi_p : (-\epsilon, \epsilon)^n \to M$ which satisfies:

  (a) $\Phi_p(0, \dots, 0) = p$.

  (b) For each i let $x_i \in (-\epsilon, \epsilon)^{n_i}$ and for $x \in (-\epsilon, \epsilon)^n$ let $x = (x_1, \dots, x_r)$. Then letting x_i vary and keeping the other coordinates fixed, one obtains a local C^k-chart in the leaf

$$W_i(\Phi_p(x_1, \dots, x_i = 0, \dots, x_r)).$$

  (c) For each $1 \le i \le r$ let dv_i denote the $(n - n_i)$-volume form on $\mathbb{R}^{n-n_i}$ lifted to $\mathbb{R}^n$ via the product structure given by (b). Then the push forward density $\omega_i^\Phi := \Phi_{p,*}(dv_i)$ is a continuous $(n - n_i)$-form on the image of Φ_p.

  (d) The global form ω_i^Φ is C^k when restricted to the leaves of W_i.

The following theorem is the main result of this section.

Theorem 3.5.2 *Let $W_1, W_2, \ldots, W_r$ be a C^k-regular web of foliations on a smooth manifold M. Suppose that $f : M \to \mathbb{R}$ is continuous, and for each $1 \leq i \leq r$ the restriction of f to the leaves of W_i is C^k, with the Jacobian along W_i being C^k along the leaves of W_i and depending transversally continuous on $p \in M$. Then f is C^{k-n-1} on M.*

Proof Since the proof of regularity is local, one can restrict the study to Euclidean open sets. In particular, one can assume that f has compact support in a common foliation chart U for all of the foliations W_i, which is centered in the origin.

The loss of regularity in the statement of the theorem appears due to the Sobolev lemma. We present the proof only for $k = \infty$. The modifications for the general case are straightforward.

If $f : \mathbb{R}^n \to \mathbb{R}$ is a continuous function with compact support, one can define the Fourier transform $\hat{f}$ of f by

$$\hat{f}(\xi) = (2\pi)^{-n/2} \int_{\mathbb{R}^n} \exp(i\xi \cdot x) f(x) dx, \quad \xi \in \mathbb{R}^n. \tag{3.5.1}$$

The proof of the theorem follows from the next fact, which we prove in the sequel. Here f is the function from the theorem after a change of coordinates that is specified below.

For each integer $m > 0$ there are constants $C(m) > 0$, $T(m) > 0$ such that for all $\xi \in \mathbb{R}^n$ with $\|\xi\| = 1$

$$|\hat{f}(t\xi)| < C(m)t^{-m} \quad \textit{for } t > T(m). \tag{3.5.2}$$

Indeed, the estimate (3.5.2) implies that for each $s > 0$ the function f belongs to the s-Sobolev space on $\mathbb{R}^n$. Since f has compact support and is continuous, one can apply the Sobolev lemma to conclude that f is C^∞.

We show now the proof of (3.5.2). The first step is to make a change of coordinates that forces the leaves of the foliations passing through the origin to become appropriate coordinate planes. This is possible because the individual leaves are C^∞ submanifolds. Since the tangential distributions are continuous fields, one can also arrange that, on the support of f, the above coordinate planes are C^0-close to TW_i near the origin.

More precisely, let $x = (x_1, \ldots, x_r) \in \mathbb{R}^n = \mathbb{R}^{n_1} \oplus \cdots \oplus \mathbb{R}^{n_r}$, with $x_i \in \mathbb{R}^{n_i}$, $1 \leq i \leq r$. Let $x^i := x - x_i$ and let $\tilde{W}_i$ be the foliation W_i in the above local coordinates. Then one can assume that there exists $\epsilon > 0$ such that:

(i) The leaf $\tilde{W}_i(0)$ contains the disk $\{x | x_j = 0,\ j \neq i,\ \|x_i\| < \epsilon\}$.
(ii) For each $1 \leq i \leq r$ there exists a function $\psi^i : (-\epsilon, \epsilon)^n \to \mathbb{R}^n \ominus \mathbb{R}^{n_i}$ such that each leaf $\tilde{W}_i(x^i)$ is locally given by the graph

$$\{x_i + \psi^i\left(x_i + x^i\right) | x_i \in \mathbb{R}^{n_i}\}, \tag{3.5.3}$$

and $0 < \delta < r^{-2}$ such that $\|\psi^i(x))\| < \delta$ for all $x \in \mathbb{R}^n$ with $\|x\| < \epsilon$.

(iii) For each x^i the function $\psi^i_{x^i} : (-\epsilon, \epsilon)^{n_i} \to \mathbb{R}^n \ominus \mathbb{R}^{n_i}$ given by $x_i \to \psi^i\left(x_i + x^i\right)$ is C^∞ and depends continuously on x^i in C^∞-topology. Moreover, the first derivative of $\psi^i_{x^i}$ is uniformly bounded by δ.

(iv) For each x_i the function $\psi^i_{x_i} : (-\epsilon, \epsilon)^{n-n_i} \to \mathbb{R}^n \ominus \mathbb{R}^{n_i}$ given by $x^i \to \psi^i\left(x_i + x^i\right)$ is absolutely continuous, with absolutely continuous inverse, and pushes forward the standard Lebesgue measure dx^i on $\mathbb{R}^{n-n_i}$ to a continuous measure.

For any vector $\xi = (\xi_1, \ldots, \xi_r) \in \mathbb{R}^n$, with $\xi_i \in \mathbb{R}^{n_i}$, there exists an index $1 \le i \le r$, such that $r\|\xi_i\| \ge \|\xi\|$. This is equivalent to saying that ξ lies in a cone centered on the coordinate plane $\mathbb{R}^{n_i} \subset \mathbb{R}^n$. Fix now a value i with this property.

Denote $\hat{F}(t) = \hat{f}(t\xi)$. Then make a change of coordinates in the integral (3.5.1) using $\left(x_i, x^i\right) = \left(x_i, \psi^i\left(x_i, v^i\right)\right)$ and separate the variable v_i to obtain

$$\hat{F}(t) = (2\pi)^{-n/2} \int_{\mathbb{R}^{n-n_i}} \Phi\left(v^i\right) dv^i,$$

where

$$\begin{aligned}\Phi(t, v^i) &\\ = \int_{\mathbb{R}^{n_i}} &\exp\left[it\left(\xi_i x_i + \xi^i \psi^i\left(x_i, v^i\right)\right)\right] f\left(x_i, \psi^i\left(x_i, v^i\right)\right) \\ &\times \operatorname{Jac}\left(\psi^i\right)\left(x_i, v^i\right) dx_i,\end{aligned} \tag{3.5.4}$$

where $\operatorname{Jac}(\psi^i)(x_i, v^i)$ is the standard Jacobian. By the hypotheses, the function $x_i \to \operatorname{Jac}(\psi^i)(x_i, v^i)$ is C^∞ in x_i and depends continuously on v^i in the C^∞-topology.

The second step of the proof is to make a second change of variable in (3.5.4), which allows to show super-polynomial decay for this function in the variable t, uniform with respect to v^i. One writes (3.5.4) as a convolution integral along the leaves of W_i, with the integrand being the product between f, restricted to the leaves, and other terms that appear from the change of variables. The other terms are uniformly smooth due to the regularity assumptions on the web of foliations. Then the decay in the variable t is a consequence of the standard properties of the Fourier transform of smooth functions. Then the proof of the fact above follows as $\hat{F}(t)$ is obtained from $\Phi(t, v^i)$ by integrating with respect to the second variable over a compact set.

Let A be an invertible $n_i \times n_i$ matrix, with first row the vector $\xi_i \in \mathbb{R}^{n_i}$, and subsequent rows an orthogonal basis to the orthogonal space to ξ_i. Let B be a $n_i \times (n - n_i)$ matrix, with first row the vector ξ^i, and all the other entries equal to zero. Introduce a new variable:

$$\begin{aligned} z_i &= z_i\left(\cdot, v^i\right) : (-\epsilon, \epsilon)^{n_i} \to \mathbb{R}^{n_i}, \\ z_i\left(x_i, v^i\right) &= x_i + A^{-1}B\psi^i\left(x_i, v^i\right). \end{aligned} \tag{3.5.5}$$

The choice of A and B guarantees that

$$\xi_i z_i(x_i, v^i) = \xi_i x_i + \xi^i \psi^i\left(x_i, v^i\right),$$

and, moreover, one has the operator norm estimate

$$\|A^{-1}B\| \leq \|\xi^i\| \|\xi_i\|^{-1} \leq r.$$

By the assumptions made on ψ^i, the function defined by (3.5.5) is injective in x_i and the matrix differential $\partial z_i / \partial x_i$ is invertible, uniformly in v^i. Denote the inverse function $x_i = \alpha(z_i, v^i)$ and observe that it is C^∞ in z_i, uniformly in v^i. After substitution, formula (3.5.4) becomes

$$\Phi(t, v^i) = \int_{\mathbb{R}^{n_i}} \exp(it\xi_i z_i) F\left(z_i, v^i\right) dz_i,$$

with integrand

$$\begin{aligned} F\left(z_i, v^i\right) =& f\left(\alpha\left(z_i, v^i\right), \psi^i\left(\alpha\left(z_i, v^i\right)\right)\right) \operatorname{Jac}\left(\psi^i\right)\left(\alpha\left(z_i, v^i\right), v^i\right) \\ & \times \left|\frac{\partial \alpha}{\partial z_i}\right|\left(z_i, v^i\right). \end{aligned} \tag{3.5.6}$$

The function F has compact support and all the terms in (3.5.6) are C^∞ in z^i, uniformly with respect to v^i. So its Fourier transform in z_i has subexponential decay in the variable z_i uniformly in v^i. □

Remark 3.5.3 There are two fundamental examples to which Theorem 3.5.2 can be applied. One appears as the pair of stable/unstable foliations of a C^k Anosov diffeomorphism or the pair stable/strong unstable foliations of a C^k Anosov flow. These foliations are continuous with C^k leaves, and, according to Theorem 3.4.4, have the Jacobians along the leaves uniformly C^k. The second example appears in the rigidity theory of higher rank abelian actions on tori that are Cartan (see Section 2.2.5). In this case the web of foliations consists of one-dimensional foliations that are intersections of stable foliations for various Anosov elements of the abelian action.

3.6 Real analytic regularity by Fourier method

In this section we present a real analytic regularity result from [100]. The statement shows the overall real analytic regularity of a real valued function once we know the real analytic regularity for it along the leaves of several foliations. The proof builds upon the method developed in the previous section.

The properties needed for the web of foliations are similar to those that appear in Definition 3.5.1. C^k-regularity is replace by real analytic regularity. Thus, the foliations need to be continuous, to have real analytic-leaves and have the Jacobian along the leaves uniformly real analytic along the leaves. The real analytic structure is defined by coordinate charts as in Definition 3.4.2.

Definition 3.6.1 Let $W_1, W_2, \ldots, W_r$ be a family of continuous foliations on a real analytic n-dimensional manifold M with leaves of dimensions n_i, $n_1 + n_2 + \cdots + n_r = n$. The family is called a *real analytic regular web of foliations* if the following are true:

(i) For each $1 \leq i \leq r$ and $p \in M$ the leaf $W_i(p)$ is a real analytic immersed submanifold of M, which depends continuously in C^k topology on the point p.
(ii) The tangential distributions TW_i are pairwise transversal and $TM = TW_1 \oplus \cdots \oplus TW_r$.
(iii) For each $1 \leq i \leq r$ there exists an $\epsilon > 0$ such that for each $p \in M$ there exists a coordinate system $\Phi_p : (-\epsilon, \epsilon)^n \to M$ which satisfies:
  (a) $\Phi_p(0, \ldots, 0) = p$.
  (b) For each i let $x_i \in (-\epsilon, \epsilon)^{n_i}$ and for x a typical point in $(-\epsilon, \epsilon)^n$ let $x = (x_1, \ldots, x_r)$. Then letting x_i vary and keeping the other coordinates fixed, one obtains a local real analytic chart in the leaf $W_i(\Phi_p(x_1, \ldots, x_i = 0, \ldots, x_r))$.
  (c) For each $1 \leq i \leq r$ let dv_i denote the $(n - n_i)$-volume form on $\mathbb{R}^{n-n_i}$ lifted to $\mathbb{R}^n$ via the product structure given by (b). Then the push forward density $\omega_i^{\Phi} := \Phi_{p,*}(dv_i)$ is a continuous $(n - n_i)$-form on the image of Φ_p.
  (d) The global form ω_i^{Φ} is real analytic when restricted to the leaves of W_i.

The following theorem is the main result of this section.

Theorem 3.6.2 *Let $W_1, W_2, \ldots, W_r$ be a real analytic regular web of foliations on a smooth manifold M, with the Jacobians along W_i uniformly real analytic in the direction of W_i and depending transversally continuous on $p \in M$. Suppose that $f : M \to \mathbb{R}$ is continuous, and for each $1 \leq i \leq r$ the restriction of f to the leaves of W_i is real analytic. Then f is real analytic on M.*

Proof Using real analytic coordinates everything reduces to proving the theorem in an open set in $\mathbb{R}^n$, $n = \dim M$. For $x_0 \in \mathbb{R}^n$ let $v_1, \ldots, v_n$ be a basis of $\mathbb{R}^n$ such that $v_{n_i+1}, \ldots, v_{n_{i+1}}$ span the tangent space to $W_i(x_0)$, $1 \le i \le r$.

For $\mathbb{T}^n$ n-dimensional torus and $\delta > 0$ that will be determined later, define the functions $\tau : \mathbb{T}^n \to \mathbb{R}^n$ and $\Omega : \mathbb{T}^n \to \mathbb{R}$ by

$$\tau(\theta) = x_0 + \delta \sum_{i=1}^{n} v_i \sin \theta_1,$$

$$\Omega(\theta) = (\cos \theta_1 \cdots \cos \theta_n)^n,$$

where $\theta = (\theta_1, \ldots, \theta_n) \in \mathbb{T}^n$.

One shows that $\tilde{f} = (f \circ \tau)\Omega$ is real analytic, which implies that f itself is real analytic on the range of τ. One studies the decay of the Fourier coefficients

$$\hat{\tilde{f}}_k = \int_{\mathbb{T}^n} e^{ik\theta} \tilde{f}(\theta) d\theta,$$

where $k = (k_1, \ldots, k_n) \in \mathbb{Z}^n$.

Fix $1 \le i \le n$ and consider $\psi = (\psi_1, \ldots, \psi_n)$ a system of coordinates in the range of τ for which $(\psi_1, \ldots, \psi_{n_i})$ gives a parameterization of the W_i direction.

Making a change of variable that "straighten" the coordinates system ψ and taking into account that the multiplicity of τ is 4^n except on a set of measure zero, one has

$$\hat{\tilde{f}}_k = 4^n \int_{\text{range}(\tau)} e^{ik\theta(\psi)} f(\psi(\theta)) \Omega(\psi(\theta)) \frac{\partial(\theta)}{\partial(\psi)} d\psi.$$

Note that according to the assumptions, the Jacobian exists and is real analytic. Moreover, θ and f are real analytic functions in the variables $(\psi_1, \ldots, \psi_{n_i})$. By taking the constant δ small enough, one can also assume that $\theta_1, \ldots, \theta_s$ are sufficiently close to $\psi_1, \ldots, \psi_s$ respectively. All of these happens uniform in the remaining variables.

One performs now the integration in the variables $\psi_1, \ldots, \psi_s$ using a deformation of the contour of integration. Since for sufficiently small δ one has $\theta(\psi_i)$ close to ψ_i, $1 \le i \le s$, on an analytic extension of the domain one has

$$|\hat{\tilde{f}}_k| = M e^{-\xi(|k_1| + \cdots + |k_s|) + \epsilon(\delta)\|k\|}, \tag{3.6.1}$$

where ξ is the width of the strip of analyticity, M is a constant times a uniform bound for the integrand in this complex strip, and ϵ is a constant that can be made as small as possible by taking δ small enough.

Similar estimates with (3.6.1) can be obtained for all foliations W_i, $1 \le i \le r$, so overall

$$|\hat{\tilde{f}}_k| = M'e^{-(\xi/2-\epsilon(\delta))\|k\|},$$

and consequently $\tilde{f}$ is real analytic. □

3.7 Smooth regularity via hypoelliptic theory

3.7.1 Vector fields, Hörmander condition, and hypoellipticity

Let X, Y be two C^∞ vector fields in an open set U in $\mathbb{R}^n$. We denote by $[X, Y]$ or $\mathrm{ad}(X, Y)$ the bracket of X and Y defined by $[X, Y]f = X(Yf) - Y(Xf)$. The bracket is a new vector field.

Let $\{X_1, \dots, X_k\}$ be a family of C^∞ vector fields in an open set U of $\mathbb{R}^n$. One says that the family satisfies *the Hörmander condition* if there exists an integer $r > 0$ such that the vector space generated by the iterated brackets of the vectors X_i, of length less or equal to r, at any point $x_0 \in U$, coincides with $\mathbb{R}^n$.

A differential operator L on U is called *hypoelliptic* if $Lf \in C^\infty$ for some distribution f implies that f is a C^∞ function. If $\{X_1, \dots, X_k\}$ is a family of C^∞ vector fields which satisfies Hörmander condition and if P is a non-commutative polynomial of degree m, that is,

$$P = \sum_{|\alpha| \le m} a_\alpha(x) X^\alpha, \ |\alpha| = k,$$

one says that P is *maximally hypoelliptic* if

$$\sum_{|\alpha| \le m} \|a_\alpha(x) X^\alpha u\| \le C\left(\|Pu\|_0^2 + \|u\|_0^2\right), \tag{3.7.1}$$

for all $u \in C_0^\infty(U)$, where $\|\cdot\|_0$ denotes L^2-norm and $\|\cdot\|$ is the usual Euclidean norm.

One can show that maximal hypoellipticity, together with the Hörmander condition, implies hypoellipticity.

Theorem 3.7.1 (Hörmander) *[52]. Let $\{X_1, \dots, X_k\}$ be a family of C^∞ vector fields in an open set $U \subset \mathbb{R}^n$. If the Hörmander condition is satisfied for some r, then the operator $L = \sum_{i=1}^k X_i^2$ is hypoelliptic on U.*

3.7.2 A smooth regularity result

Let M be a compact manifold and $\mathcal{D}$ be a proper distribution in TM. The distribution $\mathcal{D}$ is called *totally non-integrable* if the tangent space T_pM at any point $p \in M$ is spanned by brackets of vectors fields tangent to $\mathcal{D}$.

The following regularity result can be found in [79, Theorem 2.1].

Theorem 3.7.2 *Let $X_1, X_2, ..., X_k$ be C^∞ vector fields on a manifold M of dimension N. Assume that the sum $\sum_{i=1}^k X_i$ is a totally non-integrable distribution and for each $j \leq r$, the dimension of the space spanned by the commutators of length at most j at each point is constant in a neighborhood.*

Let P be a distribution on M. Assume that for any positive integer $p \leq n$ the p's partial derivatives $X_i^p(P)$ exist as continuous or local L^2 functions. Then P is a $C^{[n/r-N/2]}$ function on M. Moreover, if all partial derivative $X_i^p(P)$ exist as continuous or local L^2 function, then P is a C^∞ function on M.

Proof The proof of the finite regularity version is similar to the proof of the C^∞ version, so we discuss only the later.

Since $\sum_{i=1}^k X_i^2(P)$ is continuous or L^2_{loc}, Höermander's square theorem guarantees that any distribution P is an L^2_{loc}-function.

In order to show higher regularity, one considers the higher order polynomial differential operator $L = \sum_{i=1}^k X_i^m(P)$, where m is an even integer m. To a set of vector fields $X_1, X_2, ..., X_k$ as above, Metivier [114] attached a nilpotent Lie algebra $\mathfrak{g}_x$ to any point $x \in M$ that is generated by the elements $\hat{X}_1, \hat{X}_2, ..., \hat{X}_k$. Let G_x be the simply connected Lie group with Lie algebra $\mathfrak{g}_x$. If π is a unitary representation of G_x, let S_π denote the space of C^∞ vectors of π. By [150, Proposition 7.1] the operator $\pi(\sum_{i=1}^k X_i^m(P))$ is injective as an operator on S_π for any non-trivial irreducible representation π. Using this fact, it is proved in [152, Theorem 0.7] that the operator L is maximally hypoelliptic and (3.7.1) is satisfied.

Denote by $H^\alpha(V)$ the α's Sobolev space of V with norm $\|\cdot\|_\alpha$. By the above and [153, §16] there is a constant $C > 0$ such that for all $f \in C_0^\infty(V)$

$$\|f\|_{m/r} \leq C(\|Lf\|_0^2 + \|f\|_0^2)^{1/2}.$$

In [168] there are introduced mollifiers of the form

$$Op(\phi)u(x) = \int \phi(x, \eta, y)e^{-2i\pi(x-y)\cdot\eta}u(y)dydx,$$

where $\phi(x, \eta, y)$ is a C^∞ function in $\mathbb{R}^{3n}$ all of whose partial derivatives decay exponentially in η. These mollifiers are infinitely smoothing.

Let $\phi_t(x, \eta, y) = \phi(x, t\eta, y)$. Then for any $0 < t \leq 1$ and $s \in \mathbb{R}$, $\{Op(\phi_t)\}_t$ is a bounded set of continuous linear endomorphisms of $H^s(\mathbb{R}^n)$ in the operator norm topology, and for every $u \in H^s(\mathbb{R}^n)$, it follows from [168, Theorem 1.2] that $\{Op(\phi_t)u\}_t$ converges in $H^s(\mathbb{R}^n)$ as $t \to \infty$ to the product of u by the function $\phi(x, 0, x)$. Moreover, if X is a vector field such that for every y there is a compact set that contains the support of $\phi(\cdot, \eta, y)$ for all η, then there is another symbol ψ with super-polynomial decay such that $[X, Op(\phi_t)] = Op(\psi_t)$ [168, Theorem 1.1]. The sequence $\{Op(\psi_t)\}_t$ also satisfies the above convergence property as one can see from the explicit formula given in [168].

Let ψ be a mollifier as above, where V is identified to $\mathbb{R}^n$. By above one can find inductively mollifiers $\phi^{i,j}$ with $\phi = \phi^{0,0}$ such that $[X_j, Op(\phi^{i,j})] = Op(\phi^{i+1,j})$. Then

$$LOp(\phi_t)f = Op(\phi_t)Lf + \sum_{j=1}^{l}\sum_{i=1}^{m}\binom{m}{i} Op(\phi_t^{m-i,j})(X_j)^i f.$$

By the a priori estimates one has

$$\|Op(\phi_t)f\|_{\frac{m}{r}}^2 \leq C\left(\|LOp(\phi_t)f\|_0^2 + \|Op(\phi_t)f\|_0^2\right)$$

$$= C\left(\left\|Op(\phi_t)Lf + \sum_{j=1}^{l}\sum_{i=1}^{m}\binom{m}{i} Op(\phi_t^{m-i,j})(X_j)^i f\right\|_0^2 + \|Op(\phi_t)f\|_0^2\right).$$

By the assumptions all $(X_j)^i f$ are continuous or locally integrable and hence belong to $L^2(V)$ for small enough V. Hence, by the properties of the mollifiers chosen above, $\|Op(\phi_t)f\|_{m/r}$ is bounded as $t \to 0$ and hence $f \in H^{m/r-\epsilon}$ for every $\epsilon > 0$ by Rellich's lemma. Since this is true for all m, the Sobolev lemma implies that f is C^∞. □

Remark 3.7.3 An instance when condition (*) in the statement of Theorem 3.7.2 is satisfied is when the manifold is a quotient of a Lie group by a Lie subgroup and the vector fields have lifts to vector fields in the Lie group.

Part II

Cocycles, cohomology, and rigidity

4

First cohomology and rigidity for vector-valued cocycles

4.1 Cocycles over general group actions: an overview

4.1.1 The relevance of cocycles for dynamical systems questions

In this chapter we study the first cohomology of a group action. Higher order cohomology will be discussed in Chapter 6. First cohomology comes in two flavors: ordinary (or untwisted) and twisted. Most of the chapter deals with ordinary cohomology, which has been extensively studied and for which a more developed theory is available. The twisted version is discussed in Section 4.6.

Cohomology is an equivalence relation on a class of functions called *cocycles*. Cocycles lie at the center of many questions about the rigidity of various smooth actions, existence of invariant structures, and other important properties of the action. A cocycle over a group action allows for a lift of the action in the base to an extended action of a fibered space. Information about the extended action is used for analyzing the action in the base. Another direction is to use this construction in order to built new examples of actions of fibered spaces that inherit the properties that we have in the base. More generally, one can consider extensions to principal bundles. We refer to Section 1.4, Section 5.1, and Section 5.4.4 for more details about extensions to principal bundles.

Another appearance of cocycles is in connection to the orbit equivalence of dynamical systems, in particular time change. Consider two actions $\alpha : G \times X \to X$ and $\alpha' : G' \times Y \to Y$, which have a similar structure (measurable, topological, smooth, etc.), and a bijection $H : X \to Y$, which preserves the particular structure. In addition, H takes orbits of α into orbits of α'.

Since H takes the α-orbit of the point $x \in X$ into the α'-orbit of the point $H(x) \in Y$, the last condition is equivalent to the existence of a map $\beta : G \times X \to G'$ such that

$$H(\alpha(g,x)) = \alpha'(\beta(g,x), H(x)). \qquad (4.1.1)$$

One can show now that (4.1.1) implies that β is a cocycle. Indeed:

$$\begin{aligned}
&\alpha'(\beta(g_1g_2,x), H(x))\\
&\quad = H(\alpha(g_1g_2,x)) = H(\alpha(g_1,\alpha(g_2,x)))\\
&\quad = H(\alpha(g_1, H^{-1}(\alpha'(\beta(g_2,x), H(x)))))\\
&\quad = \alpha'(\beta(g_1, H^{-1}(\alpha'(\beta(g_2,x), H(x)))), \alpha'(\beta(g_2,x), H(x)))\\
&\quad = \alpha'(\beta(g_1,\alpha(g_2,x)), \alpha'(\beta(g_2,x), H(x)))\\
&\quad = \alpha'(\beta(g_1,\alpha(g_2,x))\beta(g_2,x), H(x)),
\end{aligned} \qquad (4.1.2)$$

which gives

$$\beta(g_1g_2,x) = \beta(g_1,\alpha(g_2,x))\beta(g_2,x).$$

If $G = G'$, the question of whether a time change of the G action α' is isomorphic to the action α reduces to the question of whether the cocycle β is cohomologous to the identity cocycle $i(g,x) \stackrel{\text{def}}{=} g$. Indeed, if $P : X \to Y$ is the corresponding transfer map, then $(g,x) \to (P(g),x)$ gives a conjugacy between α and α'.

In another direction, cohomological equations (1.4.5) for real-valued cocycles appear as linear parts in the conjugacy problem for small perturbations of integrable systems. Solving the cocycle equations is a crucial step in finding the conjugacy map. The method was initially introduced by Kolmogorov in his ICM address in 1954, and later developed by Arnold and Moser. It is known today as KAM theory. One starts by linearizing a non-linear problem about an approximate solution. Then the linearized equation, given by a cocycle equation, is solved. Here one may encounter small denominators. Then one inductively improves the approximate solution by using the solution of the linearized equation as the basis of a Newton's method argument.

An example to which the method can be applied is the problem of finding the analytic conjugacy between an irrational rotation of the circle $R_\alpha(x)$, which has a Diophantine rotation number, and a small analytic perturbation that has the same rotation number. Assume that the lift of the rotation to the real line is given by $R_\alpha(x) = x + \alpha$, and that the small perturbation is given $\phi(x) = x + \alpha + \eta(x)$, where $\eta'(x) > -1$ and $\eta(x+1) = \eta(x)$. Moreover, let $H(x) = x + h(x)$, where $h(x+1) = h(x)$, be a candidate for a conjugacy between R_α and ϕ, that is,

$$\phi \circ H = H \circ R_\alpha. \qquad (4.1.3)$$

Then (4.1.3) is equivalent to

$$h(x+\alpha) - h(x) = \eta(x + h(x)). \qquad (4.1.4)$$

If one expands now both sides of this equation, retaining only the first order terms in the (small) quantities h and η, one has

$$h(x+\alpha) - h(x) = \eta(x), \qquad (4.1.5)$$

which is a cocycle equation with the action given by the rotation R_α.

For applications like this, important questions are whether any given cocycle over the action can be trivialized and whether the transfer map is of high regularity.

The set of cohomologous cocycles makes up a *cohomology class* in the space of all cocycles over an action. If the range H of a class of cocycles over a fixed action $\alpha : G \times X \to X$ is an abelian group, then one can define the sum of two cocycles. In this case the set of coboundaries form a subgroup of the abelian group of all cocycles, and one can introduce the structure of an abelian group on the set of cohomology classes. Formally, this coincides with the abstract first cohomology group of G acting on X with coefficients in H. One should note, however, that due to the presence of nontrivial asymptotic behavior of a generic dynamical system, the computation of the cohomology groups only rarely reduces to formal algebraic manipulations.

If the range H is not abelian, the set of cohomology classes does not posses any group structure.

Describing cohomology classes for cocycles with values in an abelian group over an action naturally reduces to two problems. First one is to determine sufficient conditions for the trivialization of a cocycle and provide some reasonable description of these conditions so that they reflect the dynamics of the action. For example, for systems with hyperbolic behavior, closing conditions (see Section 4.2.1) are necessary and sufficient for the trivialization of a cocycle. Equivalently, cohomology classes are determined by averages with respect to all invariant measures. This result was discovered by Livshitz [96, 97], and was immediately applied to thermodynamical formalism [9].

On the other hand, for uniquely ergodic systems it may happen that the obstructions to triviality are given by a finite or even infinite set of invariant distributions that are not signed measures or it may happen that the only obstruction is a zero average with respect to the unique invariant measure. The key notion here is cocycle stability (see Definition 1.4.4), which asserts that the vanishing of all obstructions implies cocycle trivialization. As a simple example, one may mention the Diophantine–Liouville dichotomy for the equation (4.1.5) over the circle rotation. If the rotation number is Diophantine, then the equation has a solution for any C^∞ function η. In contrast, if the rotation number is Liouville, there are always C^∞ functions η for which the equation has no solution.

The second problem is to determine how regular is the transfer map for two cohomological cocycles of certain regularity. There are two directions that have been pursued: showing that a C^r regularity for a cocycle forces some $C^{r'}$ regularity on a Hölder transfer map, and showing that a measurable solution to a Hölder cocycle has to be Hölder. These questions were mostly studied for cocycles over hyperbolic actions.

4.1.2 Principal methods for classification of cocycles

There are three basic techniques for solving cohomological questions. Two of them are geometric in nature: extension along the orbit of a transitive point, and extension along the stable and unstable leaves. The other technique is analytic in nature and involves decomposing the appropriate function spaces into direct sums (or direct integrals) of subspaces invariant under the action, solving the cohomological equations separately and then glueing the solutions together. We will call this method the harmonic analysis method. Before proceeding to a detailed description, let us provide a quick overview of these methods.

Extension along the orbits is introduced in the seminal papers of Livshitz [96, 97] and can be applied to Livshitz' cohomological equation (1.4.5). As mentioned before, closing conditions are natural obstructions to solving Livshitz' cohomological equation. Livshitz proved that these are the only obstructions. To find the solution, one iterates the cohomological equation along the orbit of a transitive point in order to find a candidate for the transfer map P, and then, under the assumption that closing conditions are satisfied, shows that the resulting solution exists everywhere as a continuous function. This method requires only Hölder regularity or even a weaker Dini condition for the cocycles, and can be applied to cocycles with either abelian and non-abelian ranges.

Extension along the stable and unstable manifolds appears in Livshitz' proof of the lift of regularity for the transfer map from Hölder to C^1. The idea of the method is to construct a conjectural solution along the stable and unstable manifolds of various points and see when this construction is consistent. It can also be interpreted as viewing the extended action defined by the cocycle as a partially hyperbolic transformation, and trying to use as much as possible the existence and the properties of the invariant stable and unstable foliations that usually appear for such maps. This method was much extended by several authors and it constitutes now a powerful tool for analyzing cocycles with non-abelian range. Among the first papers emphasizing the method we mention [127], in which optimal results for regularity of the transfer maps in the case of non-abelian cocycles over $\mathbb{Z}$ or $\mathbb{R}$ hyperbolic actions are obtained.

The results show that if two cohomologous cocycles are smooth, a Hölder transfer map with sufficiently large Hölder exponent is actually smooth. The idea of the proof is to show that a proper Hölder assumption implies that the stable and unstable leaves of one extension are carried into stable, respectively unstable, leaves of the other by the natural conjugacy given by the transfer map. Other papers in which this method is developed and used are dedicated to the study of rigidity for cocycles over higher rank abelian actions. See, for example, [21, 71–73]. The basic idea in [73] is to construct an invariant foliation that is sub-foliated by both the stable and unstable foliations of the extension. This foliation is then shown to have compact leaves. In the case of real valued cocycles, an alternative approach is to construct the solution as a differential form, by finding a candidate working along the stable and unstable directions, and then show that it is closed and exact using the commutativity of the action.

The *harmonic analysis method* also appears briefly in Livshitz' papers, in his proof of existence of smooth solutions to cohomological equations for smooth cocycles over a hyperbolic automorphism of a torus. This method is naturally restricted to spaces and actions that arise from certain algebraic constructions which implies more information about the structure of the dual space. Decompositions arising from representation theory along with careful study of the properties of the dual map in many cases give cocycle stability and rigidity results. This is particularly useful in the higher rank setting since most higher rank actions considered so far come from algebraic constructions. One can look at Section 4.4 for a collection of examples.

The harmonic analysis method appears in a more sophisticated way in the work of Veech [169]. Veech investigated cohomological equations over ergodic partially hyperbolic, but not necessarily hyperbolic automorphisms of a torus. His results give necessary and sufficient dual obstructions for solving the cohomological equation using Fourier analysis. In this setup the obstruction to cocycle triviality is the sum of the Fourier coefficients along the individual $\mathbb{Z}^d$-orbits of the dual action. Using number theory, Veech showed that the collection of dual obstructions is equivalent to the usual periodic obstructions that come from the closing conditions. We note that, so far, there is no geometric proof of Veech's result. These results are non-trivial even for the case of a hyperbolic automorphism of a torus.

Katok and Spatzier in [78], using Veech description of the dual obstructions, proved cocycle rigidity for higher rank actions by automorphisms of a torus. They also proved cocycle rigidity for higher rank abelian continuous actions on locally symmetric spaces. The proof of cocycle rigidity for actions on locally symmetric spaces uses the decay of matrix coefficients of irreducible representations of semisimple Lie groups, and the fast decay of

Fourier coefficients along the orbits of the dual action is used in the case of actions by automorphisms of a torus. The method of Fourier analysis is also used in [68] and [69], which investigate the higher order cohomology of higher rank actions of toral automorphisms. We will discuss these results in Chapter 6.

Harmonic analysis appears in a different way in the proofs of the regularity of the transfer map. Here it is used locally. As we mentioned before, smoothness of the solution along the stable and unstable directions of a hyperbolic system can be obtained using the extension along these foliations. Showing that this implies smoothness overall was done initially using the theory of elliptic operators [104] and Fourier transform [56]. The state of art right now, in obtaining regularity of the transfer map for hyperbolic rank-one actions, is to instead use a convenient result of Journé that implies $C^{K+\epsilon}$ overall if we know that the solution is $C^{K+\epsilon}$ along the stable and unstable directions. Nevertheless, for higher rank abelian actions this is not enough due to the fact that often the span of stable/unstable directions that appear in the phase space is not the whole tangent space of the manifold supporting the action. Here one still needs to use techniques from elliptic theory. See, for example, Section 3.7, which summarizes [79].

Let us also mention the paper [37], where the cohomological equation is considered in the prototype non-hyperbolic algebraic situation, namely for horocycle flows on surfaces of constant negative curvature. The main tool used in the paper is the theory of infinite-dimensional unitary representations of $SL(2, \mathbb{R})$. It is shown that there are infinitely many obstructions to the existence of a smooth solution of the cohomological equation over the horocycle flow. If these obstructions vanish, one can obtain Sobolev estimates for the solution. In [38] it is shown a similar result for cocycles over area preserving flows on compact higher genus surfaces under certain assumptions that hold generically.

Mieczkowski [113] extended the techniques from [37] to the study of cohomology for parabolic higher rank actions and discovered remarkable rigidity phenomena in this parabolic situation.

4.2 Vector-valued cocycles in rank-one hyperbolic case

4.2.1 Extension along orbits: the Livshitz theorem

In this section we consider abelian cocycles over rank-one Anosov actions, i.e., Anosov diffeomorphisms and flows. We show that for Hölder cocycles closing conditions are necessary and sufficient for trivialization. We use the notations and results from Section 1.8.1.

Any differentiable $\mathbb{Z}$-action on a compact manifold M is generated by a single diffeomorphism $f : M \to M$, and each cocycle $\beta : \mathbb{Z} \times M \to \mathbb{R}$ is determined by the function $\beta(x) := \beta(1, x)$. Abusing the notation we call the map $\beta(\cdot) : M \to \mathbb{R}$ cocycle as well. Thus, after switching to the additive notation, the Livshitz cohomological equation (1.4.5) becomes:

$$P(fx) - P(x) = \beta(x), \tag{4.2.1}$$

and the closing conditions (1.4.6) become

$$\sum_{k=0}^{n-1} \beta\left(f^k x\right) = 0 \qquad \text{whenever} \qquad f^n x = x. \tag{4.2.2}$$

The following lemma gives a uniform estimate for the difference between the sum of the values of the cocycle along a pseudo-orbit and the sum of the values of the cocycle over a corresponding shadowing orbit, independent of the length of the orbit.

Lemma 4.2.1 *Let $M, f, c, \lambda, x, y, z$ as in Lemma 1.8.7. Let $\beta : M \to \mathbb{R}$ an α-Hölder cocycle. Then there exists a constant $K_1 > 0$ depending only on f and β such that:*

(i) $\sum_{k=0}^{N-1} |\beta\left(f^k x\right) - \beta\left(f^k z\right)| \le K_1 d_M\left(f^N x, x\right)$; *and*
(ii) $\sum_{k=0}^{N-1} |\beta\left(f^k z\right) - \beta\left(f^k y\right)| \le K_1 d_M\left(f^N x, x\right)$.

Proof We prove only (i). The proof of (ii) is similar.

$$\begin{aligned}\sum_{k=0}^{N-1} |\beta(f^k x) - \beta(f^k z)| &\le \sum_{k=0}^{N-1} \|\beta\|_{\text{Hölder}} d_M\left(f^k x, f^k z\right)^\alpha \\ &\le \|\beta\|_{\text{Hölder}} \sum_{k=0}^{N-1} c^\alpha \lambda^{\alpha k} d_M\left(f^N x, x\right)^\alpha \\ &\le \|\beta\|_{\text{Hölder}} c^\alpha \frac{1}{1-\lambda^\alpha} d_M\left(f^N x, x\right)^\alpha \le K_1 d_M\left(f^N x, x\right)^\alpha,\end{aligned}$$

where $\|\beta\|_{\text{Hölder}}$ is the usual Hölder norm:

$$\|\beta\|_{\text{Hölder}} = \sup_{x \ne y} \frac{|\beta(x) - \beta(y)|}{d_M(x, y)^\alpha}.$$

□

Theorem 4.2.2 (Livshitz [97]) *Let M be a compact Riemannian manifold, f a topologically transitive C^1 Anosov diffeomorphism of M, and $0 < \alpha < 1$. Then f is $C^\alpha_{\mathbb{R}}$-cocycle stable i.e. for any α-Hölder real-valued function β satisfying* *condition* (4.2.2) *there exists an* α*-Hölder function* P *on* M *such that* $P(fx) - P(x) = \beta(x)$ *holds.*

Proof Let x be a transitive point of f and set

$$\phi_n(x) = \sum_{k=0}^{n-1} \beta\left(f^k x\right).$$

We will prove first that there is a constant C' such that

$$|\phi_n(x)| < C' d_M(f^n x, x)^\alpha.$$

This needs to be proved only for small $d_M(f^n x, x)$, say for

$$d_M(f^n x, x) < \eta < \Delta/K,$$

where Δ, K are as in Theorem 1.8.4 and the proof of Lemma 1.8.7.

Let $y, z \in M$ be provided by Lemma 1.8.7. Now Lemma 4.2.1 yields

$$\left|\phi_n(x) - \sum_{k=0}^{n-1} \beta\left(f^k z\right)\right| < K_1 d_M(f^n x, x)^\alpha,$$

and

$$\left|\sum_{k=0}^{n-1} \beta\left(f^k z\right)\right| = \left|\sum_{k=0}^{n-1} \beta\left(f^k z\right) - \sum_{k=0}^{n-1} \beta(f^k y)\right| \leq K_1 d_M\left(f^N x, x\right)^\alpha,$$

since y is a periodic point for f and by the closing conditions

$$\sum_{k=0}^{n-1} \beta(f^k y) = 0.$$

Summing up two of the previous inequalities yields

$$|\phi_n(x)| \leq K' d_M(f^n x, x)^\alpha,$$

with $K' = 2K_1$. Observe that the constant K' can be chosen to depend only on C, μ, c, and the diameter of M. Thus the same argument applied to $f^k x$ in place of x will show that

$$|\phi_n(f^k x)| \leq K' d_m(f^{n+k} x, f^k x)^\alpha.$$

Since the orbit of x is dense in M we obtain a Hölder function $P : M \to \mathbb{R}$ such that $P(f^n x) = \phi_n(x)$ for $n = 0, 1, 2, \ldots$. Finally, we have $\beta(x) = P(fx) - P(x)$ on a dense set, and hence everywhere. □

4.2.2 Regularity of the transfer function: the first appearance of the extension along stable manifolds method

The first results about the lift of regularity for the transfer map were proved by Livshitz [96, 97], who showed $C^{\infty}_{\mathbb{R}}$ and $C^{\omega}_{\mathbb{R}}$-cocycle stability for hyperbolic automorphisms of a torus. Guillemin and Kazhdan [43] proved $C^{\infty}_{\mathbb{R}}$-cocycle stability for geodesic flows on negatively curved surfaces, and Collet *et al.* [17] proved $C^{\omega}_{\mathbb{R}}$-cocycle stability for geodesic flows on surfaces of constant negative curvature.

In the general case of Anosov diffeomorphisms and flows the $C^{\infty}_{\mathbb{R}}$-cocycle stability was proved by de la Llave *et al.* [104].

Theorem 4.2.3 *Let M be a compact Riemannian manifold, and f a C^{∞} Anosov diffeomorphism. Let $\beta : M \to \mathbb{R}$ be a C^{∞} cocycle, and $P : M \to \mathbb{R}$ is continuous map such that $\beta(x) = P(fx) - P(x)$. Then P is C^{∞}.*

The proof of Theorem 4.2.3 is done in two steps. The first step shows the regularity along the stable and unstable foliations, and the second step shows that the regularity of P along stable and unstable foliations implies the regularity of P.

To prove the existence of the first derivative along the stable leaves one finds a formula for the derivative of P along the stable leaf $W^s(x_*)$ for $x_* \in M$ a periodic point. The formula will be valid for any stable leaf. Thus P is differentiable along any stable leaf and the derivatives are continuous transversally.

Once there is a formula for the first derivative along the stable leaves, one can introduce nice coordinates and differentiate the formula further to show differentiability of higher order along the stable leaves. This method works in a similar fashion for the unstable leaves.

Let $x_* \in M$ periodic point and let $x \in W^s(x_*)$. Define

$$\begin{aligned} f_n^{x_*}(x) &= P(x_*) + \beta(f^{n-1}x) + \cdots + \beta(fx) + \beta(x) \\ &= P(x_*) + P(x) - P(f^n x). \end{aligned} \tag{4.2.3}$$

For $x \in W^s(x_*)$, $f^n x$ converges to x_*, and hence, from the continuity of P and (4.2.3), it follows that

$$\lim_{n\to\infty} f_n^{x_*}(x) = P(x) \text{ for all } x \in W^s(x_*). \tag{4.2.4}$$

One calculates now $D^s f_n^{x_*}$, where D^s is the derivative along the stable direction. Start with

$$f_n^{x_*} = f_{n-1}^{x_*}(fx) + \beta(x),$$

and differentiate both sides:

$$D^s f_n^{x_*}|_x = D^s f_{n-1}^{x_*}|_{fx} D^s f|_x + D^s \beta|_x. \tag{4.2.5}$$

Iterating (4.2.5) follows that

$$D^s f_n^{x_*}|_x = \sum_{l=0}^{n-1} D^s \beta|_{f^l x} D^s f^l|_x. \tag{4.2.6}$$

The first multiple derivative $D^s \beta|_{f^l x}$ that appears in the general term of the sum is uniformly bounded, and the second multiple derivative $D^s f^l|_x$ can be bounded by λ^l, where λ is the contraction constant of the Anosov diffeomorphism. Thus

$$\lim_{n\to\infty} \sum_{l=0}^{n-1} D^s \beta|_{f^l x} D^s f^l|_x = \sum_{l=0}^{\infty} D^s \beta|_{f^l x} D^s f^l|_x. \tag{4.2.7}$$

In conclusion, the sequence $\{f_n^{x_*}|_{W(x_*)}\}_n$ converges pointwise to $P|_{W(x_*)}$ and the sequence $\{D^s f_n^{x_*}|_{W(x_*)}\}_n$ is uniformly convergent on compact sets. Hence the limit $P|_{W(x_*)}$ is differentiable and

$$D^s P|_x = \sum_{l=0}^{\infty} D^s \beta|_{f^l x} D^s f^l|_x. \tag{4.2.8}$$

Since β and F are smooth, one can differentiate further in formula (4.2.8). In order to write down explicit formulas for higher derivatives, one introduces a convenient system of coordinates. If E^s is the stable bundle and $\delta > 0$, denote by E^s_δ the bundle of balls of radius δ centered in the origin. The proof of the existence of the stable foliation as in [51] guarantees the existence of a $\delta > 0$ and a continuous map $w^s : E^s_\delta \to M$ such that for each $x \in M$, $w^s_x = w^s|_{E^s_{\delta,x}}$ is a C^∞-embedding onto a neighborhood of x in $W^s(x)$ and $w^s_x(0) = x$. Moreover, the functions w^s_x depend continuously on x in C^∞-topology. Alternatively, one can use the fact that the stable foliation has C^∞-leaves with continuous dependence transversely and define w^s_x to be the exponential map of the $W^s(x)$ with respect to a metric on M. Choosing the metric to be the adapted metric (see Remark 1.8.9) induced by f further simplifies the computations.

Define $\phi_x := (w^s_x)^{-1} \circ f \circ w^s_x$, $\phi_x : E^s_{\delta,x} \to E^s_{\delta,x}$. Then ϕ_x is a C^∞ map and satisfies

$$\|\phi_x(v^s)\| \le \lambda \|v^s\|, \; v^s \in E^s_{\delta,x}. \tag{4.2.9}$$

We need the higher derivative of the composition:

$$\Phi_{x,n} = \phi_{\phi^{n-1}(x)} \circ \cdots \circ \phi_{\phi^2(x)} \circ \phi_{\phi(x)} \circ \phi_x. \tag{4.2.10}$$

Differentiating (4.2.10) gives

$$\left[\Phi_{x,n}\right]' = \left[(\phi_{\phi^{n-1}x})' \circ \Phi_{x,n}\right]\left[(\phi_{\phi^{n-2}x})' \circ \Phi_{x,n}\right] \cdots \left[(\phi_x^s)'\right], \tag{4.2.11}$$

and since each first derivative is a contraction by λ one has

$$\|(\Phi_{x,n})'\| \leq \lambda^n. \tag{4.2.12}$$

To estimate higher derivatives we differentiate (4.2.11) further. Note that the higher derivatives are multilinear functions, and obey the chain and product rules. Using these, the kth derivative of $\Phi_{x,n}$ is an expression with $a_{n,k}$ terms, each one has $b_{n,k}$ factors. The product rule implies that

$$a_{n,k+1} \leq a_{n,k} b_{n,k}. \tag{4.2.13}$$

Since the number of factors in a term can increase only by the appearance of new first derivatives, and since there are only n of them, we also have

$$b_{n,k+1} \leq b_{n,k} + n. \tag{4.2.14}$$

Note also that the number of times the first derivatives appear in a term is at least $n - k$. This is because each terms contains at least n factors, and only k of them can be derivatives of higher that first term.

One chooses now a finite number of coordinate charts. The first derivatives are estimated by λ, and the higher derivatives by their supremum. The supremum of the higher derivatives appearing in the kth derivative is denoted by C_k. Since the recurrence formulas (4.2.13) and (4.2.14) can be easily solved, one has

$$\|(\Phi_{x,n})^{(k)}\| \leq (k+1)!(1 + C_k)n^{k+1}\lambda^{n-k}. \tag{4.2.15}$$

It follows now from (4.2.15) that, due to the powers of λ that appear any time when we differentiate f, and due to the fact that only a finite number of derivatives of $\beta \in C^\infty$ appear in the derivatives, that the series that appear after further differentiation of (4.2.8) are still uniformly convergent on compact sets. Consequently P is C^∞ along the stable leaves. Moreover, the formula that we have for the first and subsequent derivatives shows that the derivatives along the leaves are continuous overall.

For the second step in the proof of Theorem 4.2.3, which can be viewed as a regularity result independent of the theory of hyperbolic systems, there are several proofs. The original proof in [104] uses the theory of elliptic operators. Necessary assumptions are the smoothness of the function along transverse absolutely continuous foliations and certain regularity conditions on Jacobians. A fact that is very important in some applications involving iterative schemes is that the proof in [104] produces not only a smooth solution to the cohomology equation but also tame estimates for the C^r norm of the transfer function.

Same assumptions are used by Hurder and Katok in [56], but their proof, which uses also Fourier analysis, is simplified. We presented their proof in Chapter 3, as Theorem 3.5.2. This approach, as well as de la Llave approach, has a loss of regularity if one works with finite regularity.

Yet other proof was found by Journé in [58]. His proof uses the Hölder continuity of the foliations, but makes no assumption on the regularity of the Jacobian, and gives finite regularity results without any loss of regularity. Journé's result is restricted to pairs of transverse foliations. A proof of the Journé's result is given in Chapter 3, as Theorem 3.3.1.

In [100] de la Llave proved a C^ω version of the Livshitz theorem for Anosov systems using an approach inspired by [56] based on changes of coordinates and estimates of Fourier coefficients. The Fourier analysis approach, in general, does not produce tame estimates for the transfer function. A proof of de la Llave analyticity result is given in Chapter 3, Theorem 3.6.2.

In [101] de la Llave proves a regularity result for Anosov systems for cocycles belonging to certain Sobolev spaces.

A related direction that was studied in the literature is the lift of regularity for the transfer map from measurable to Hölder. The main idea in the proof of the following theorem is also due to Livshitz.

Theorem 4.2.4 *Let M be a compact Riemannian manifold, and f a C^2 Anosov diffeomorphism. Let μ be an ergodic invariant volume. Assume that $\beta : M \to \mathbb{R}$ is α-Hölder, and there is a μ measurable function $P : M \to \mathbb{R}$ such that*

$$\beta(x) = P(fx) - P(x). \tag{4.2.16}$$

Then there is an α-Hölder function $P' : M \to \mathbb{R}$ such that $P' = P$ μ a.e.

Proof For $x \in M$ set

$$\phi_n(x) = \beta(f^{n-1}x) + \cdots + \beta(fx) + \beta(x).$$

If $y \in W^s(x)$ then estimates as in Lemma 4.2.1 give

$$|\phi_n(x) - \phi_n(y)| \leq \sum_{i=0}^{n-1} |\beta(f^i x) - \beta(f^i y)| \leq C d_M(x, y)^\alpha,$$

which together with (4.2.16) gives

$$|P(x) - P(y)| \leq C d_M(x, y)^\alpha + |P(f^n x) - P(f^n y)|. \tag{4.2.17}$$

By Luzin's theorem there exists a measurable set $S \subset M$ such that $\mu(S) > 1/2$ and P restricted to S is uniformly continuous. Since f is ergodic with respect to μ [67, Theorem 20.4.1], the Birkhoff ergodic theorem applied for the characteristic function of S implies that

$$\lim_{n\to\infty} \frac{1}{n}\mathrm{Card}\{i \,|\, f^i(x) \in S, 0 \leq i \leq n-1\} = \mu(S) > \frac{1}{2}. \tag{4.2.18}$$

The ergodic measure μ is a product measure, that is, for μ a.e. $x \in M$, in a small neighborhood of x, the measure μ is locally equivalent to a product measure $\mu_x^s \times \mu_x^u$, where μ_x^s and μ_x^u are the conditional measures of μ along the local stable and unstable leaves $W^s_{\mathrm{loc}}(x)$, respectively $W^u_{\mathrm{loc}}(x)$ (see [161]).

It follows from Fubini's theorem that for μ a.e. $x \in M$ and for μ_x^s a.e. $y \in W^s_{\mathrm{loc}}(x)$ equation (4.2.18) holds for x and y. Hence for μ a.e. $x \in M$ and for μ_x^s a.e. $y \in W^s_{\mathrm{loc}}(x)$ one can apply repeatedly pigeon-hole principle and choose a subsequence n_i such that $f^{n_i}(x)$, $f^{n_i}(y)$ belong to the set S. Due to the uniform continuity of $P|_S$ this implies that for μ a.e. $x \in M$ and for μ_x^s a.e. $y \in W^s_{\mathrm{loc}}(x)$ one has $|P(x) - P(y)| \leq C d_M(x, y)^\alpha$. Similar considerations can be done along the unstable foliation. The local product for μ implies now that P coincides μ a.e. to a Hölder function (see [67, Proposition 19.1.1]). □

Since we are primarily concerned in this book with cocycles of higher regularity, we refer the reader to [103, 123, 136, 138, 139] for more discussion on the lift of regularity for measurable cocycles.

4.2.3 Invariant foliations for vector-valued extensions

Throughout this section $\mathcal{A}$ denotes $\mathbb{Z}$ or $\mathbb{R}$.

Let M be a compact manifold, $\alpha : \mathcal{A} \times M \to M$ a smooth action, and $\beta : \mathcal{A} \times M \to \mathbb{R}$ a Hölder cocycle. Assume that M is foliated by a continuous, α-invariant, and contracting foliation W, with C^1 leaves and contraction constant λ. A basic result proved in this section is the existence of a lifted contracting invariant foliation for the extension of α by β. The proof is constructive and gives explicit formulas for the leaves of the lifted foliation as graphs of invariant functions over the leaves of W. If the action is a higher rank abelian, we also show that the lifted foliation is independent of the particular map used to build it.

These results can be applied to extensions of Anosov diffeomorphisms and flows, in which case the contracting foliation is the stable one, as well as to extensions of higher rank partially hyperbolic actions, in which case the contracting foliation is an intersection of the stable foliations of various hyperbolic elements of the action. For the sake of completeness we begin with detailed definitions.

Definition 4.2.5 Let M be a compact manifold, and $\alpha : \mathcal{A} \times M \to M$ be a C^1 action on M. Let W be a continuous foliation of M with C^1 leaves $W(x)$, $x \in M$. The foliation W is called *α-invariant* if

$$\alpha(n, W(x)) \subset W(\alpha(n, x)), \quad x \in M, \quad n \in \mathcal{A}.$$

Definition 4.2.6 Let M be a compact manifold and $\alpha : \mathcal{A} \times M \to M$ is a C^1 action on M. Let $0 < \lambda < 1$. An α-invariant foliation W is called *contracting with contraction constant* λ, if there exists a constant $C > 0$, such that

$$\text{dist}_{W(\alpha(n,x))}(\alpha(n, x), \alpha(n, y)) \leq C\lambda^n \text{dist}_{W(x)}(x, y), \tag{4.2.19}$$

for all $x, y \in M, n \geq 0$.

An α-invariant foliation W is called *expanding* if there exists a constant $C > 0$, such that

$$\text{dist}_{W(\alpha(-n,x))}(\alpha(-n, x), \alpha(-n, y)) \leq C_2\lambda^n \text{dist}_{W(x)}(x, y), \tag{4.2.20}$$

for all $x, y \in M, n \geq 0$.

Remark 4.2.7 The contracting/expanding foliations we will use in the future are intersections of stable or unstable foliations of a partially hyperbolic element of an $\mathbb{R}^k$- or $\mathbb{Z}^k$-action, and the $\mathcal{A}$-action α is the restriction of that action to the one-parameter subgroup generated by the element. These foliations have the property that the distance between pairs of points in the same local leaf is equivalent to the distance between points on the manifold. This will allow to replace the induced metric on the leaves $d_{W(x)}(x, y)$ by $d_M(x, y)$ in future arguments.

Recall that the extension of an action $\alpha : \mathcal{A} \times M \to M$ via a cocycle $\beta : \mathcal{A} \times M \to \mathbb{R}$ is the action $\alpha_\beta : \mathcal{A} \times M \times \mathbb{R} \to M \times \mathbb{R}$ defined by

$$\alpha_\beta(n, x, g) = (\alpha(n, x), \beta(n, x) + g). \tag{4.2.21}$$

If n is a positive integer and $f(\cdot) := \alpha(1, \cdot)$, then

$$\beta(n, x) = \beta(f^{n-1}x) + \cdots + \beta(fx) + \beta(x). \tag{4.2.22}$$

Let W be an α-invariant contracting foliation. Our goal is to find a contracting foliation $\{\mathbf{W}(x, h)\}_{(x,h)\in M\times\mathbb{R}}$ of the product space $M \times \mathbb{R}$ that is invariant under the extended action α_β. In addition, we want the lifted foliation $\mathbf{W}$ to have the property that the projections of its leaves into M coincide with leaves of W.

The last condition follows if there exists a family of continuous functions $\{\gamma_x | \gamma_x : W(x) \to \mathbb{R}\}_x$ such that each leaf $\mathbf{W}(x,h)$ is given as the graph of a function γ_x:

$$\mathbf{W}(x,h) := \{(t, \gamma_x(t) + h) | t \in W(x)\}, x \in M, g \in \mathbb{R}.$$

The α_β-invariance of the foliation $\{\mathbf{W}(x,h)\}_{(x,h)}$ amounts to the relation

$$\beta(n,t) + \gamma_x(t) = \gamma_{\alpha(n)(x)}(\alpha(n)(t)) + \beta(n,x),$$

or

$$\gamma_x(t) = -\beta(n,t) + \gamma_{\alpha(n)(x)}(\alpha(n)(t)) + \beta(n,x), \tag{4.2.23}$$

for $n \in \mathcal{A},\ t \in W(x)$.

Since we want the functions γ_x to be continuous and satisfy $\gamma_x(x) = 0$, (4.2.23) suggests to define γ_x by $\gamma_x(t) = \lim_{n\to\infty} [\beta(n,x) - \beta(n,t)]$.

Proposition 4.2.8 *Let M be a compact manifold. Let $\alpha : \mathcal{A} \times M \to M$ be a smooth action. Let W be a contracting α-invariant foliation with contraction constant λ. Let $\beta : M \to \mathbb{R}$ be a Hölder cocycle. For any $x \in M$ and $n \in \mathcal{A}$ define the family of functions $\gamma_{x,n} : W(x) \to \mathbb{R}$ by*

$$\gamma_{x,n}(t) = \beta(n,x) - \beta(n,t) \tag{4.2.24}$$

Then the following statements are true:

(i) The family of functions $\{\gamma_{x,n}\}_n$ converges pointwise as $n \to \infty$.

(ii) The map $\gamma_x : (W(x), d_{W^s(x)}) \to \mathbb{R}$ given by

$$\gamma_x(t) = \lim_{n\to\infty} \gamma_{x,n}(t)$$

is uniformly θ-Hölder.

(iii) $\gamma_x(x) = 0$.

(iv) The family of graphs

$$\mathbf{W}(x,h) := \{(t, \gamma_x(t) + h) | t \in W^s(x)\}, x \in M, h \in \mathbb{R},$$

gives an α_β-invariant foliation of $M \times \mathbb{R}$. This is equivalent to

$$\beta(n,t) + \gamma_x(t) = \gamma_{\alpha(n)(x)}(\alpha(n)(t)) + \beta(n,x),\ n \in \mathcal{A},\ t \in W^s(x).$$

(v) If $y \in W(x)$ and $\nu > \lambda^\theta$ then

$$\lim_{n\to\infty} \nu^{-n} |\beta(n,x) - \beta(n,t) - \gamma_x(t)| = 0.$$

In particular, the foliation $\mathbf{W}$ is α_β contracting.

(vi) The family of functions $\{\gamma_x\}_x$ is uniquely determined by the properties (ii), (iii), and (iv).

(vii) *The family of functions* $\{\gamma_x\}_x$ *is uniquely determined by the property (v), satisfied for a value* $0 < \nu < 1$.

(viii) *The foliation* **W** *depends continuously on the cocycle* β, *that is, the application*

$$\{\beta(1, \cdot) : M \to \mathbb{R}\} \to \{\gamma_x : W(x) \to \mathbb{R}\},\ x \in M,$$

is continuous from the topology of uniform convergence of maps from M into $\mathbb{R}$, *to the topology of uniform convergence on compact sets of maps from the leaves of W to* $\mathbb{R}$.

Proof (i) We claim that the sequence $\{\gamma_{x,n}(t)\}_n$ is uniformly Cauchy, so the limit $\gamma_x(t) = \lim_{n\to\infty} \gamma_{x,n}(t)$ exists uniformly for $t \in W_{loc}(x)$ and defines a continuous function $\gamma_x : W_{loc}(x) \to \mathbb{R}$.

Let $m > n$ and $t \in W_{loc}(x)$. Then one has

$$\begin{aligned} |\gamma_{x,m}(t) - \gamma_{x,n}(t)| &\le \sum_{k=n}^{m-1} |\gamma_{x,k+1}(t) - \gamma_{x,k}(t)| \\ &= \sum_{k=n}^{m-1} |\beta(f^{k+1}x) - \beta(f^{k+1}t)| \qquad (4.2.25) \\ &\le \sum_{k=n}^{m-1} \|\beta\|_{Holder} \lambda^{(k+1)\theta} d_M(x,t)^\theta \le C\lambda^n d_M(x,t)^\theta, \end{aligned}$$

where $C > 0$ is a constant independent of n, x, and t. Notice now that the identity $\gamma_{x,n}(t) = \gamma_{x',n}(t) + \gamma_{x,n}(x')$ implies $\gamma_x(t) = \gamma_{x'}(t) + \gamma_x(x')$, which allows to extend the function γ_x to a whole leaf $W(x)$.

(ii) We show that the functions $\gamma_x : W(x) \to \mathbb{R}$ are θ-Hölder and their Hölder norm is bounded by a constant $C > 0$ independent of $x \in M$.

Let $t, t' \in W(x)$ and n positive integer. Then one has

$$\begin{aligned} |\gamma_{x,n+1}(t) - \gamma_{x,n+1}(t')| &\le \sum_{k=0}^{n} |\beta^{-1}(f^k t) - \beta^{-1}(f^k t')| \\ &\le \sum_{k=0}^{n} \lambda^{k\theta} \|\beta^{-1}\| d_M(t,t')^\theta \le C d_M(t,t')^\theta. \end{aligned}$$

(iii) This follows from the definition of γ_x.

(iv) This follows from $\gamma_{x,n+1}(t) = -\beta(t) + \gamma_{fx,n}(ft) + \beta(x)$, and then taking limit as $n \to \infty$.

(v) From (4.2.21) follows that

$$f_\beta^n(t, \gamma_x(t) + g) = (f^n t, \beta(n,x) + \gamma_x(t) + g).$$

Since f is contracting along W with contraction constant λ it remains to show that

$$\lim_{n\to\infty} \nu^{-n}|\beta(n,x)-\beta(n,t)-\gamma_x(t)|=0.$$

By (iii), (iv), and (ii) one has

$$\begin{aligned}
&|\beta(n,x)-\beta(n,t)-\gamma_x(t)| \\
&\quad = |\beta(n,x)-\gamma_{f^n x}(f^n t)-\beta(n,x)| \\
&\quad = |\gamma_{f^n x}(f^n t)-\gamma_{f^n x}(f^n t)| \\
&\quad = C\lambda^{\theta n} d_M(x,t)^\theta .
\end{aligned}$$

(vi) Let $\{\omega_x : W(x) \to \mathbb{R}\}_{x\in M}$ be a family of functions that satisfies statements (ii), (iii), and (iv). From (iv) we have

$$\omega_x(t) = -\beta(n,t) + \omega_{f^n x}(f^n t) + \beta(n,x), \quad t \in W(x).$$

Then using (iii) and (ii) follows that

$$\begin{aligned}
&|\beta_x(t)-\gamma_{x,n}(t)| \\
&\quad = |-\beta(n,t)+\omega_{f^n x}(f^n t)+\beta(n,t)+\beta(n,t)-\beta(n,x)| \\
&\quad = |\omega_{f^n x}(f^n t)-\omega_{f^n x}(f^n x)| \\
&\quad \le C\lambda^{n\theta} d_M(x,t)^\theta ,
\end{aligned}$$

where $C > 0$ is a constant independent of x, n, t. Now $n \to \infty$ implies $\omega_x(t) = \gamma_x(t)$.

(vii) Let $\{\omega_x : W(x) \to \mathbb{R}\}_{x\in M}$ be a family of functions that satisfies statement (v). Then

$$\begin{aligned}
&|\gamma_{x,n}(t)-\omega_x(t)| \\
&\quad = |\beta(n,x)-\beta(n,t)-\omega_x(t)| \\
&\quad \le \nu^n \nu^{-n}|\beta(n,x)-\beta(n,t)-\omega_x(t)|.
\end{aligned}$$

Now $n \to \infty$ and (5) satisfied by ω_x implies that $\omega_x(t) = \gamma_x(t)$.

(viii) This statement is a consequence of the fact that the application $\beta \to \gamma_x$ is the "uniform limit of a sequence of continuous functions."

Let $\epsilon > 0$ and $x \in M$ be fixed. Let $t \in K$, $K \subset W(x)$ be compact. Let $\beta, \beta' : M \to \mathbb{R}$ be cocycles, with the corresponding families of functions γ_x, γ'_x. If n is large enough and fixed, then statement (ii) implies that

$$|\gamma_{x,n}(t)-\gamma_x(t)| < \frac{\epsilon}{4} \text{ and } |\gamma'_{x,n}(t)-\gamma'_x(t)| < \frac{\epsilon}{4}, \tag{4.2.26}$$

for all $t \in K$. Moreover, if the cocycles β, β' are close in uniform topology, then, for the same n, one can show that

$$|\gamma_{x,n}(t) - \gamma'_{x,n}(t)| < \frac{\epsilon}{2}. \tag{4.2.27}$$

Combining (4.2.26) and (4.2.27) gives

$$|\gamma_x(t) - \gamma'_x(t)| < \epsilon, \tag{4.2.28}$$

and the statement follows. □

Assume now that the foliation W is contracting/expanding under the several $\mathcal{A}$-actions that are parts of a higher rank abelian $\mathcal{A}^k$-action. The following proposition shows that the "height" $\gamma_x(y)$ introduced in Proposition 4.2.8 does not depend on the particular $\mathcal{A}$-subflow used to build it. In order to mark the dependence of γ_x on α, we write γ_x^a if $a \in \mathcal{A}^k$ is a generator of the $\mathcal{A}$-subaction α used to construct the flow.

Proposition 4.2.9 *Let M be compact manifold, $\alpha : \mathcal{A}^k \times M \to M$ a smooth action, and W an α-invariant foliation of M. Let $0 < \lambda < 1$ and $\beta : \mathcal{A}^k \times M \to \mathbb{R}$, be a θ-Holder cocycle over α. Let $a, b \in \mathcal{A}^k$. Assume that W is contracting for the sub-actions induced by $\mathcal{A}a$ and $\mathcal{A}b$. Then:*

(i) $\beta(b, y) + \gamma_x^a(y) = \gamma_{bx}^a(by) + \beta(b, x)$, for $y \in W(x)$; and
(ii) $\gamma_x^a = \gamma_x^b$, for all $x \in M$.

Proof We start by proving (i). For $x \in M$, define the function $\Gamma_x : W(x) \to \mathbb{R}$ by

$$\Gamma_x(y) = \gamma_{bx}^a(by) + \beta(b, y) - \beta(b, y). \tag{4.2.29}$$

Clearly $\Gamma_x(x) = Id_N$, and since the family $\{\gamma_x\}_x$ is uniformly θ-Hölder, the family $\{\Gamma_x\}_x$ is also uniformly θ-Hölder. We will show that Γ_x satisfies condition (iv) in Proposition 4.2.8, and then Proposition 4.2.8(v), implies that $\Gamma_x = \gamma_x$.

Since $a+b = b+a$, the cocycle equation (5.4.26) gives $\beta(b, ax)+\beta(a, x) = \beta(a, bx) + \beta(b, x)$. Together with (5.4.32) this gives:

$$\begin{aligned} \beta(a, y) + \Gamma_x(y) &= \beta(a, y) - \beta(b, y) + \gamma_{bx}^a(by) + \beta(b, x) \\ &= \gamma_{(a+b)x}^a((a+b)y) - \beta(b, ay) + \beta(a, bx) + \beta(b, x) \\ &= \gamma_{(a+b)x}^a((a+b)y) - \beta(b, ay) + \beta(b, ax) + \beta(a, x) \\ &= \Gamma_{ax}(ay) + \beta(a, x). \end{aligned} \tag{4.2.30}$$

To prove (ii), observe that γ_x^a satisfies Proposition 4.2.8(iv), as applied to γ_x^b: this is exactly (i) in this proposition. So, using Proposition 4.2.8(v) again, it follows that $\gamma_x^a = \gamma_x^b$. □

4.2.4 The first appearance of the harmonic analysis argument

The following result appears in [97]. We can view it as a regularity result. Its historical importance should be emphasized since it is the first time when certain harmonic analysis obstructions are exhibited. As we will see later, in Section 4.4, and in Chapter 6, understanding these obstructions is crucial to the generalization of Livshitz' cohomological result, Theorem 4.2.2, to smooth cocycles over ergodic partially hyperbolic automorphisms of a torus.

Theorem 4.2.10 *Hyperbolic automorphisms of the torus $\mathbb{T}^N$ are $C_{\mathbb{R}}^{a,a-\kappa}$ ($\kappa > N$) and $C_{\mathbb{R}}^{\infty}$-cocycle stable.*

Proof Let $A \in SL(n, \mathbb{Z})$ be hyperbolic, and let f be the toral automorphism induced by A. Let $\beta : \mathbb{T}^n \to \mathbb{R}$ be a C^∞ cocycle, and let

$$\beta(x) = \sum_{n \in \mathbb{Z}^n} \beta_n e_n$$

be the Fourier series of β.

One starts by showing that $\beta_0 = 0$ for a cocycle that satisfies the closing conditions. Then define $P_n^+ = \sum_{i=0}^{\infty} \beta_{A^i n}$, $n \in \mathbb{Z}^n$, and

$$P^+(x) = \sum_{n \in \mathbb{Z}^n} P_n^+ e_n.$$

Now, formally, $\beta(x) = P^+(Ax) - P^+(x)$ and P^+ is a distribution. Similarly, the coefficients $P_n^- = -\sum_{-\infty}^{-1} \beta_{A^i n}$ define a distribution P^- which formally satisfies $\beta(x) = P^-(Ax) - P^-(x)$. One needs to show that P^+ is C^∞.

A crucial point of the proof is that if β satisfies the closing conditions then $P^+ = P^-$. Indeed, the closing conditions and Theorem 4.2.2 imply the existence of a Hölder transfer map Q such that $\beta(x) = Q(Ax) - Q(x)$. Since the solution of the cohomological equation is unique and both P^+ and P^- are solutions, they have to coincide with Q.

Thus one can define $P = P^+ = P^-$ and the proof can be finished by showing that P_n decreases faster than any negative power of $\|n\|$, which implies that P is C^∞. Since A is hyperbolic $\mathbb{R}^n = E^s \oplus E^u$ splits into contracting and expanding components. There exist $\lambda > 0$, $C_0 > 0$ such that $\|A^n x\| > C_0 \lambda^l \|x\|$, $x \in E^u$, $n \in \mathbb{N}$. Assume now that n is such that $\|n\| \geq \|n^+\| \geq \|n^-\|$. Then

$$
\begin{aligned}
\|P_n\| &\le \sum_0^\infty \|\beta_{A^i n}\| \\
&\le C\sum_i^\infty \|A^i n\|^{-m} \le C\sum_i^\infty \|A^i n^+\|^{-m} \\
&\le C_1 \sum_i^\infty \lambda^{-im}\|n^+\|^{-m} \le C_2(m)\|n\|^{-m}.
\end{aligned}
$$

Similarly, since $P = P^+ = P^-$ one can show that P_n decreases faster than any negative power of $\|n\|$ for $\|n^-\| \ge \|n^+\|$.

The finite regularity result follows due to standard Sobolev estimates. □

Remark 4.2.11 Veech showed in [169] that the $C^1_{\mathbb{R}}$-cocycle stability for ergodic non-hyperbolic automorphisms of a torus fails. Namely, there exists $f \in C^1(\mathbb{T}^N)$ such that f satisfies closing conditions over some ergodic non-hyperbolic endomorphism of a torus, but the Livshitz equation has no solution of class C^1.

4.3 Cocycles over partially hyperbolic systems

4.3.1 An overview

In Section 4.1.2 we mentioned the result of Veech for ergodic automorphisms of a torus [169]. We will not present a proof of this result here since one can obtain it is a corollary of more general results about (higher) cohomology of higher rank abelian partially hyperbolic actions presented in Chapter 6. Let us note that Veech's proof uses certain arithmetic properties characteristic for ergodic automorphisms of a torus and thus is was difficult to generalize so far to other more general partially hyperbolic systems, like smooth perturbations of partially hyperbolic toral automorphisms or for algebraic actions on manifolds other than the torus..

A more geometric result, that holds for a large class of partially hyperbolic systems, disjoint from the class of ergodic toral automorphisms, appeared in [71]. A simple model result there asserts that partially hyperbolic diffeomorphisms of a compact manifold with C^∞ totally non-integrable (see Section 3.7.2) stable and unstable foliations are $C^\infty_{\mathbb{R}}$-cocycle stable. The obstructions are given by an infinite set of invariant distributions called periodic cycle functionals. The $C^{a,0}_{\mathbb{R}}$-cocycle stability ($0 < a < 1$) is actually proven for a larger class of systems, namely systems satisfying a certain version of the uniform accessibility property. See Definition 4.3.3.

The C^0 solution to the cohomology equation is actually C^∞ along stable and unstable foliations, by the same argument as the one used in the proof of Theorem 4.2.3, which, providing those foliations are smooth and totally non-integrable, implies global smoothness of the solution by a general elliptic operator theory argument. See Section 3.7.2. We note that, in particular, this result implies $\mathcal{C}^\infty_{\mathbb{R}}$-cocycle stability for actions by individual elements of a Cartan subgroup of a semisimple Lie group G on the locally symmetric space $M \setminus G/\Gamma$ as described in Section 2.3.3.

4.3.2 Accessibility of foliations

Let M be a Riemannian manifold and $S \subset M$ a submanifold. For $x, y \in M$ (correspondingly $x, y \in S$) we denote by $d_M(x, y)$ ($d_S(x, y)$) the infimum of the lengths of the smooth curves in M (respectively S) connecting x and y.

Definition 4.3.1 Let $\mathcal{F}_1, \ldots, \mathcal{F}_r$ be a family of foliations of M with smooth leaves. An ordered set of points

$$(x_1, \ldots, x_l, x_{l+1}), x_i \in M, \ 1 \le i \le l+1,$$

is called an $\mathcal{F}_{1,\ldots,r}$*-path* of length l if for every $i = 1, \ldots, l$ there exists $j(i) \in \{1, \ldots, r\}$ such that $x_{i+1} \in \mathcal{F}_{j(i)}(x_i)$. If $x_{l+1} = x_1$, the path is called $\mathcal{F}_{1,\ldots,r}$-*cycle*.

Note that since the leaves of a foliation are connected, and hence path-wise connected, condition $x_{i+1} \in \mathcal{F}_{j(i)}(x_i)$ implies the existence of a smooth path starting in x_i and ending in x_{i+1} included in the leaf $\mathcal{F}_j(x_i)$.

Definition 4.3.2 Let $\mathcal{F}_1, \ldots, \mathcal{F}_r$ be a family of foliations of M with smooth leaves. The family is called *transitive*, or *accessible*, if for any $x, y \in M$ there exists $(x, x_2, \ldots, x_l, y)$ an $\mathcal{F}_{1,\ldots,r}$-path joining x and y.

The family is called *locally transitive* if there exists an integer $N \ge 1$, called the *transitivity index*, such that for any $\epsilon > 0$ there exists $\delta > 0$ such that for any $x, y \in M$, $d_M(x, y) < \delta$, there is an $\mathcal{F}_{1,\ldots,r}$-path $(x = x_1, \ldots, x_{l+1} = y)$, $l \le N$, such that

$$d_{\mathcal{F}_{j(i)}(x_i)}(x_{i+1}, x_i) < \epsilon, \ i = 1, \ldots, l, \ j(i) \in \{1, \ldots, r\}. \tag{4.3.1}$$

Note that local transitivity implies transitivity. These notions can be defined in the finite differentiability setup as well. For our applications it is enough to consider the case when the foliations have smooth leaves.

It is a consequence of Theorem 1.8.4 (local product structure) that the pair of stable/unstable foliations of an Anosov diffeomorphism has local transitivity with a transitivity index of 2. In general, this is not the case for the pair of stable/unstable foliations of a partially hyperbolic diffeomorphism. For example, the stable and unstable foliations of a partially hyperbolic automorphism of a torus are jointly integrable.

The notions of transitivity/local transitivity of a pair of foliations were introduced by Brin and Pesin in [12]. They studied extensively the local transitivity property for the pair of stable and unstable foliations for partially hyperbolic dynamical systems, its relations to other properties of the system, and its stability under perturbations.

Several important classes of partially hyperbolic dynamical systems generically possess the transitivity property. Those include compact Lie group extensions of an Anosov diffeomorphism and the frame flows on manifolds of negative curvature (see [11]).

Later the transitivity condition appeared in the work of Pugh and Shub [144] on stable ergodicity. It is one of the conditions which they require in order for a partially hyperbolic system to be stably ergodic. It is conjectured that for $k \geq 1$, in the class of C^k partially hyperbolic diffeomorphisms those that are transitive are open and dense. This is a subject of extensive current research. See, for example, [27, 49, 129].

The following definition is a Hölder version of local transitivity.

Definition 4.3.3 Let $\mathcal{F}_1, \ldots, \mathcal{F}_r$ be a family of foliations of M with smooth leaves. The family is called *locally θ-Hölder transitive* if there exist $N \in \mathbb{N}$, $\delta > 0$, $C > 0$ such that for every $x, y \in M$ with $d_M(x, y) < \delta$ there is an $\mathcal{F}_{1,\ldots,r}$-path $(x = x_1, \ldots, x_{l+1} = y)$, $l \leq N$, such that

$$d_{\mathcal{F}_{j(i)}(x_i)}(x_{i+1}, x_i) < C d_M(x, y)^\theta, \; i = 1, \ldots, l, \; j(i) \in \{1, ..., r\}. \quad (4.3.2)$$

We recall now an instance of accessibility for foliations generated by smooth vector fields that was widely used in control theory. See [106]. If X is a smooth vector field on a compact smooth manifold M, denote by $e^{Xt}(x)$ the flow of X through the point $x \in M$ after time t. Let $\mathcal{F}$ be a family of smooth vector fields on M. The *orbit* of $x \in M$ under $\mathcal{F}$ is the set of all points $e^{t_1 X_1} \ldots e^{t_k X_k}(x)$ for all $X_i \in \mathcal{F}$ and $t_i \in \mathbb{R}$.

Proposition 4.3.4 *Let M be a compact smooth manifold and $\mathcal{F}$ be a family of smooth vector fields for which the corresponding family of distributions generates the tangent space at each $x \in M$. Then the orbit of any point $x \in M$ under the family $\mathcal{F}$ coincides with M.*

Proof We can assume that $\mathcal{F}$ is the largest possible family of vectors fields that generates the orbit of $\mathcal{F}$. Let $k = \dim M$. If $X_1, \ldots, X_k$ is a family of vector fields that generates the tangent space at the point $x \in M$. For U an open neighborhood of the origin in $\mathbb{R}^k$, consider the map

$$\psi_x : U \to M, \quad \psi_x(t_1, \ldots, t_k) = e^{t_1 X_1} \cdots e^{t_k X_k}(x).$$

The tangent space at each point $e^{t_1 X_1} \ldots e^{t_k X_k}(x)$ is spanned by the linearly independent vector fields from $\mathcal{F}$:

$$X_1, e_*^{t_1 X_1} X_2, \ldots, e^{t_1 X_1} \cdots e_*^{t_{k-1} X_{k-1}} X_k,$$

where for $X, Y \in \mathcal{F}$ we define $e^{t(e_*^X Y)} = e^X e^{tY}$. Thus ψ_x is a local diffeomorphism, and its image covers an open neighborhood of x.

Now one can finish the proof using a standard covering argument. □

The following notion already appeared in Section 3.7.2.

Definition 4.3.5 A family $\mathcal{F}$ of vector fields on a compact manifold M is called *totally non-integrable with index* $p \in \mathbb{N}$ if for any $x \in M$ the Lie brackets of degree at most p of the vector fields in $\mathcal{F}$ span the whole $T_x M$.

A family of smooth foliations of M is called *totally non-integrable with index* $p \in \mathbb{N}$ if there is a family of vector fields, tangent to the foliations, that is totally non-integrable with index p.

The proof of the following theorem is similar to that of [12, Theorem 4.2].

Proposition 4.3.6 *If a pair of smooth foliations* $\mathcal{F}_1, \mathcal{F}_2$ *of a compact manifold* M *is totally non-integrable with index* p*, then the pair is locally* $(1/2^p)$*-Hölder transitive.*

Proof Let $n = \dim M, k_1 = \dim \mathcal{F}_1, k_2 = \dim \mathcal{F}_2$. From the total non-integrability of the foliations, there exist smooth vector fields $v_1, \ldots, v_{k_1}$ tangent to $\mathcal{F}_1$ and $v_{k_1+1}, \ldots, v_{k_1+k_2}$ tangent to $\mathcal{F}_2$, called basis vector fields, such that for any $x \in M$ the vectors $v_1(x), \ldots, v_{k_1}(x), v_{k_1+1}(x), \ldots, v_{k_1+k_2}(x)$ and their brackets of length at most p form a basis of the tangent space $T_x M$. Since this condition is open, there exists an open neighborhood $U(x)$ for which the vectors $v_1(y), \ldots, v_{k_1}(y), v_{k_1+1}(y), \ldots, v_{k_1+k_2}(y)$ and their brackets of length at most p form a basis of the tangent space $T_y M$ for each $y \in U(x)$.

Denote by $T(t, x, v)$ the image of x under the time t map of the flow generated by the vector field v, and by $G(t, x, v)$ the translation by time t of a point $x \in M$ along the geodesic determined by the vector field $v(x)$. Then there is a constant C independent of x and t such that

$$d_M(T(t, x, v), G(t, x, v)) \leq Ct^2. \tag{4.3.3}$$

For $w = [u, v]$ the bracket of two basis vector fields u, v, define the map

$$Q_1(t, x, w) = T(t^{1/2}, T(t^{1/2}, T(t^{1/2}, T(t^{1/2}, x, u), v), -u), -v).$$

If w is the Lie bracket of length m of basis vector fields such that $w = [v, w_1]$, where w_1 is a Lie bracket of length $m - 1$ of basis vector fields, define Q_m by

$$\begin{aligned} &Q_m(t, x, w) \\ &= Q_{m-1}(t^{1/2^m}, T(t^{1/2^m}, Q_{m-1}(t^{1/2^m}, T(t^{1/2^m}, x, v), w_1), -v), -w_1). \end{aligned} \tag{4.3.4}$$

It is standard that (see for example [48, Lemma 8.1, Chapter II])

$$d_M(T(t, x, w), Q_1(t, x, w)) \leq Ct^2, \tag{4.3.5}$$

where C is a constant independent of x and t. By (4.3.5) and induction it follows that

$$d_M(T(t, x, w), Q_m(t, x, w)) \leq Ct^2, \tag{4.3.6}$$

where C is a constant independent of x and t.

Let $\exp_x : B(\epsilon) \to U(x)$ be the exponential map defined in a neighborhood of the origin. Each vector $v \in B(\epsilon)$ can be written in terms of the basis $\{v_i(x)\}_i$ in T_xM as $v = \sum_i^n a_i v_i(x)$. One defines the map $\phi_x : B(\epsilon) \to U(x)$ as

$$\begin{aligned} \phi_x(v) =& Q_{m_n}(a_n, \ldots, v_n) \circ \cdots \circ Q_{m_{k_1+k_2+1}}(a_{k_1+k_2+1}, \ldots, v_{k_1+k_2+1}) \\ &\circ T(a_{k_1+k_2}, \ldots, v_{k_1+k_2}) \circ T(a_1, \ldots, v_1). \end{aligned}$$

From (4.3.3) and (4.3.6) follows that for any $v \in B(\epsilon)$ one has

$$d_M(\exp_x v, \phi_x(v)) \leq C\epsilon^2.$$

Since $\exp_x$ is a local diffeomorphism and ϕ is continuous, the proposition follows. □

Corollary 4.3.7 *Let G be a real semisimple Lie group of non-compact type, A its maximal split Cartan subgroup, Γ an irreducible co-compact lattice, and K any compact subgroup of G that commutes with A. Then for any regular element $a \in A$ acting on $M = K \backslash G / \Gamma$, the family of vector fields corresponding to the hyperbolic directions of a is $(1/2^p)$-Hölder transitive for some $p \in \mathbb{N}$.*

Proof It follows from the general Lie group theory presented in Section 2.3.3 that the family of vector fields corresponding to the hyperbolic directions of the regular element a is totally non-integrable of index p for some $p \in \mathbb{N}$. □

The following stability result for the transitivity property can be found in [12]. See also [26] for a stable transitivity result where the foliations are not smooth.

Theorem 4.3.8 *Let $f_0 : M \to M$ be a partially hyperbolic diffeomorphism with a pair of smooth stable/unstable foliations that is locally transitive. Then there exists a neighborhood U of f_0 in $\mathrm{Diff}^2(M)$ such that any perturbation $f \in U$ is partially hyperbolic and its pair of stable/unstable foliations is locally transitive.*

Proof It follows from Theorem 1.8.14 that f is partially hyperbolic.

Let $x \in M$, $R > 0$, and let $B^s(x, R)$ ($B^u(x, R)$) be the open ball of radius R centered in x inside the stable (unstable) leaf of x. If k_1 (k_2) is the dimension of the stable (unstable) foliation, then there exist open balls $B^s \subset \mathbb{R}^{k_1}$ and $B^u \subset \mathbb{R}^{k_2}$ and smooth injective embeddings $\phi^s_x : B^s \to M$ and $\phi^u_x : B^u \to M$ such that $\phi^s(B^s) = B^s(x, R)$ and $\phi^u(B^u) = B^u(x, R)$. Since the foliations are smooth, one can select ϕ^s_x, ϕ^u_x to depend smoothly on x.

Consider now the smooth maps $\phi^s : M \times B^s \to M$ and $\phi^u : M \times B^u \to M$ given by

$$\phi^s(x, y_1) = \phi^s_x(y_1), \qquad \phi^u(x, y_2) = \phi^s_x(y_2).$$

We construct a sequence of maps $\theta^1, \dots, \theta^N$, where N is the transitivity index of f_0. Pick $x_0 \in M$ and define $\theta^1 = \phi^s_{x_0}$. The map $\theta^2 : B^s \times B^u \to M$ is defined by

$$\theta^2(y_1, y_2) = \phi^u_{\phi^s(y_1)}(y_2).$$

Assume now that the maps $\theta^{2k} : (B^s \times B^u)^k \to M$ are already defined. Then the maps $\theta^{2k+1} : (B^s \times B^u)^k \times B^s \to M$ and $\theta^{2k+2} : (B^s \times B^u)^{k+1} \to M$ are defined by the formulas

$$\theta^{2k+1}(\xi, y_1) = \phi^s_{\theta^{2k}(\xi)}(y_1),$$
$$\theta^{2k+2}(\xi, y_1, y_2) = \phi^u_{\theta^{2k+1}(\xi, y_1)}(y_2),$$

where $\xi \in (B^s \times B^u)^k$, $y_1 \in B^s$, $y_2 \in B^u$. It follows from the local transitivity of f_0 that the image of the map $\theta^N : D^N \to M$ is the whole manifold M, where D^N is the domain of θ^N. Due to Sard's theorem (see [91]), the set of regular values of θ^N is a set of second category and of full measure in M. Let $x_1 \in M$ be a regular point and $y \in D^N$ be a point in its preimage. Since the differential $D\theta^N_y$ is onto, there exists an n-dimensional subspace $E \subset T_y D^N$ such that $D\theta^N_y(E) = T_{x_1}M$. Let $S^{n-1}_\delta \subset E$ be the sphere of radius δ centered in the origin. If δ is sufficiently small, the sphere S^{n-1}_δ can be assumed to be embedded in D^N. Moreover, $\theta^N(S_\delta^{n-1})$ is diffeomorphic to S_δ^{n-1}, while the set bounded by it is diffeomorphic to a ball in M that contains x_1.

From Theorem 1.8.14 follows that the stable and unstable foliations of f depend continuously on the perturbation. This construction can be carried out for any sufficiently small perturbation f of f_0. Let us denote the corresponding maps by $\hat{\phi}^s, \hat{\phi}^u, \hat{\theta}^N$.

Thus for any $\epsilon > 0$ there exists a neighborhood $U(x_0) \subset \mathrm{Diff}^2(M)$ of f_0 for which the corresponding maps $\hat{\phi}^s, \hat{\phi}^u, \hat{\theta}^N$ can be chosen such that

$$\sup_{y \in D^N} d_M(\hat{\theta}^N(y), \theta^N(y)) < \epsilon.$$

Choose $\epsilon = 1/2 \max_{x \in \theta^N(S_\delta^{n-1})} d_M(x, x_1)$. Then for any $f \in U(x_0)$, $(\hat{\theta}^N)^{-1}(x_1)$ is non-empty and, in addition, for each point $x \in M$ with $d_M(x_1, x) < \epsilon/2$ the preimage $(\hat{\theta}^N)^{-1}(x)$ is not empty. Thus there exist points $y_1, \ldots, y_n \in M$ such that $y_1 = x_0, y_N = x_1, y_{i+1} \in W^\nu(y_i), i = 1, \ldots, N-1, \nu = s$ or u with $d_{W^\nu(y_i)}(y_i, y_{i+1}) < 2R$.

For $f \in U(x_0)$ consider the map $\hat{\theta}^N$ constructed starting with the balls $B_{2R}^s(x)$ and $B_{2R}^u(x)$. Let $M'(x_0)$ be the range of this map. Since $M'(x_0)$ contains the range of $\hat{\theta}^N$, and hence a neighborhood of x_1, it also contains a neighborhood of y_{N-1}. By recursively considering the points $y_{N-1}, y_{N-2}, \ldots, y_1 = x_1$, one can show that $M'(x_0)$ contains a neighborhood $M''(x_0)$ of x_0.

Construct now the sets $U(x_0)$ and $M''(x_0)$ for each $x_0 \in M$. The sets $M''(x_0)$ are open and cover M. Choose a finite cover $M''(x_0^1), M''(x_0^2), \ldots, M''(x_0^m)$. Then the intersection $U = \cap_{i=1}^m U(x_0^i)$ is the desired neighborhood of f_0. □

4.3.3 Periodic cycle functionals

In what follows $\mathcal{A} = \mathbb{R}$ or $\mathbb{Z}$, $\alpha : \mathcal{A}^k \times M \to M, k \geq 1$, is a smooth action on M, and $\beta : \mathcal{A}^k \times M \to \mathbb{R}$ a θ-Hölder cocycle over α. All the foliations that appear in this section are assumed to be α-invariant, continuous, and with smooth leaves, and contracting or expanding under the action of $\mathcal{A}a$, $a \in S$ a compact set of generators for $\mathcal{A}$. Therefore the construction of the functions γ_x introduced in Section 4.2.3 can be carried over. In order to emphasize the dependence of the function γ_x on a certain contracting/expanding foliation W in M, we introduce the notation γ_x^W.

Definition 4.3.9 Let $\mathcal{F}_1, \ldots, \mathcal{F}_r$ be a family of foliations of M, each $\mathcal{F}_i$ either contracting or expanding under the action of $\mathcal{A}a_i \subset \mathcal{A}^k, a_i \in Q(\mathcal{A}^k)$, and

$\mathcal{P} = (x_1, \ldots, x_l, x_{l+1})$ an $\mathcal{F}_{1,\ldots,r}$-*path*. We define *the height of* β *over the path* $\mathcal{P}$ to be

$$H(\beta, \mathcal{P}) = \gamma_{x_l}^{\mathcal{F}_{j(l)}}(x_{l+1}) + \cdots + \gamma_{x_2}^{\mathcal{F}_{j(2)}}(x_3) + \gamma_{x_1}^{\mathcal{F}_{j(1)}}(x_2). \tag{4.3.7}$$

Remark 4.3.10 It follows from Proposition 4.2.9 that the height $H(\beta, \mathcal{P})$ does not depend on the particular actions of $\mathcal{A}a_i$. A different choice of the flows for which the foliations are still contracting/expanding gives the same height.

Remark 4.3.11 If the path $\mathcal{P}$ is a cycle, the height was introduced in [71] as *periodic cycle functional.*

The following proposition shows a necessary condition for the triviality of a cocycle.

Proposition 4.3.12 *Let* $\mathcal{F}_1, \ldots, \mathcal{F}_r$ *be a family of foliations of* M*, each* $\mathcal{F}_i$ *either contracting or expanding under the action of* $\mathcal{A}a_i \subset \mathcal{A}^k$*,* $a_i \in \mathcal{Q}(\mathcal{A}^k)$*. Assume that the cocycle* β *is cohomologous to a constant cocycle. Then all the heights of* β *over* $\mathcal{F}_{1,\ldots,r}$*-cycles are equal to zero.*

Proof Let $\pi : \mathcal{A}^k \to \mathbb{R}$ be a homomorphism. Assume that β is cohomologous to π via a continuous transfer map $h : M \to \mathbb{R}$. Let $\mathcal{C} = (x_1, \ldots, x_l, x_{l+1})$, $x_{l+1} = x_1$, be a $\mathcal{F}_{1,\ldots,r}$-cycle. Assume that the foliation $\mathcal{F}_{j(i)}$ is contracting (the proof for expanding is similar) under the action of $\mathcal{A}a \subset \mathcal{A}^k$. Then

$$\begin{aligned}\gamma_{x_i}^{\mathcal{F}_{j(i)}}(x_{i+1}) &= \lim_{t\to\infty}(\beta(ta, x_i) - \beta(ta, x_{i+1}))\\ &= \lim_{t\to\infty}[(h(tax_i) + \pi(ta) - h(x_i)) - (h(tax_{i+1}) + \pi(ta) - h(x_{i+1}))]\\ &= h(x_{i+1}) - h(x_i),\end{aligned} \tag{4.3.8}$$

where for the last equality we use the continuity of h and that

$$\lim_{t\to\infty} d_M(tax_i, tax_{i+1}) = 0.$$

Thus

$$\begin{aligned}H(\beta, \mathcal{C}) &= \gamma_{x_l}^{\mathcal{F}_{j(l)}}(x_1) + \gamma_{x_{l-1}}^{\mathcal{F}_{j(l-1)}}(x_l) + \cdots + \gamma_{x_2}^{\mathcal{F}_{j(2)}}(x_3) + \gamma_{x_1}^{\mathcal{F}_{j(1)}}(x_2)\\ &= h(x_1) - h(x_l) + h(x_l) - h(x_{l-1}) - \cdots\\ &\quad + h(x_3) - h(x_2) + h(x_2) - h(x_1) = 0.\end{aligned} \tag{4.3.9}$$

□

Another instance when all the heights are trivial appears when we work with only one foliation.

Proposition 4.3.13 *Let $\mathcal{F}$ be a contracting or expanding foliation under the action of $\mathcal{A}a \subset \mathcal{A}^k$. Then the heights of β over all $\mathcal{F}$-cycles are trivial.*

Proof We assume that $\mathcal{F}$ is contracting. Let $\mathcal{C} = (x_1, \ldots, x_l, x_{l+1}), x_{l+1} = x_1$, be a $\mathcal{F}$-cycle. Then

$$
\begin{aligned}
H(\beta, \mathcal{C}) &= \gamma_{x_l}^{\mathcal{F}}(x_1) + \gamma_{x_{l-1}}^{\mathcal{F}}(x_l) + \cdots + \gamma_{x_2}^{\mathcal{F}}(x_3) + \gamma_{x_1}^{\mathcal{F}}(x_2) \\
&= \lim_{t\to\infty} \Big(\beta(ta, x_l) - \beta(ta, x_1) + \beta(ta, x_{l-1}) - \beta(ta, x_l) + \cdots \\
&\quad + \beta(ta, x_2) - \beta(ta, x_3) + \beta(ta, x_1) - \beta(ta, x_2) \Big) = 0.
\end{aligned} \tag{4.3.10}
$$

□

If the family of foliations is locally transitive, then the necessary condition presented in Proposition 4.3.12 is also sufficient for the cocycle to be cohomologous to a constant.

The next result proves a criterion for the vanishing of the cohomology in terms of the heights over the cycles (or the periodic cycle functionals). We formulate the result for the higher rank case.

Proposition 4.3.14 *Let $\mathcal{F}_1, \ldots, \mathcal{F}_r$ be a family of locally transitive foliations, each foliation $\mathcal{F}_i$ either contracting or expanding under the action of $\mathcal{A}a_i \subset \mathcal{A}^k$, $a_i \in S$. Assume that $H(\beta, \mathcal{C}) = 0$ for all cycles $\mathcal{C}$ determined by the family. Then β is cohomologous to a constant cocycle via a transfer map h.*

Moreover, if the family of foliations is locally θ'-Hölder transitive, then h is $\theta\theta'$-Hölder.

Proof (i) *Definition of h.* Fix $x \in M$ and let $y \in M$ be an arbitrary point. Since the family of foliations is transitive, there is a $\mathcal{F}_{1,\ldots,r}$-path $\mathcal{C}$ connecting x and y. Define the function $h : M \to \mathbb{R}$ by

$$
h(y) = H(\beta, \mathcal{C}). \tag{4.3.11}
$$

Since $H(\beta, \mathcal{C}) = 0$ for all cycles $\mathcal{C}$ it follows that the function h is well defined. Indeed, if $\mathcal{C}'$ is another path connecting x and y, then the concatenation of $\mathcal{C}$, listed from x to y, and $\mathcal{C}'$, listed from y to x, gives a cycle. Thus $H(\beta, \mathcal{C}) - H(\beta, \mathcal{C}') = 0$, and $H(\beta, \mathcal{C}) = H(\beta, \mathcal{C}')$.

(ii) *Continuity of h.* Continuity of h is a consequence of the local transitivity of the family of foliations. Fix $\epsilon > 0$ and let $\delta > 0$, $N > 0$ be as in Definition 4.3.2. Let $y_1, y_2 \in M$ such that $d_M(x, y_1) < \delta$, $d_M(x, y_2) < \delta$, and consequently $d_M(y_1, y_2) < 2\delta$. Let $(x = x_1, \ldots, x_{l+1} = y_1)$, $l \leq N$, and $(x = x'_1, \ldots, x'_{l'+1} = y_2)$, $l' \leq N$, be paths for which (4.3.1) holds. Recall that $h(x) = 0$. Then

$$
\begin{aligned}
|h(y_1) - h(y_2)| &\le |h(y_1) - h(x)| + |h(x) - h(y_2)| \\
&\le |\gamma_{x_l}^{\mathcal{F}_{j(l)}}(x_{l+1}) + \cdots + \gamma_{x_2}^{\mathcal{F}_{j(2)}}(x_3) + \gamma_{x_1}^{\mathcal{F}_{j(1)}}(x_2)| \\
&\quad + |\gamma_{x'_{l'}}^{\mathcal{F}_{j'(l')}}(x'_{l'+1}) + \cdots + \gamma_{x'_2}^{\mathcal{F}_{j'(2)}}(x'_3) + \gamma_{x'_1}^{\mathcal{F}_{j'(1)}}(x'_2)| \\
&\le \sum_{i=1}^{l} C\|\beta\|_{Holder} d_{\mathcal{F}_{j(i)}(x_i)}(x_i, x_{i+1})^\theta \qquad (4.3.12) \\
&\quad + \sum_{i=1}^{l'} C\|\beta\|_{Holder} d_{\mathcal{F}_{j'(i)}(x'_i)}(x'_i, x'_{i+1})^\theta \\
&\le 2NC\|\beta\|_{Holder}\epsilon^\theta,
\end{aligned}
$$

where the second inequality follows from (4.2.25).

(iii) *h is a transfer map.* Let $a \in \mathcal{A}^k$. Note that if $\mathcal{C} = (x = x_1, \ldots, x_l = y)$ is an $\mathcal{F}_{1,\ldots,r}$-path connecting x and y, then it follows from α-invariance of the family of foliations that $a\mathcal{C} = (ax_1, \ldots, ax_l)$ is an $\mathcal{F}_{1,\ldots,r}$-path connecting ax and ay. Hence by Proposition 4.2.8

$$
\begin{aligned}
h(ay) &= H(\beta, a\mathcal{C}) + h(ax) \\
&= \gamma_{ax_{l-1}}^{\mathcal{F}_{j(l-1)}}(ax_l) + \cdots + \gamma_{ax_2}^{\mathcal{F}_{j(2)}}(ax_3) + \gamma_{ax_1}^{\mathcal{F}_{j(1)}}(ax_2)h(ax) \\
&= \lim_{t\to\infty}(\beta(\epsilon_{l-1}ta_{j(l-1)}, ax_{l-1}) - \beta(\epsilon_{l-1}ta_{j(l-1)}, ax_l)) + \cdots \qquad (4.3.13) \\
&\quad + (\beta(\epsilon_2 ta_{j(2)}, ax_2) - \beta(\epsilon_2 ta_{j(2)}, ax_3)) \\
&\quad + (\beta(\epsilon_1 ta_{j(1)}, ax_1)h(ax_1) - \beta(\epsilon_1 ta_{j(1)}, ax_2)),
\end{aligned}
$$

where $\epsilon_i \in \{\pm 1\}$, depending on the foliation $\mathcal{F}_{j(i)}$ being contracting or expanding.

Observe now that

$$
\begin{aligned}
&\beta(ta_{j(i)}, ax_i) - \beta(ta_{j(i)}, ax_{i+1}) \\
&\quad = \beta(a, x_{i+1}) - \beta(ta_{j(i)} + a, x_{i+1}) + \beta(ta_{j(i)} + a, x_i) - \beta(a, x_i),
\end{aligned} \qquad (4.3.14)
$$

for $1 \le i \le l-1$. So (4.3.13) becomes

$$
\begin{aligned}
&h(ay) \\
&\quad = \beta(a, x_l) + \lim_{t\to\infty}\Big(\beta(\epsilon_{l-1}ta_{j(l-1)} + a, x_{l-1}) - \beta(\epsilon_{l-1}ta_{j(l-1)} + a, x_l) \cdots \\
&\qquad \beta(\epsilon_2 ta_{j(2)} + a, x_2) - \beta(\epsilon_2 ta_{j(2)} + a, x_3) + \beta(\epsilon_1 ta_{j(1)} + a, x_1) \\
&\qquad - \beta(\epsilon_1 ta_{j(1)} + a, x_2)\Big) - \beta(a, x_1) + h(ax_1). \qquad (4.3.15)
\end{aligned}
$$

Note that

$$\begin{aligned}
&\lim_{t\to\infty}\Big(\beta(\epsilon_{l-1}ta_{j(l-1)}+a, x_{l-1}) - \beta(\epsilon_{l-1}ta_{j(l-1)}+a, x_l)\Big)\\
&\quad = \lim_{t\to\infty}\Big(\beta(a, \epsilon_{l-1}ta_{j(l-1)}x_l) + \beta(\epsilon_{l-1}ta_{j(l-1)}, x_{l-1})\\
&\qquad -\beta(\epsilon_{l-1}ta_{j(l-1)}, x_l) - \beta(a, \epsilon_{l-1}ta_{j(l-1)}x_l)\Big)\\
&\quad = \lim_{t\to\infty}\Big(\beta(\epsilon_{l-1}ta_{j(l-1)}, x_{l-1}) - \beta(\epsilon_{l-1}ta_{j(l-1)}, x_l)\Big),
\end{aligned}$$

because β θ-Hölder implies that

$$\lim_{t\to\infty}\Big(\beta(a, \epsilon_{l-1}ta_{j(l-1)}x_{l-1}) - \beta(a, \epsilon_{l-1}ta_{j(l-1)}x_l)\Big) = 0.$$

Similar identities hold for the other products on the right-hand side of (4.3.15), so (4.3.15) becomes

$$\begin{aligned}
h(ay) &= \beta(a, x_l) + h(y) - \beta(a, x_1) + h(ax_1)\\
&= \beta(a, y) + h(y) - \beta(a, x) + h(ax).
\end{aligned} \tag{4.3.16}$$

Define $\pi : \mathcal{A}^k \to \mathbb{R}$ by

$$\pi(a) = \beta(a, x) - h(ax). \tag{4.3.17}$$

Note that π is well defined because x is fixed. We show that π is a representation, that is,

$$\pi(a + b) = \pi(a) + \pi(b). \tag{4.3.18}$$

Formula (4.3.18) is equivalent to

$$\beta(a + b, x) - h((a + b)x) = \beta(a, x) - h(a) + \beta(b, x) - h(b), \tag{4.3.19}$$

which follows immediately from (4.3.16) if we replace y by bx and take into account that $\beta(a + b, x) - \beta(b, x) = \beta(a, b)$.

To show that β is cohomologous to a constant cocycle, observe that (4.3.16) is equivalent to

$$\beta(a, y) = h(ay) + \pi(a) - h(y). \tag{4.3.20}$$

(iv) *h is Hölder.* It remains to show that h is $\theta\theta'$-Hölder if the family of foliations is locally θ'-Hölder. Let $\delta > 0, N > 0, C > 0$ be as in Definition 4.3.3. Let $y_1, y_2 \in M$ such that $d_M(x, y_1) < \delta, d_M(x, y_2) < \delta$, and consequently $d_M(y_1, y_2) < 2\delta$. Let $(x = x_1, \ldots, x_{l+1} = y_1), l \leq N$, and $(x = x'_1, \ldots, x'_{l'+1} = y_2), l' \leq N$, be paths for which (4.3.2) holds. Then, using estimates similar to those in (4.2.25), (4.3.11), and (4.3.7), and $h(x) = 0$, one has

$$\begin{aligned}
|h(y_1) - h(y_2)| &\le |h(y_1) - h(x)| + |h(x) - h(y_2)| \\
&\le |\gamma_{x_l}^{\mathcal{F}_{j(l)}}(x_{l+1}) + \cdots + \gamma_{x_2}^{\mathcal{F}_{j(2)}}(x_3) + \gamma_{x_1}^{\mathcal{F}_{j(1)}}(x_2)| \\
&\quad + |\gamma_{x'_{l'}}^{\mathcal{F}_{j'(l')}}(x'_{l'+1}) + \cdots + \gamma_{x'_2}^{\mathcal{F}_{j'(2)}}(x'_3) + \gamma_{x'_1}^{\mathcal{F}_{j'(1)}}(x'_2)| \\
&\le \sum_{i=1}^{l} C\|\beta\|_{\text{Holder}} d_{\mathcal{F}_{j(i)}(x_i)}(x_i, x_{i+1})^{\theta} \qquad (4.3.21)\\
&\quad + \sum_{i=1}^{l'} C\|\beta\|_{\text{Holder}} d_{\mathcal{F}_{j'(i)}(x'_i)}(x'_i, x'_{i+1})^{\theta} \\
&\le 2NC\|\beta\|_{\text{Holder}} d_M(y_1, y_2)^{\theta\theta'},
\end{aligned}$$

and h is $\theta\theta'$-Hölder. □

The following results give general criteria for cocycle's stability for partially hyperbolic diffeomorphisms. They are both immediate corollaries of Proposition 4.3.14.

Theorem 4.3.15 *If f is a partially hyperbolic diffeomorphism such that the pair of stable/unstable foliations $\mathcal{F} = (W^s, W^u)$ is locally transitive, then, for any $\theta \in (0, 1]$, the space of θ-Hölder cocycles over f is C^0-stable, and the subspace of cocycles cohomologous to a constant is the common zero set of the cycles functionals, i.e., a cocycle β is cohomologous to a constant, with C^0 transfer function, if and only if $H(\beta, C) = 0$ for all cycles C determined by $\mathcal{F}$.*

Theorem 4.3.16 *If f is a partially hyperbolic diffeomorphism such that the pair of stable/unstable foliations (W^s, W^u) is locally θ-Hölder transitive, then, for any $\theta \in (0, 1]$, the space of θ'-Hölder cocycles is both $\theta\theta'$-Hölder stable and C^0-stable. Moreover, in all cases, the subspaces of cocycles cohomologous to a constant are the common zero set of the cycles functionals.*

4.4 Higher rank results for vector-valued cocycles

For cocycles over actions of higher rank abelian groups the cohomological picture may be very different from that in the rank-one case. For the classes of genuinely higher rank abelian Anosov actions described in Chapter 2, including actions by toral automorphisms and Weyl chamber flows, Katok and Spatzier proved that *every* real-valued cocycle is cohomologous to a constant cocycle and that the transfer map is C^∞ if the cocycle is C^∞ [78]. This extends to some partially hyperbolic actions, including actions by ergodic toral automorphisms and compact group extensions of the Weyl chamber flow described in Section 2.3.3. The reasons are of quite a different nature: for the first kind of actions global Fourier analysis can be carried out in non-Anosov situation due to the specific properties of the dual action, while in the latter case the complete non-integrability of stable and unstable foliations, along with general elliptic theory arguments, allows the extension of the cocycle rigidity result to the partially hyperbolic case.

Related results for expansive $\mathbb{Z}^k$-actions by automorphisms of compact abelian groups were established by Katok and Schmidt [77] and for higher dimensional shifts of finite type by Schmidt [155, 156].

Remark 4.4.1 In the case of $\mathbb{Z}^k$ ergodic actions by toral automorphisms, the higher rank assumption is necessary and sufficient for the trivialization of the first cohomology. Namely, in [164] it is proven that any cocycle over such an action is cohomologous to a constant if and only if the action has no non-trivial rank-one subactions which is in turn equivalent to the existence of a $\mathbb{Z}^2$ subaction consisting of ergodic toral automorphisms. It is conjectured that similar results holds for other examples of higher rank abelian actions.

4.4.1 Higher rank trick and vanishing of the first cohomology

Before stating the results and giving the main ideas of the proofs, we note that, as in the rank-one case, there are in general two ways of approaching the cocycle rigidity problem.

The first is based on harmonic analysis and, as we mentioned before, is only applicable to actions of algebraic nature. An essential ingredient in the proof of cocycle rigidity is:

- the fast decay of Fourier coefficients along the orbits of the dual action in the case of actions by toral automorphisms;
- the decay of matrix coefficients of irreducible representations of semisimple Lie groups for actions on locally symmetric spaces.

The role of the fast decay of coefficients is to show that the obstructions to the trivialization of a cocycle vanish providing the action is *genuinely* higher rank. Specifically, one shows that obstructions to the solution of the coboundary equation for one generator is invariant under the rest of the action and the fast decay guarantees that after applying sufficiently large elements of the action the obstructions become arbitrarily small. This argument is sometimes called *higher rank trick*.

Once the obstructions vanish, hyperbolicity of the action allows to establish the smoothness of the transfer map. Estimates for the C^r-norm of the solution may be obtained although in the case of partially hyperbolic actions on locally symmetric spaces the estimate obtained from the general elliptic operator theory argument is not tame. In various cases tame estimates can be obtained using the structure of the neutral direction.

The second approach is geometric, it is not restricted to algebraic actions, and can be applied to non-commutative cocycles as well. However, it requires a special structure of the stable and unstable manifolds for different elements of the action and, in addition, in the non-commutative case, the smallness of the cocycle. These conditions are satisfied for the TNS actions [73]. The main idea is that, given a cocycle, one can assign to it a differential form which is exact along the stable and unstable directions for a regular element of the action. Showing that the cocycle is a coboundary reduces to showing that the form is exact overall and that it agrees for various elements of the action. The geometric approach also applies to cocycles over Weyl chamber flows where a particular structure of the tangent space can be used [33]. A further development of this method uses the structure of the web of stable foliations at a deeper level and draws on techniques from algebraic K-theory to show the vanishing of the periodic cycle functionals. We give a brief introduction to this approach in the next section.

We present now in detail various proofs of cocycle rigidity for several classes of higher rank abelian actions in order to illustrate both methods.

4.4.2 Harmonic analysis method

Rigidity results for real-valued cocycles over higher rank abelian actions are proven by Katok and Spatzier in [78] for Anosov case and are extended to some partially hyperbolic situations in [79]. In this section we present a complete proof for the Weyl chamber flow and sketch the proof for the case of toral automorphisms. The latter case follows from results proved in full detail in Chapter 6, in Theorem 6.2.2.

In both examples rigidity appears due to the exponential mixing property of the action. This fact implies the existence of a solution to the cohomology equation in distributions that have continuous derivatives of all orders in both stable and unstable directions. In the case of Weyl chamber flows, the complete non-integrability of the stable and unstable foliations allows us to apply the general elliptic operator theory argument to establish the smoothness of the solution. In the toral automorphisms case we do not have this geometric property. Nevertheless, the fast decay of Fourier coefficients that exists for this type of actions is sufficient to imply smoothness.

Theorem 4.4.2 *Let G be a semisimple connected real Lie group of non-compact type and of $\mathbb{R}$-rank at least 2. Assume that the Lie algebra $\mathfrak{g}$ of G does not have any factors isomorphic to $so(n, 1)$ or $su(n, 1)$. Let A be the connected component of a split Cartan subgroup of G. Suppose Γ is an irreducible torsion-free co-compact lattice in G. Let $Z(A) = MA$ be the centralizer of A, where M is compact. Let α be the Weyl chamber flow (see Section 2.3.3) of A on $M \setminus G/\Gamma$.*

Then α is $C^{\infty}_{\mathbb{R}}$-cocycle rigid.

The same result holds for the partially hyperbolic action given by the corresponding compact extension.

We start the proof by recalling a few facts on matrix coefficients. Estimates on the decay of matrix coefficients of semisimple Lie groups play an important role in representation theory, and already appear in the work of Harish–Chandra. Among other results in this field, both Ratner and Moore prove exponential decay for Hölder vectors in the real rank-one case [117, 147]. Their results are not directly applicable, so here, following [78], we give a standard treatment based on the notion of a K-finite C^{∞} vector.

Let G be a connected semisimple Lie group with finite center. For π an irreducible unitary representation of G on a Hilbert space $(\mathcal{H}, < \cdot >)$ define the *matrix coefficient* of $v, w \in \mathcal{H}$ as the function $\phi_{v,w} : G \to \mathbb{R}$ given by $g \to < \pi(g)v, w >$. Let K be a maximal compact subgroup of G. A vector $v \in \mathcal{H}$ is called *K-finite* if the K-orbit of v spans a finite dimensional space. Let $\hat{K}$ denote the unitary dual of K. Then one has the decomposition

$$\mathcal{H} = \oplus_{\mu \in \hat{K}} \mathcal{H}_{\mu},$$

where $\mathcal{H}_{\mu}$ is $\pi(K)$-invariant and the action of K on $\mathcal{H}_{\mu}$ is equivalent to $n\mu$ where n is an integer or ∞, and is called the *multiplicity* of μ in $\mathcal{H}$. Note that the subset of K-finite vectors is dense in $\mathcal{H}$.

The representation π is called *strongly L^p* if there is a dense subset of $\mathcal{H}$ such that for v, w in this subspace $\phi_{v,w} \in L^p(G)$. Let A be a maximal split Cartan subgroup of G, with Lie algebra $\mathfrak{a}$. Fix an order on the roots, let $\mathcal{C}$ be the positive Weyl chamber, and let $\rho : \mathfrak{a} \to \mathbb{R}$ be half the sum of the positive roots on $\mathcal{C}$. Howe [53, Corollary 7.2 and §7] obtained the following estimates for the matrix coefficients of $v \in \mathcal{H}_{\mu}$ and $w \in \mathcal{H}_{\nu}$ of a strongly L^p-representation of G:

$$|\phi_{v,w}(\exp(tH))| \leq C_1 \|v\| \, \|w\| \dim \nu \, \dim \mu \, e^{-\frac{t}{2p}\rho(H)},$$

where $H \in \bar{\mathcal{C}}$ and $C_1 > 0$ is a universal constant. Cowling [18] shows that any irreducible unitary representation of G with discrete kernel is strongly L^p for some p. Moreover, if the Lie algebra $\mathfrak{g}$ of G does not have factors isomorphic to $\mathfrak{so}(n, 1)$ or $\mathfrak{su}(n, 1)$ then p can be chosen independent of π.

A vector $v \in \mathcal{H}$ is called C^∞ if the map $g \in G \to \pi(g)v$ is C^∞. We will use now classical estimates on the size of Fourier coefficients of C^∞-vectors.

Let $m = \dim K$ and $X_1, X_2, \ldots, X_m$ be an orthonormal basis of $\mathfrak{k}$, the Lie algebra of K. Let $\Omega = 1 - \sum_{i=1}^m X_i^2$. Then Ω belongs to the center of the universal enveloping algebra of $\mathfrak{k}$, and acts on the set of K-finite vectors $\mathcal{H}$ since K-finite vectors are smooth.

Theorem 4.4.3 *Let v and w be C^∞-vectors in an irreducible unitary representation π of G with discrete kernel. Then there is a universal constant $E > 0$ and an integer $p > 0$ such that for all $H \in \bar{\mathcal{C}}$ and large enough m one has*

$$| < \exp(tH)v, w > | \leq E\, e^{-\frac{t}{2p}\rho H} \|\Omega^m(v)\|\ \|\Omega^m(w)\|.$$

In fact, p can be any number for which π is strongly L^p, and, if $\mathfrak{g}$ does not have factors isomorphic to $\mathfrak{so}(n, 1)$ or $\mathfrak{su}(n, 1)$, p only depends on G.

Proof It follows from Schur's Lemma that Ω acts as a multiple $c(\mu) id_{\mathcal{H}_\mu}$ on $\mathcal{H}_\mu$. Let $v = \sum_{\mu \in \hat{K}} v_\mu$. By [173, Lemma 4.4.2.2], one has for all integers $m > 0$

$$\|v_\mu\| \leq c(\mu)^{-m} \dim^2 \mu \|\Omega^m(v)\|.$$

Following [173, Lemma 4.4.2.3] one sees that for m large enough

$$\sum_{\mu \in \hat{K}} c(\mu)^{-2m} \dim^6 \mu < \infty.$$

Similar estimates hold for $w = \sum_{\mu \in \hat{K}} w_\mu$. Pick $p > 0$ such that π is L^p. Then for m large enough one has

$$\begin{aligned}
| < \exp(tH)v, w > | &= \left| < \sum_{\mu \in \hat{K}} \exp(tH)v_\mu, \sum_{\nu \in \hat{K}} w_\nu \right| \\
&\leq De^{-\frac{t}{2p}\rho H} \sum_{\mu,\nu \in \hat{K}} \|v_\mu\|\ \|w_\nu\| \dim \mu \dim \nu \\
&\leq De^{-\frac{t}{2p}\rho H} \left(\sum_{\mu \in \hat{K}} \|v_\mu\|^2 \dim^2 \mu \right)^{\frac{1}{2}} \left(\sum_{\nu \in \hat{K}} \|w_\nu\|^2 \dim^2 \nu \right)^{\frac{1}{2}} \\
&\leq De^{-\frac{t}{2p}\rho H} \|\Omega^m(v)\|\ \|\Omega^m(w)\| \sum_{\mu \in \hat{K}} c(\mu)^{-2m} \dim^6(\mu),
\end{aligned}$$

and the theorem is proved. □

Lemma 4.4.4 *Let G be a semisimple connected Lie group with finite center. Assume that its Lie algebra $\mathfrak{g}$ does not have factors isomorphic to $\mathfrak{so}(n, 1)$ or $\mathfrak{su}(n, 1)$. Let Γ be a co-compact lattice in G. Let $f, h \in L^2(G/\Gamma)$ be C^∞-functions orthogonal to the constants. Let $\mathcal{C}$ be a positive Weyl chamber in a maximal split Cartan D. Then there is an integer $p > 0$ which only depends on G and a constant $E > 0$ such that for all $H \in \overline{\mathcal{C}}$*

$$\langle(\exp(tH)_*(f), h\rangle \leq E\, e^{-\frac{t}{2p}\rho(H)} \|f\|_m \|h\|_m,$$

where $\|f\|_m$ is the Sobolev norm of f.

Proof Since the lattice Γ is irreducible it follows from Moore's theorem [179] that there are no L^2-functions on G/Γ orthogonal to the constants which are also invariant under any non-compact element of G. Consequently any non-trivial irreducible component of $L^2(G/\Gamma)$ has a discrete kernel. By the previous theorem it is sufficient to show that any non-trivial irreducible component is strongly L^p for a p that only depends on G. But this is Cowling's result as $\mathfrak{g}$ does not have factors isomorphic to $\mathfrak{so}(n, 1)$ or $\mathfrak{su}(n, 1)$. □

We return to the proof of Theorem 4.4.2.

Proof Let a and b be two distinct generators of the action α and let β be a C^∞ real-valued cocycle over α. Then the cocycle equation implies that C^∞ functions $f(a) = \beta(a, x)$ and $g(x) = \beta(b, x)$ satisfy the following equation:

$$f(\alpha(b, x)) - f(x) = g(\alpha(a, x)) - g(x), \tag{4.4.1}$$

or

$$bf - f = ag - g, \tag{4.4.2}$$

where a and b will also denote the corresponding induced maps on $L^2(G/\Gamma)$. In this notation, the cocycle equation for a,

$$aP - P = f,$$

has two formal solutions:

$$P_a^+ = \sum_{k=0}^{\infty} a^k f \quad \text{and} \quad P_a^- = -\sum_{k=-\infty}^{-1} a^k f.$$

Both are distributions. Moreover, they coincide. Both facts are consequences of Lemma 4.4.4. Indeed,

$$|\langle a^k f, h\rangle| \leq Ee^{-k\rho} \| f \|_m \| h \|_m .$$

Hence $\sum_{k=0}^{\infty} \langle a^k f, h \rangle$ converges absolutely, and there is a constant $C > 0$ such that $| \sum_{k=0}^{\infty} \langle a^k f, h \rangle | \leq C \parallel h \parallel_m$. Thus P_+ and similarly P_- are distributions. In fact, they are elements of the Sobolev space H^{-m}.

The fact that they coincide is a consequence of the cocycle equation (4.4.2) and the estimates from Lemma 4.4.4. Namely, from (4.4.2),

$$\sum_{k=-l}^{l} a^k bf - \sum_{k=-l}^{l} a^k f = \sum_{k=-l}^{l} a^{k+1} g - a^k g = a^{l+1} g - a^{-l} g.$$

Since Γ is an irreducible lattice the matrix coefficients of elements in $L^2(G/\Gamma)$ orthogonal to the constants vanish [179, Chapter 2]. Hence we see that, for $g \in C^\infty(G/\Gamma)$,

$$\sum_{k=-\infty}^{\infty} \langle a^k f, b^{-1} h \rangle - \sum_{k=-\infty}^{\infty} \langle a^k f, h \rangle = \lim_{l \to \infty} \langle a^{l+1} g - a^{-l} g, h \rangle = 0.$$

Since $a^k b^m \to \infty$ as $(k, m) \to \infty$ and the matrix coefficients decay exponentially, the sum

$$\sum_{m=-\infty}^{\infty} \sum_{k=-\infty}^{\infty} \langle a^k f, b^m h \rangle = \lim_{m \to \infty} 2m \sum_{k=-\infty}^{\infty} \langle a^k f, h \rangle$$

converges absolutely. Thus we get $\sum_{k=-\infty}^{\infty} \langle a^k f, h \rangle = 0$. Henceforth we will denote $P_a^+ = P_a^-$ by P.

It follows by a similar computation to that of Section 4.2.2 that P_a^+ has continuous derivatives of all orders along the strong stable manifold of a and P_a^- has continuous derivatives of all orders along the strong unstable manifold of a. A similar conclusion holds for any other regular element b. This follows once we show that P is a transfer map for all $b \in A$.

The cocycle identity implies that

$$\begin{aligned} bP - P &= \sum_{k=0}^{\infty} ba^n \beta(a, \cdot) - a^n \beta(a, \cdot) \\ &= \sum_{k=0}^{\infty} a^{n+1} b\beta(b, \cdot) - a^n \beta(b, \cdot) = \beta(b, \cdot). \end{aligned}$$

Since the stable and unstable directions of the elements in A along with their Lie brackets generate the tangent space at any point, by the general elliptic theory argument (see Section 3.7) P is C^∞. □

We finish the proof of Theorem 4.4.2. Recall that the Weyl chamber flow is the action on $X = M \setminus G/\Gamma$ induced from the action of A on G/Γ by left translations. If $\bar{\beta}$ is a C^∞-cocycle on X, then it lifts to a cocycle β on G/Γ. Since $\beta(a,x)$ is M-invariant so is P. Thus P projects to a distribution $\bar{P}$ on $M \setminus G/\Gamma$ which solves the cohomological equation $\bar{\beta}(a,x) = \bar{P}(x) - \bar{P}(ax)$ for all $a \in A$. A general elliptic theory argument (see Section 3.7) implies now that $\bar{P}$ is C^∞. □

Remark 4.4.5 Recently Kanai revealed a link between the classical vanishing theorems of Matsushima and Weil and the rigidity of Weyl chamber flow. See [65].

The counterpart of Theorem 4.4.2 for actions by automorphisms of a torus is shown below.

Theorem 4.4.6 *An action of $\mathbb{Z}^k$ $(k \geq 2)$ by automorphisms of a torus with no rank one algebraic factors is $C^\infty_{\mathbb{R}}$-cocycle rigid.*

Outline of the proof The following lemma is proved later, as a part of Theorem 6.2.12. Its proof follows from the higher rank assumption for the action α. The lemma gives an exponential estimate which is essential for the proof of Theorem 4.4.6.

Lemma 4.4.7 *For any non-zero $n \in \mathbb{Z}^N$ and any $m = (m_1, .., m_k) \in \mathbb{Z}^k$ the dual action α^* satisfies the following:*

$$|(\alpha^*)^k n| \geq C|n|^{-N} \exp\{\tau \|k\|\},$$

for some $\tau > 0$ that depends on the action only.

Proof Ergodicity and commutativity assumptions imply that it is enough to show the existence of a smooth solution of the cohomology equation corresponding to some element of the action. However, for individual elements of the action there are infinitely many obstructions to cocycle trivialization. This is where one needs to take into account the higher rank assumption. Since the action α has no rank-one factors there exist two multiplicatively independent generators A and B such that every non-trivial $A^l B^k$ is ergodic. See Proposition 2.2.3.

Let β be a cocycle over α and denote $\beta(A,x) = f(x)$ and $\beta(B,x) = g(x)$. To simplify further notation, we assume that β has average zero. Showing that β trivializes reduces then to solving the following equations:

$$P \circ A - P = f,$$

$$P \circ B - P = g.$$

Veech [169] showed that individual equations above have smooth solutions providing the dual obstructions $\mathcal{O}_n^A(f) = \sum_{l=-\infty}^{l=\infty} \hat{f}_{(A^*)^l n}$ and $\mathcal{O}_n^B(g) = \sum_{k=-\infty}^{k=\infty} \hat{g}_{(B^*)^k n}$ for f and g vanish. We show below that cocycle equation implies that all the obstructions vanish and that the smooth solutions to the equations above coincide. Indeed, β being a cocycle over a commutative action implies

$$f \circ B - f = g \circ A - g.$$

Passing to Fourier coefficients implies for every $n \in \mathbb{Z}^N$

$$\hat{f}_{B^*n} - \hat{f}_n = \hat{g}_{A^*n} - \hat{g}_n. \tag{4.4.3}$$

Iterating (4.4.3) and adding all the iterates gives

$$\sum_{k=-\infty}^{k=\infty} \hat{f}_{(B^*)^{k+1}n} - \hat{f}_{(B^*)^k n} = \sum_{k=-\infty}^{k=\infty} \hat{g}_{A^*(B^*)^k n} - \hat{g}_{(B^*)^k n}.$$

Since f and g are C^∞ and because of the exponential decay from Lemma 4.4.7, all the sums above converge absolutely. Thus the left-hand side in the expression above is zero which implies that the sum

$$\mathcal{O}_n^B(g) = \sum_{k=-\infty}^{k=\infty} \hat{g}_{(B^*)^k n}$$

is invariant not only under the action of B^* but also under the action of A^*. Therefore, the series $\sum_{l=-\infty}^{l=\infty} \mathcal{O}^B_{(A^*)^l n}(g)$ does not converge, since all summands are the same, unless they are all zero, i.e., unless $\mathcal{O}_n^B(g) = 0$.

But the series above is equal to

$$\sum_{l=-\infty}^{l=\infty} \sum_{k=-\infty}^{k=\infty} \hat{g}_{(A^*)^l (B^*)^k n},$$

and it converges due to Lemma 4.4.7.

The argument we just presented is the simplest and most transparent instance of the higher rank trick.

Thus the obstructions vanish and the existence of smooth solution follows as in [169]. Namely the solution may be constructed by defining $\hat{P}_n$ to be $\hat{P}_n^+$, i.e., the "positive" part of the sum $\mathcal{O}_n^B(g)$, if n is largest in the expanding direction, and defining $\hat{P}_n$ to be $\hat{P}_n^-$, i.e., the "negative" part of $\mathcal{O}_n^B(g)$ if n is largest in the contracting direction. Of course, n may be largest in the neutral direction as well, in which case either the positive or negative sum may be used. As it turns out, this only results in fixed loss of regularity for the solution P since no integer vector can stay large in the neutral direction for "too long" due to the ergodicity assumption (according to Lemma 3 in [83]). □

Remark 4.4.8 We note here that the same method implies $C^{\omega}_{\mathbb{R}}$-cocycle rigidity as well, and also gives results in finite regularity, namely $C^{a,a-\kappa}_{\mathbb{R}}$-cocycle rigidity for any $a > \kappa > N$.

4.4.3 Geometric approach and TNS actions

In this subsection we consider Anosov $\mathbb{Z}^k$-actions on infranilmanifolds.

Let M be an infranilmanifold, and $\alpha : \mathbb{Z}^k \times M \to M$ be a C^K-action of $\mathbb{Z}^k$. View α as a homomorphism from $\mathbb{Z}^k$ into $\mathrm{Diff}^K(M)$ and denote by $\mathcal{A} \subset \mathrm{Diff}^K(M)$ its image.

The following notion was introduced in [73].

Definition 4.4.9 The action α is said to be *totally non-symplectic*, or *TNS*, if there exists a family S of Anosov elements in $\mathcal{A}$ and a continuous splitting of the tangent bundle $TM = \oplus_{i=1}^{m} E_i$ into $\mathcal{A}$-invariant distributions such that:

(i) the stable and unstable distributions of any element in S are direct sums of sub-families of the E_is;
(ii) any two distributions E_i and E_j, $1 \le i, j \le m$, are included in the stable distribution of some element in S.

If, moreover, the action α is C^∞ and each distribution E_i is smooth, we say that the action is *smoothly TNS*.

Note that any linear TNS action on an infranilmanifold is actually smoothly TNS. Moreover, if the linear action is on a torus, one can assume that the distributions E_i are constant, that is, given by translates of some fixed vector subspaces.

Consider a TNS $\mathbb{Z}^k$-action α on an infranilmanifold M. Since it contains Anosov elements, there is a subgroup $G \subset \mathbb{Z}^k$ of finite index acting with a fixed point, say x_0, and by the Franks–Manning classification $\alpha|_G$ is conjugated to the linear action $\bar{\alpha} := (\alpha|_G)_*$ induced on $\pi_1(M, x_0)$. Using the fact that the elements of $S \subset \mathbb{Z}^k$ can be replaced by their powers, one can assume that $S \subset G$. The action $\bar{\alpha}$ is TNS as well, because the TNS property can be described in terms of the intersections of the stable and unstable foliations of the elements of S.

We recall (see Sections 1.5 and 1.6) now the notions of Lyapunov exponent and Weyl chamber in order to show an equivalent definition of the TNS action.

The action α_* of the derivative on the tangent bundle of the universal cover of M is determined by commuting invertible matrices. Let $L_j : \mathbb{Z}^k \to \mathbb{R}$ be the *Lyapunov exponents* of α_*. Each Lyapunov exponent can be extended to a linear map $L_j : \mathbb{R}^k \to \mathbb{R}$, called also Lyapunov exponent. There is a splitting of the tangent bundle into $\mathbb{Z}^k$-invariant sub-bundles $TM = \oplus_j F_j$ such that the Lyapunov exponent of $v \in F_j$ with respect to $\alpha(a)$ is given by $L_j(a)$. We call F_j a *Lyapunov space* or *Lyapunov distribution* for the action. The kernel of each Lyapunov exponent is a hyperplane $\mathcal{H}_j$ in $\mathbb{R}^k$. We denote by $\mathcal{H}_j^-$ the half-space where L_j is negative. The connected components of $\mathbb{R}^k - \cup\mathcal{H}_j$ are the *Weyl chambers*.

Note that, using Lyapunov exponents, the TNS property can be characterized by

$$L_j = cL_i \text{ for some constant } c \implies c > 0.$$

We present now a rigidity result for abelian cocycles over TNS actions. The following theorem refers to the special case when the action is linear and the cocycle is real-valued. The main geometric idea of the proof is to construct a C^∞ 1-form on M which is closed and determines a $\mathbb{Z}^k$-invariant class in cohomology. Since the action induced in cohomology is hyperbolic, the form has to be exact. This allows to recover the constant cocycle and the transfer map.

Note that a closed form determines a foliation of $M \times \mathbb{R}$ with leaves of dimension $m = \dim M$. A related argument, which constructs directly an invariant foliation, will be shown later, and will be used to prove a rigidity result for cocycles with non-abelian range. Considering the holonomy of the invariant foliation, one can show that the leaves are closed and cover M simply. This fact allows to recover the constant cocycle and the transfer map.

Theorem 4.4.10 *Let M be an infranilmanifold. Let $\alpha : \mathbb{Z}^k \times M \to M$ be a linear TNS $\mathbb{Z}^k$-action. Let $\beta : \mathbb{Z}^k \times M \to \mathbb{R}$ be a C^∞ cocycle over α. Then β is cohomologous with a constant cocycle and the transfer map $P : M \to \mathbb{R}$ is C^∞.*

In the rest of the subsection we prove Theorem 4.4.10.

Let $x \in M$ and let $a \in \mathbb{Z}^k$ act hyperbolic on M. Let $W_a^s(x)$ be the stable leaf of a though x, and $y \in W_a^s(x)$. Then it follows from Proposition 4.2.8 that the sum below is convergent:

$$\gamma_x^a(y) = \sum_{n=0}^{\infty} [\beta(a, (na)x) - \beta(a, (na)y)]. \tag{4.4.4}$$

Moreover, arguing as in the proof of Theorem 4.2.3, it follows that the function $\gamma_x^a(y)$, $y \in W_a^s(x)$, is actually C^∞. One can define now a 1-form ω_a^- on $E_a^s(x)$ by taking the derivative of γ_x^a in the y-variable along the stable leaf $W_a^s(x)$.

Similarly, for $x \in M$ and $z \in W^u(x; a) = W^s(x; -a)$, let the 1-form ω_a^+ on $E_x^u(a)$ be defined as the z-differential of $\gamma_x^{-a}(z)$ along the unstable leaf of $W_a^u(x)$. Consider the form $\omega_a = \omega_a^+ \oplus \omega_a^-$ on $T_x M = E_x^u(a) \oplus E_x^s(a)$.

We show that for a large set of hyperbolic elements in $\mathbb{Z}^k$ the above construction leads to the same form. Moreover, this form is smooth and closed.

Lemma 4.4.11 *Consider a linear $\mathbb{Z}^k$-action α which contains an Anosov element. Then there is a subset $S \subset \mathbb{Z}^k$ of hyperbolic generators of $\mathbb{Z}^k$ which contains elements from each Weyl chamber, and with the property that if $a, b \in S$ then*

$$\omega_a = \omega_b.$$

Proof Let $L_j : \mathbb{R}^k \to \mathbb{R}, 1 \leq j \leq m$ be the Lyapunov exponents and $TM = \oplus_j F_j$ the decomposition in Lyapunov distributions. Let $\lambda_j := \exp \circ L_j : \mathbb{Z}^k \to [0, \infty)$, and denote by $\mathcal{F}_j$ the foliation corresponding to F_j.

Assume first that $a, b \in \mathbb{Z}^k$ are partially hyperbolic, $F_j \subset E^s(a) \cap E^s(b)$ and $\lambda_j(b)$, the contraction coefficient along F_j, is smaller than the inverse of the Lipschitz norm of $\alpha(a-b)$. Let $z \in \mathcal{F}_j(x) \subset W^s(x; a) \cap W^s(x; b)$. Using the cocycle relation we find that

$$\begin{gathered}\sum_{k=0}^{n-1} \beta(a, (ka)z) = \beta(na, z), \\ \beta(na, z) - \beta(nb, z) = \beta(n(a-b), (nb)z),\end{gathered} \tag{4.4.5}$$

and similarly for x instead of z. Therefore, in order to show that $\gamma_x^a(z) = \gamma_x^b(z)$, and consequently that $\omega_a|_{F_j} = \omega_b|_{F_j}$, it is enough to show that

$$\lim_{n\to\infty} \big(\beta(n(a-b), (nb)z) - \beta(n(a-b), (nb)x)\big) = 0.$$

But

$$\begin{aligned} &|\beta(n(a-b), (nb)z) - \beta(n(a-b), (nb)x)| \\ &\quad = \left| \sum_{k=0}^{n-1} \big(\beta(a-b, [nb + k(a-b)]z) \right. \\ &\qquad \left. - \beta(a-b, [nb+k(a-b)]x)\big) \right| \\ &\quad \leq \|\beta(a-b, \cdot)\|_{\text{Holder}} \\ &\qquad \cdot \left[\sum_{k=0}^{n-1} \text{dist}_M(\alpha(nb + k(a-b))(z), \alpha(nb+k(a-b))(x))^\delta \right] \\ &\quad \leq \|\beta(a-b, \cdot)\|_{\text{Holder}} \cdot \lambda_j(b)^{n\delta} \cdot \\ &\qquad \cdot C \cdot (\text{dist}_M(z, x))^\delta \sum_{k=0}^{n-1} \|\alpha(a-b)\|_{\text{Lip}}^{k\delta}, \end{aligned} \tag{4.4.6}$$

where C is a constant independent of n. Since $\lambda_j(b) < 1$ and $\lambda_j(b) \cdot \|\alpha(a - b)\|_{\mathrm{Lip}} < 1$, the conclusion follows.

We construct now the set $S \subset \mathbb{Z}^k$. Consider first a finite set F of elements in $\mathbb{Z}^k$ close to the origin, which contains a $\mathbb{Z}$-basis of $\mathbb{Z}^k$. There is a constant $M > 1$ such that $\|\alpha(c)\|_{\mathrm{Lip}} \leq M$, for all $c \in F$. Let $L_j : \mathbb{R}^k \to \mathbb{R}$ be the Lyapunov exponent for j and $\mathcal{H}_j$ the hyperplane in $\mathbb{R}^k$ determined by the kernel of λ_j. Then there exist a ball B around the origin and cones $C(\mathcal{H}_j) \subset \mathcal{H}_j^-$ intersecting all Weyl chambers in $\mathcal{H}_j^-$, such that for each j and any element $b \in C(\mathcal{H}_j) \cap (\mathbb{Z}^k - B)$ we have

$$L_j < -\log M,$$

and therefore

$$\lambda_j(b) < M^{-1}. \tag{4.4.7}$$

Consider two elements $a, b \in C(\mathcal{H}_j) \cap (\mathbb{Z}^k - B)$. We can join a and b by a sequence of elements in $C(\mathcal{H}_j) \cap (\mathbb{Z}^k - B)$ adding at each step an element from F. Formula (4.4.7) allows us to apply the first part of the proof repeatedly and deduce that

$$\omega_a|_{F_j} = \omega_b|_{F_j}. \tag{4.4.8}$$

By the construction of the 1-form, formula (4.4.8) still holds if a and b are in the union of $C(\mathcal{H}_j)$ with the opposite cone, $-C(\mathcal{H}_j)$.

Define the set S to be

$$S = \left[\cap_{j=1}^m \left(C(\mathcal{H}_j) \cup (-C(\mathcal{H}_j))\right)\right] \cap (\mathbb{Z}^k - B). \qquad \square$$

Lemma 4.4.12 *If the linear $\mathbb{Z}^k$-action is TNS then the form $\omega \equiv \omega_a$, $a \in S$ constructed above is smooth and closed.*

Proof Denote $m = \dim M$. Let $U \subset M$ be a open set included in a coordinate chart. Since the distributions E_i are smooth, one can find a frame of smooth vector fields $\{X_j\}_{j=1,m}$ over U such that each field X_j is contained in some E_i. Let $\{\eta_j\}_{j=1,m}$ be the dual frame of one-forms over U, and write

$$\omega|_U = \sum_{j=1}^m f_j \eta_j, \text{ where } f_j = \omega(X_j).$$

We will show that each function f_j is smooth along all the distributions E_i and the derivatives are continuous on U. Since the $\oplus E_i = TM$ this implies that each f_j is smooth on U (see [56], Theorem 2.6).

To show that f_j is smooth along E_i, pick an Anosov element $a \in S$ such that $X_i, E_j \subset E^s(a)$. This is possible due to the TNS condition. Since $P_a^-(\cdot, x)$ is smooth along $W^s(x; a)$ and varies continuously in the C^∞-topology with $x \in M$, one concludes that $\omega_a^-|_{E^s(a)}$ is continuously C^∞ along $W^s(a)$. By Lemma 4.4.11, this proves our assertion.

To show that ω is closed, use the TNS condition and Lemma 4.4.11. Clearly $\omega_a^-|_{W^s(x;a)}$ is exact, hence, using the fact that pull-back and exterior differentiation commute,

$$(d\omega)|_{W^s(x;a)} = d(\omega|_{W^s(x;a)}) = 0 \quad \text{for} \quad a \in S.$$

Since over U any two directions X_i and X_j are included in the stable subspace of some hyperbolic element $a \in S$, we obtain that $(d\omega)|_U = 0$. □

Lemma 4.4.13 *The cohomology class of ω in $H^1(M, \mathbb{R})$ is $\mathbb{Z}^k$-invariant, hence it has to be zero, i.e., ω is exact.*

Proof Let $a \in \mathbb{Z}^k$ be hyperbolic and $\omega_a = \omega_a^+ \oplus \omega_a^-$ on $TM = E^u(a) \oplus E^s(a)$. Then

$$b^*\omega_a = \omega_a + d\beta(b, \cdot), \tag{4.4.9}$$

for any diffeomorphism $b \in \mathbb{Z}^k$. Indeed, since $ab = ba$, the cocycle equation implies that

$$\beta(a, bt) = \beta(a, t) + \beta(b, at) - \beta(b, t),$$

and therefore

$$\begin{aligned} P_a^-(by, bx) &= P_a^-(y, x) + [\beta(b, y) - \beta(b, x)], \\ P_{-a}^-(bz, bx) &= P_{-a}^-(z, x) + [\beta(b, z) - \beta(b, x)], \end{aligned} \tag{4.4.10}$$

for $y \in W^s(x; a)$ and $z \in W^u(x; a)$. Hence, for $\xi \in E_x^s(a)$,

$$\begin{aligned} \left(b^*\omega_a^-\right)_x(\xi) &= \left(\omega_a^-\right)_{bx}(Db(\xi)) \\ &= d_- P_a^-(\cdot, bx)(Db(\xi)) = d_- P_a^-(b\,\cdot, bx)(\xi) \qquad (4.4.11) \\ &= d_-\left[P_a^-(\cdot, x) + \beta(b, \cdot) - \beta(b, x)\right](\xi) = \omega_a^-(\xi) + d_-\beta(b, \cdot)(\xi), \end{aligned}$$

where d_- denotes the differential along $E^s(a)$. A similar computation for ω_a^+ completes the proof of (4.4.9). This shows that the class $\bar{\omega} \in H^1(M, \mathbb{R})$ corresponding to ω is $\mathbb{Z}^k$-invariant.

That ω is exact (i.e., that $\bar{\omega} = 0$) now follows from the fact that any linear hyperbolic automorphism of an infranilmanifold induces a hyperbolic map of the first cohomology group, and therefore the only invariant class is the trivial one.

Indeed, let the infranilmanifold be $M = N/G$, where $G \subset NC$ is a lattice, and let $\bar{A} : NC \to NC$ be an automorphism which preserves both N and G, is hyperbolic on N, and induces the infranilmanifold automorphism $A : M \to M$. Then $\pi_1(M) = G$, $H_1(M, \mathbb{Z}) = G/[G, G]$, $H_1(M, \mathbb{R}) = H_1(M, \mathbb{Z}) \otimes_{\mathbb{Z}} \mathbb{R}$ and $H^1(M, \mathbb{R})$ is the dual of $H_1(M, \mathbb{R})$ in a natural way, where $[G, G]$ is the commutator subgroup of G. Note that $[G, G]$ is invariant under $\bar{A}$, hence it defines a map on $G/[G, G]$, which induces the action of A on $H_1(M, \mathbb{R})$.

Let $G_0 := G \cap N$, which has finite index in G and is an $\bar{A}$-invariant lattice of N. Recall that a lattice in a simply connected nilpotent Lie group and any subgroup of such a lattice are finitely generated ([146], Theorems 2.10 and 2.7).

Since $G_0/(G_0 \cap [G, G]) \hookrightarrow G/[G, G]$ is of finite index and both are finitely generated abelian groups,

$$(G_0/(G_0 \cap [G, G])) \otimes_{\mathbb{Z}} \mathbb{R} \cong (G/[G, G]) \otimes_{\mathbb{Z}} \mathbb{R}$$

in a way that identifies the natural actions of $\bar{A}$. Therefore, it is enough to show that the action of $\bar{A}$ on $(G_0/(G_0 \cap [G, G])) \otimes_{\mathbb{Z}} \mathbb{R}$ is hyperbolic. Since $[G_0, G_0] \subset G_0 \cap [G, G]$, the above statement follows once we show that $\bar{A}$ acts hyperbolically on $(G_0/[G_0, G_0]) \otimes_{\mathbb{Z}} \mathbb{R}$, because

$$G_0/(G_0 \cap [G, G]) \cong \frac{G_0/[G_0, G_0]}{(G_0 \cap [G, G])/[G_0, G_0]},$$

and all of the above quotient groups are finitely generated abelian.

Consider the short exact sequence of finitely generated abelian groups

$$\{1\} \to \frac{G_0 \cap [N, N]}{[G_0, G_0]} \to \frac{G_0}{[G_0, G_0]} \to \frac{G_0}{G_0 \cap [N, N]} \to \{1\}.$$

Since both $G_0 \cap [N, N]$ and $[G_0, G_0]$ are co-compact in $[N, N]$ ([146], Corollary 1 of Theorem 2.3 and proof of Theorem 2.1), the left group in the above sequence is finite. On the other hand, $G_0/\,(G_0 \cap [N, N]) \hookrightarrow N/[N, N]$ and the action of $\bar{A}$ on the abelian group $N/[N, N]$ is hyperbolic because the derivative of $\bar{A}$ is hyperbolic at the origin of N. These two observations complete the proof of the fact that A acts hyperbolically on $H_1(M, \mathbb{R})$, hence on $H^1(M, \mathbb{R})$ as well. □

End of proof of Theorem 4.4.10 Once we know that ω is exact, the conclusion of Theorem 4.4.10 follows easily. Let $P : M \to \mathbb{R}$ be a C^∞ function such that $\omega = dP$ (P can be chosen C^∞ because so is ω). From (4.4.9) we obtain that

$$d\,[\beta(b, \cdot) - P \circ b(\cdot) + P(\cdot)] = 0,$$

for each $b \in \mathbb{Z}^k$. Now the proof is complete because this means that the cocycle cohomologous to β given by

$$\widetilde{\beta}(b, \cdot) := \beta(b, \cdot) - P \circ b(\cdot) + P(\cdot) : M \to \mathbb{R}$$

is constant for all $b \in \mathbb{Z}^k$. □

4.4.4 Geometric approach for Weyl chamber flows

For Weyl chamber flows several geometric methods have been found that allow us to solve the cohomological equation. The approach in this subsection was introduced by Ferleger and Katok in the unpublished manuscript [33]. A different approach, based on the work of Damjanovic and Katok [21], is described later in Section 4.4.5.

Let G be a simple connected real Lie group of non-compact type and of $\mathbb{R}$-rank at least two. Consider a maximal split Cartan subgroup A of G. The centralizer $Z(A)$ of the Cartan subgroup splits as a product $Z(A) = MA$, where M is a compact subgroup of G. Pick an irreducible torsion-free cocompact lattice Γ in G. Let $N := M\backslash G/\Gamma$.

Let $\rho : A \times N \to N$ be the chamber Weyl flow of A (see Section 2.3.3.2). For an arbitrary $x \in N$, and a hyperbolic (regular) element $a \in A$ the tangent space at x splits as $T_x N = E^c_a(x) \oplus E^s_a(x) \oplus E^u_a(x)$, where $E^c_a(x)$, $E^s_a(x)$ and $E^u_a(x)$ are the distribution tangent to the A-flow orbit, and respectively the stable and unstable directions of a. The distributions $E^c_a(x)$, $E^s_a(x)$, $E^u_a(x)$ are C^∞.

Consider the Lie algebra $\mathfrak{g}$ of G, endowed with a norm $\|\cdot\|$. Let A be a Cartan subgroup with (real) Lie algebra $\mathfrak{a}$. Let R be the complex root system of $\mathfrak{g}$ associated with A. According to the general semisimple real Lie algebra structure theory (see, for example, [48]), the root system splits as $R = K \cup \Sigma_{\mathbb{C}}$, where K is the subset of compact roots, that is, $K = \{\delta \in R | \mathrm{Re}\,\delta = \delta + \bar{\delta} = 0\}$, and $\Sigma_{\mathbb{C}}$ is the set of non-compact roots. We have the following Cartan decomposition:

$$\mathfrak{g} = \mathfrak{a} \oplus \bigoplus_{\delta \in K} \mathfrak{m}^\delta \oplus \bigoplus_{\alpha \in \Sigma} \mathfrak{g}^\alpha. \tag{4.4.12}$$

Here $\mathfrak{m}^\delta$ is a linear space such that its complexification is the δ-root space in $\mathfrak{g}_{\mathbb{C}}$, $\Sigma = \{\mathrm{Re}\,\alpha | \alpha \in \Sigma_{\mathbb{C}}\}$ is the set of real roots with respect to $\mathfrak{a}$, and

$$\mathfrak{g}^\alpha = \{X \in \mathfrak{g} | [H, X] = (\mathrm{Re}\,\alpha)(H)X, \text{ for every } H \in \mathfrak{a}\}.$$

Let r be the real rank of $\mathfrak{g}$, $\{H_1, \ldots, H_r\}$ be a basis of $\mathfrak{a}$, and K_δ be a generator of $\mathfrak{m}^\delta$. It follows from the definitions and the general structural theory that one can pick up generators $X_{\alpha+\delta} \in \mathfrak{g}^{\alpha}$, $\delta \in K \cup \{0\}$ in such a way that there exist constants $c_{k\alpha}, c_{\delta_1\delta_2}, c_{\delta\alpha}, c_{\alpha\beta} \in \mathbb{R}$ satisfying:

$$\begin{aligned} [H_k, H_l] &= 0, \\ [H_k, K_\delta] &= 0, \\ [H_k, X_\alpha] &= c_{k\alpha} X_\alpha, \\ [K_{\delta_1}, K_{\delta_2}] &= c_{\delta_1\delta_2} K_{\delta_1+\delta_2}, \\ [K_\delta, X_\alpha] &= c_{\delta\alpha} X_{\alpha+\delta}, \\ [X_\alpha, X_\beta] &= c_{\alpha\beta} X_{\alpha+\beta} \ (\alpha \neq -\beta), \\ [X_\alpha, X_{-\alpha}] &\in \mathfrak{a}. \end{aligned} \tag{4.4.13}$$

If $\{Y_1, \ldots, Y_n\} \subset \mathfrak{g}$ is a basis, consider the corresponding basis of left invariant vector fields $\{Y_1(x), \ldots, Y_n(x)\}$ in T_*G, $Y_k(e) = Y_k, k = 1, \ldots, n$. We study the $C^\infty(N)$ module $T^*N := C^\infty(T_*N, \mathbb{R})$ of $\mathbb{R}$-valued C^∞ differential forms on N. Denote by $dY_k, k = 1, \ldots, n$, the dual basis of $\mathfrak{g}^*$, and set

$$\begin{aligned} \eta_k &= dH_k, \\ \kappa_\delta &= dK_\delta, \\ \chi_{\alpha+\beta} &= dX_{\alpha+\beta}. \end{aligned} \tag{4.4.14}$$

Let $S = \{\exp H | d_{\mathfrak{g}}(H, \bigcup_{\alpha\in\Sigma} \operatorname{Ker}\alpha) \geq 1\}$, where

$$d_{\mathfrak{g}}(X, Y) := \inf_{x \in X y \in Y} ||x - y||,$$

for $X, Y \subset \mathfrak{g}$.

The set S consists of hyperbolic elements of the action and has a non-trivial intersection with every Weyl chamber. For every hyperbolic element $a \in A$, denote by λ_a the contraction coefficient of $\rho(a)$ along the stable manifold of the action, and denote by $\operatorname{dist}_a(\cdot, \cdot)$ the Lyapunov metric corresponding to the action.

The main result of this subsection is the following theorem.

Theorem 4.4.14 *Let G be a simple real connected Lie group of non-compact type and of rank at least two, with a maximal split Cartan subgroup A, and centralizer $Z(A) = MA$. Let $\Gamma \subset G$ be an irreducible torsion-free co-compact lattice. Let $N = M\backslash G/\Gamma$ and let $\rho : A \times N \to N$ be the Weyl chamber flow.*

Then every $\mathbb{R}$-valued C^∞-cocycle $\beta : A \times N \to \mathbb{R}$ over ρ is C^∞ cohomologous to a constant cocycle.

In analogy to the geometric approach employed in Section 4.4.3 for higher rank abelian TNS actions by automorphisms of a torus, the idea of the proof is to construct an invariant differentiable form corresponding to the cocycle, and show that the form is closed and exact.

We start the proof of the theorem with some preliminary remarks. Recall from Proposition 4.2.8 that for $x \in N, a \in A$ hyperbolic, $W_a^s(x)$ the stable leaf of a, and $y \in W_a^s(x)$, the limit below converges to a C^∞ function $\gamma_x^a : W_a^s(x) \to \mathbb{R}$:

$$\gamma_x^a(y) := \lim_{n\to\infty} (\beta(na, x) - \beta(na, y)). \tag{4.4.15}$$

Moreover, arguing as in the proof of Theorem 4.2.3, it follows that the function $\gamma_x^a(y)$, $y \in W_a^s(x)$, is actually C^∞. So one can differentiate γ_x^a along the stable direction of a.

We define a form $\omega_a(x) \in T^*N$ as follows:

$$\begin{aligned} \omega_a(x)(h) &= D^c\beta(a, x)h, \ h \in E_a^c(x), \\ \omega_a(x)(h) &= D^s\gamma_x^a(y)h, \ h \in E_a^s(x), \\ \omega_a(x)(h) &= D^s\gamma_x^{-a}(y)h, \ h \in E_a^u(x), \end{aligned} \tag{4.4.16}$$

where D^c is the derivative along the orbit direction and D^s is the derivative along the stable direction.

Thus, for every cocycle β, a certain field of $\mathbb{R}$-valued forms on N has been defined. The construction, a priori, depends on the choice of the hyperbolic element $a \in A$.

The following lemma is crucial to the proof of Theorem 4.4.14 because it allows to reduce the analysis of cohomology classes of cocycles to the strictly geometric question of classifying the cohomology classes of C^∞-forms associated with the cocycles.

Lemma 4.4.15 *Under the hypothesis of Theorem 4.4.14 one has:*

(i) $\omega_a|_{E_a^s \cap E_b^s} = \omega_b|_{E_a^s \cap E_b^s}$ *for any* $a, b \in S$.
(ii) The cocycle β *is a coboundary if and only if the form* ω_a *is exact.*
(iii) The form ω_a *is* C^∞.
(iv) The differential of the form ω_a *can be represented as*

$$\begin{aligned} d\omega_a^i(x) = &\sum_{\alpha,\delta_1,\delta_2} f^i_{\alpha,\delta_1,\delta_2}(x)\chi_{\alpha+\delta_1} \wedge \chi_{-\alpha-\delta_2} \\ &+ \sum_{\alpha,\delta_1,\delta_2} g^i_{\alpha,\delta_1,\delta_2}(x)\chi_{\alpha+\delta_1} \wedge \chi_{-2\alpha-\delta_2}, \end{aligned} \tag{4.4.17}$$

for some $f^i_{\alpha,\delta_1,\delta_2}, g^i_{\alpha,\delta_1,\delta_2} \in C^\infty(N)$.

Proof (i) This follows immediately from (4.4.17), which defines the form, and from Proposition 4.2.9.

(ii) The cocycle β is a coboundary if

$$\beta(a, x) = P(ax) - P(x).$$

This implies $\gamma_x^a(y) = P(y) - P(x)$ for $y \in W_a^s(x)$, which in turn implies that

$$\omega_a(x)|_{E_a^c(x)} = \omega_a(x)|_{E_a^s(x)} = \omega_a(x)|_{E_a^u(x)} = dP(x).$$

On the other hand, if ω_a is exact, let P be a C^∞ function such that $\omega_a = dP$. From (4.4.9) we obtain that

$$d\left[\beta(b, \cdot) - P \circ b(\cdot) + P(\cdot)\right] = 0$$

for each $b \in A$. This means that the cocycle cohomologous to β given by $\widetilde{\beta}(b, \cdot) := \beta(b, \cdot) - P \circ b(\cdot) + P(\cdot) : M \to \mathbb{R}$ is zero for all $b \in \mathbb{Z}^k$, hence β is a coboundary.

(iii) Consider the distribution spanned by a family

$$\{X_{\alpha+\delta_1}(x), X_{-\alpha-\delta_2}(x)\}_{\alpha>0},$$

where the order on the set of roots is generated by a certain element $a \in S$. The corresponding form ω_a is C^∞-differentiable in any direction from the distribution since each one of the directions belongs either to the stable or to the unstable subspace of a. On the other hand, the relations (4.4.13) show that the distribution is totally non-integrable. Thus, by Theorem 3.7.2, the form ω_a is C^∞-differentiable.

(iv) We have to prove that for every $X \in \mathfrak{g}^\alpha$, $Y \in \mathfrak{g}^\beta$, where $\alpha \neq s\beta$, $s = 1, 2, 1/2$ one has $d\omega(X, Y) = 0$. For every X, Y of that kind there exists an element $a \in S$ such that $X, Y \in E_a^s(x)$ and therefore $[X, Y] \in E_a^s(x)$. Then

$$\begin{aligned}
d\omega(x)(X, Y) &\\
&= \mathcal{L}_X d\omega(x)(Y) - \mathcal{L}_Y d\omega(x)X - \omega(x)([X, Y]) \\
&= \mathcal{L}_X \mathcal{L}_Y P_a(y; x) - \mathcal{L}_Y \mathcal{L}_X P_a(y; x) - \mathcal{L}_{[X,Y]} P_a(y; x) \\
&= \mathcal{L}_{[X,Y]} P_a(y; x) - \mathcal{L}_{[X,Y]} P_a(y; x) = 0.
\end{aligned}$$

□

The next step in the proof of the theorem is to show the closeness of the form.

Lemma 4.4.16 *Under the hypothesis of Theorem 4.4.14 one has*

$$d\omega = \sum_{\alpha,\delta_1,\delta_2} c^i_{\alpha,\delta_1,\delta_2} \chi_{\alpha+\delta_1} \wedge \chi_{-\alpha-\delta_2},$$

where $c^i_{\alpha,\delta_1,\delta_2} \in \mathbb{R}$ *are some constants.*

Proof According to Lemma 4.4.15 the differential of the form can be written as

$$d\omega(x) = \sum_{\alpha,\delta_1,\delta_2} f_{\alpha,\delta_1,\delta_2}(x)\chi_{\alpha+\delta_1} \wedge \chi_{-\alpha-\delta_2} + \sum_{\alpha,\delta_1,\delta_2} g_{\alpha,\delta_1,\delta_2}(x)\chi_{\alpha+\delta_1} \wedge \chi_{-2\alpha-\delta_2},$$

for some functions $f_{\alpha,\delta_1,\delta_2}, g_{\alpha,\delta_1,\delta_2} \in C^\infty(W)$. On the other hand, the differentials of the coefficients may be represented as

$$df_{\alpha,\delta_1,\delta_2} = \sum_{\beta,\delta} q_{\alpha,\delta_1,\delta_2,\beta,\delta}\chi_{\beta+\delta} + \sum_k s_{\alpha,\delta_1,\delta_2,k}\eta_k,$$

with C^∞ smooth coefficients (by construction, the coefficients f, g are M-invariant so that there are no κs in the expression), and an analogous expression for $dg_{\alpha,\delta_1,\delta_2}$ can be obtained as well. We are going to obtain some restrictions on the coefficients by grouping those that appear in the equation $0 = dd\omega$.

First, we make the observation that if $\beta \neq \alpha, -\alpha, -2\alpha$ then $q_{\alpha,\delta_1,\delta_2,\beta,\delta}$ is the only coefficient in front of $\chi_{\beta+\delta} \wedge \chi_{\alpha+\delta_1} \wedge \chi_{-\alpha-\delta_2}$, and therefore it has to be equal to zero. Thus, $df_{\alpha,\delta_1,\delta_2}$ is an exact form of the following kind:

$$df_{\alpha,\delta_1,\delta_2} = \sum_\delta q^1_{\alpha,\delta}\chi_{\alpha+\delta} + q^2_{\alpha,\delta}\chi_{-\alpha+\delta} + q^3_{\alpha,\delta}\chi_{-2\alpha+\delta},$$

where

$$\begin{aligned} q^1_{\alpha,\delta} &= q_{\alpha,\delta_1,\delta_2,\alpha,\delta}, \\ q^2_{\alpha,\delta} &= q_{\alpha,\delta_1,\delta_2,-\alpha,\delta}, \\ q^3_{\alpha,\delta} &= q_{\alpha,\delta_1,\delta_2,-2\alpha,\delta}. \end{aligned} \tag{4.4.18}$$

But then

$$\begin{aligned} 0 &= ddf_{\alpha,\delta_1,\delta_2} \\ &= \sum_\delta dq^1_{\alpha,\delta}\chi_{\alpha+\delta} + dq^2_{\alpha,\delta}\chi_{-\alpha+\delta} + dq^3_{\alpha,\delta}\chi_{-2\alpha+\delta} + \cdots \\ &\quad + q^1_{\alpha,\delta}\chi_{\beta_1+\delta} \wedge \chi_{\gamma_1} + q^2_{\alpha,\delta}\chi_{\beta_2+\delta} \wedge \chi_{\gamma_2} + q^3_{\alpha,\delta}\chi_{\beta_3+\delta} \wedge \chi_{\gamma_3} + \cdots, \end{aligned}$$

where

$$\begin{aligned} \beta_1 + \gamma_1 &= \alpha, \\ \beta_2 + \gamma_2 &= -\alpha, \\ \beta_3 + \gamma_3 &= -2\alpha. \end{aligned} \tag{4.4.19}$$

Note that the existence of such β, γ is guaranteed by the fact that the rank of the group G is bigger than one.

Formula (4.4.19) implies that $q^j_{\alpha,\delta} = 0$, $j = 1, 2, 3$, i.e., $df_{\alpha,\delta_1,\delta_2} = 0$. By the same token, $dg_{\alpha,\delta_1,\delta_2} = 0$. Therefore, one has

$$d\omega(x) = \sum_{\alpha,\delta_1,\delta_2} c_{\alpha,\delta_1,\delta_2}\chi_{\alpha+\delta_1} \wedge \chi_{-\alpha-\delta_2} + \sum_{\alpha,\delta_1,\delta_2} b_{\alpha,\delta_1,\delta_2}\chi_{\alpha+\delta_1} \wedge \chi_{-2\alpha-\delta_2},$$

for some constants $c_{\alpha,\delta_1,\delta_2}, b_{\alpha,\delta_1,\delta_2} \in \mathbb{R}$. The last observation here is that $b_{\alpha,\delta_1,\delta_2}$ is the only coefficient in front of $\chi_{\alpha+\delta_1}\wedge\chi_{-\alpha-\delta_2}\wedge\chi_{-\alpha}$ in the expression for $0 = dd\omega$ and hence $b_{\alpha,\delta_1,\delta_2} = 0$. □

Lemma 4.4.17 *If* $\theta = \sum_{\alpha,\delta_1,\delta_2} x_{\alpha,\delta_1,\delta_2}\chi_{\alpha+\delta_1}\wedge\chi_{-\alpha-\delta_2}$, $x_{\alpha,\delta_1,\delta_2} \in \mathbb{R}$ *is a closed form then* $\theta = d\omega$, *where* $\omega = \sum_k c_k\eta_k$.

Proof We have to show that dimension of the linear space V of forms θ is equal to the rank of Lie group G. To that end, let us gather coefficients in front of terms of the form $\kappa_\delta \wedge \chi_\alpha \wedge \chi_{-\alpha-\delta_2}$ in the expression of

$$0 = d\theta = \sum_{\alpha,\delta_1,\delta_2} x_{\alpha,\delta_1,\delta_2}(d\chi_{\alpha+\delta_1} \wedge \chi_{-\alpha-\delta_2} - \chi_{\alpha+\delta_1} \wedge d\chi_{-\alpha-\delta_2}).$$

Notice that if $[K_{\delta_1}, X_\alpha] \neq 0$ and $[K_{-\delta_2}, X_{-\alpha}] \neq 0$ then $X_{-\alpha-\delta_2+\delta_1} = 0$, unless $\delta_1 = \delta_2$. Indeed, consider a complex root, the real part of which is equal to α. Because of our conditions, $\alpha+\delta_1$ and $-\alpha-\delta_2$ are complex roots as well, which implies that with respect to every order, α, δ_1, and δ_2 are of the same sign. Then $-\alpha-\delta_2+\delta_1$ is not a root, unless $\delta_1 = \delta_2$.

Thus, if $\delta_1 \neq \delta_2$, the only coefficient in front of $\kappa_{\delta_1}\wedge\chi_\alpha\wedge\chi_{-\alpha-\delta_2}$ is $x_{\alpha,\delta_1,\delta_2}$, which therefore is equal to zero. The coefficient in front of $\kappa_\delta \wedge \chi_\alpha \wedge \chi_{-\alpha-\delta}$ is $x_{\alpha,\delta} - x_{\alpha,0}$ and we conclude that the form θ is represented as

$$\theta = \sum_{\alpha,\delta} x_\alpha \chi_{\alpha+\delta} \wedge \chi_{-\alpha-\delta},$$

with $x_\alpha \in \mathbb{R}$.

But in that case, gathering the terms that are free of δ, we see that

$$\begin{aligned} 0 = d\theta &= \sum_{\alpha,\delta} x_\alpha(d\chi_{\alpha+\delta} \wedge \chi_{-\alpha-\delta} - \chi_{\alpha+\delta} \wedge d\chi_{-\alpha-\delta}) \\ &= \cdots + \sum_{\beta+\gamma=\alpha} x_\alpha c_{\beta\gamma}\chi_\beta \wedge \chi_\gamma \wedge \chi_{-\alpha} \\ &\quad - \sum_{\beta+\gamma=-\alpha} x_\alpha c_{\beta\gamma}\chi_\alpha \wedge \chi_\beta \wedge \chi_\gamma \\ &= \sum_{\beta+\gamma=\alpha} x_\alpha c_{\beta\gamma}\chi_\beta \wedge \chi_\gamma \wedge \chi_{-\alpha} \end{aligned}$$

$$- \sum_{\beta+\gamma=\alpha} x_{-\alpha} c_{\beta\gamma} \chi_{-\alpha} \wedge \chi_\beta \wedge \chi_\gamma$$
$$= 2 \sum_{\beta+\gamma=\alpha} x_\alpha c_{\beta\gamma} \chi_\beta \wedge \chi_\gamma \wedge \chi_{-\alpha}.$$

It means that for any $\alpha, \beta, \gamma \in \Sigma$ such that $\alpha = \beta+\gamma$ the following equality holds:

$$x_\alpha(c_{\beta\gamma} - c_{\gamma\beta}) + x_{-\gamma}(c_{-\alpha\beta} - c_{\beta-\alpha}) + x_{-\beta}(c_{\gamma-\alpha} - c_{-\alpha\gamma}) = 0,$$

or, which is the same,

$$x_\alpha c_{\beta\gamma} + x_\gamma c_{\beta-\alpha} + x_\beta c_{-\alpha\gamma} = 0.$$

This equality implies that for every $\beta, \gamma \in \Sigma$ such that $[X_\beta, X_\gamma] \neq 0$ (meaning $[X_\beta, X_\gamma] = c_{\beta\gamma} X_\alpha, c_{\beta\gamma} \neq 0$) the coefficient x_α is uniquely determined by x_β, x_γ. Since the set $\{X_\alpha, X_{-\alpha}\}|_{\alpha\in B}$, where $B \subset \Sigma$, is a Lie basis of the root system of $\mathbf{g}$ and thus generates $\mathbf{g}$, we see that if the coefficients $x_\alpha, \alpha \in B$ are given, then the coefficients $x_\alpha, \alpha \in \Sigma$ are uniquely determined. It means that $\dim(V) \leq \operatorname{Rank} \mathbf{g} = \operatorname{Card} B$, i.e., $\theta = d\omega$ for some ω. □

We are ready now to finish the proof of Theorem 4.4.14.

Proof For a given C^∞ cocycle β, one constructs the corresponding form $\omega \in T_1^* N$. According to Lemma 4.4.16 and Lemma 4.4.17 there exists a form $\omega_0 = \sum_{k=1}^{\operatorname{Rank} \mathbf{g}} x_k \eta_k, x_k \in \mathbb{R}$ such that $d\omega = d\omega_0$, i.e., $\omega - \omega_0$ is a closed form.

Let us show now that the first cohomology group of W is trivial. According to Margulis theorem on normal subgroups [110], either G/Γ or Γ is finite; in our case, of course, only latter is possible. Since the fiber is connected, every closed path, as well as homotopy, can be lifted from the base onto G/Γ to a closed path or a homotopy respectively. Hence the fundamental group of the base $\pi_1(N)$ is embedded into $\pi_1(G/\Gamma)$. Thus, because $\pi_1(G/\Gamma) = \Gamma$ is finite, $\pi_1(N)$ is also finite and

$$H^1(N, \mathbb{R}) = \pi_1(N)/[\pi_1(N), \pi_1(N)] \otimes_{\mathbb{Z}} \mathbb{R} = 0.$$

Therefore, $\omega - \omega_0$ is also an exact form. It follows from the first statement of Lemma 4.4.15 that there are no more than Rank $\mathfrak{g}$ different cohomology classes of cocycles. On the other hand, there are at least Rank $\mathfrak{g}$ non-cohomologous cocycles, namely, all the constant ones. It implies that there are exactly Rank $\mathfrak{g}$ cohomology classes, or, in other words, every small smooth cocycle is cohomologous to a constant cocycle. □

4.4.5 Rigidity of abelian cocycles over Cartan actions and algebraic K-theory

Review of algebraic K-theory

We start with a summary of K-theory results relevant to our exposition. Proofs for the K-theory results in this section can be found in [115]. See also [166] and [112]. Throughout this section we assume $n \geq 3$, and $\mathbb{K}$ is either the field of real numbers $\mathbb{R}$ or the field of complex numbers $\mathbb{C}$.

The abstract Steinberg group $St_n(\mathbb{K})$ is defined by generators and relations. The generators are denoted by $x_{ij}(t)$, $t \in \mathbb{K}$, $i, j \in \{1, 2, ..., n\}$, $i \neq j$, and are subject to the relations

$$x_{ij}(t)x_{ij}(s) = x_{ij}(t+s), \tag{4.4.20}$$

and

$$[x_{ij}(t), x_{kl}(s)] = \begin{cases} 1, & j \neq k, i \neq l, \\ x_{il}(st) & j = k, i \neq l, \\ x_{kj}(-st), & j \neq k, i = l, \end{cases} \tag{4.4.21}$$

Steinberg obtained the following presentation of the special linear group $SL(n, \mathbb{K})$.

Theorem 4.4.18 *The group $SL(n, \mathbb{K})$ is generated by $e_{ij}(t), i \neq j, t \in \mathbb{K}$, subject to the relations*

$$[e_{ij}(t), e_{kl}(s)] = \begin{cases} 1, & j \neq k, i \neq l, \\ e_{il}(st) & j = k, i \neq l, \\ e_{kj}(-st), & j \neq k, i = l, \end{cases} \tag{4.4.22}$$

where $[\cdot, \cdot]$ denotes the commutator,

$$e_{ij}(t)e_{ij}(s) = e_{ij}(t+s), \tag{4.4.23}$$

and

$$h_{12}(t)h_{12}(s) = h_{12}(ts), \tag{4.4.24}$$

where

$$h_{12}(t) = e_{12}(t)e_{21}(-t^{-1})e_{12}(t)e_{12}(-1)e_{21}(1)e_{12}(-1),$$

for each $t \in \mathbb{K}^*$.

The natural map $\phi : St_n(\mathbb{K}) \to SL(n, \mathbb{K})$, defined by $\phi(x_{ij}(t)) = e_{ij}(t)$, is a homomorphism. Its kernel is denoted by $K_2(\mathbb{K})$. The kernel coincides with the center of the Steinberg group. We use for it multiplicative notation, and denote the neutral element by 1.

Here is a way to construct elements in $K_2(\mathbb{K})$. Let $u, v \in \mathbb{K}^*$. Then the diagonal matrices

$$D_u = \begin{pmatrix} u & 0 & 0 \\ 0 & u^{-1} & 1 \\ 0 & 0 & 1 \end{pmatrix}, \; D'_v = \begin{pmatrix} v & 0 & 0 \\ 0 & 1 & 0 \\ 0 & 0 & v^{-1} \end{pmatrix},$$

commute and belong to $SL(3, \mathbb{K})$. Using an embedding of the $SL(3, \mathbb{K})$ in the upper left corner of $SL(n, \mathbb{K}), n \geq 3$, it follows that D_u, D'_v belong to any $SL(n, \mathbb{K}), n \geq 3$. Choose now representatives $U, V \in St_n(\mathbb{K})$, that is, $\phi(U) = u, \phi(V) = v$, and define $\{u, v\} = UVU^{-1}V^{-1}$. Then $\{u, v\}$ is an element in $K_2(\mathbb{K})$.

Alternatively, for any unit in $u \in \mathbb{K}$ and $i \neq j$ one can define

$$w_{ij}(u) = x_{ij}(u)x_{ji}(-u^{-1})x_{ij}(u),$$

and

$$h_{ij}(u) = w_{ij}(u)w_{ij}(-1).$$

Then $\{u, v\} := [h_{ij}(u), h_{ik}(v)]$ is an element in $K_2(\mathbb{K})$. The map

$$\mathbb{K}^* \times \mathbb{K}^* \ni (u, v) \to \{u, v\} \in K_2(\mathbb{K})$$

is bi-multiplicative, that is,

$$\{u_1u_2, v\} = \{u_1, v\}\{u_2, v\} \text{ and } \{u, v_1v_2\} = \{u, v_1\}\{u, v_2\}, \tag{4.4.25}$$

and skew-symmetric, that is, $\{u, 1 - u\} = 1$. Moreover, it is shown in [115] that

$$\{u, v\} = h_{ij}(uv)h_{ij}(u)^{-1}h_{ij}(v)^{-1}. \tag{4.4.26}$$

A presentation for the $K_2(\mathbb{K})$ in terms of relations and generators was found by Matsumoto [112].

Theorem 4.4.19 *Let $\mathbb{K}$ be a field. Then the kernel $K_2(\mathbb{K})$ of the natural map $\phi : St_n(\mathbb{K}) \to SL(n, \mathbb{K})$ is generated by the elements $\{u, v\}, u, v \in \mathbb{K}^*$ subject to the relations:*

(i) $\{u, 1 - u\} = 1$, *for* $u \neq 0, 1$;
(ii) $\{u_1u_2, v\} = \{u_1, v\}\{u_2, v\}$; *and*
(iii) $\{u, v_1v_2\} = \{u, v_1\}\{u, v_2\}$.

Any bi-multiplicative map $c(\cdot, \cdot) : \mathbb{K}^* \times \mathbb{K}^* \to A$ into an abelian group A satisfying $c(u, 1 - u) = 1_A$ is called a *Steinberg symbol*. If A has a structure of Hausdorff space, and the Steinberg symbol is continuous as a function $\mathbb{K}^* \times \mathbb{K}^* \to A$, then the symbol is called *continuous*.

Theorem 4.4.20 ***(Milnor)*** *(a) Every continuous Steinberg symbol on the field $\mathbb{C}$ of complex numbers is trivial.*

(b) If $c(x, y)$ is a continuous Steinberg symbol on the field $\mathbb{R}$ of real numbers, then $c(x, y) = 1$ if x or y is positive, and $c(x, y) = c(-1, -1)$ has order at most 2 if x and y are both negative.

Rigidity of abelian cocycles over Cartan actions

Let $D_n^+ \subset SL(n, \mathbb{K})$ be the group of diagonal matrices with positive elements. We parameterize D_n^+ as follows:

$$D_n^+ = \{\text{diag}(e^{t_1}, \dots, e^{t_n}) | \mathbf{t} = (t_1, \dots, t_n), \sum_{i=1}^{n} t_i = 0\}.$$

The $n - 1$ dimensional subspace of $\mathbb{R}^n$, given by

$$\mathbb{D}_n = \{(t_1, \dots, t_n) | \sum_{i=1}^{n} t_i = 0\},$$

can be viewed as the Lie algebra of D_n^+ via the inverse of the usual exponential map. So D_n^+ is isomorphic to $\mathbb{R}^{n-1}$.

Let $\Gamma \subset SL(n, \mathbb{K})$ be a torsion free co-compact lattice, that is, a discrete group of co-finite volume without elements of finite order. The quotient space $SL(n, \mathbb{K})/\Gamma$ has a structure of compact manifold. We consider the action of D_n^+ on the space $SL(n, \mathbb{K})/\Gamma$ by left translations. This type of action is called a *Cartan action*.

Let $\alpha : D_n^+ \times SL(n, \mathbb{K})/\Gamma \to SL(n, \mathbb{K})/\Gamma$ be a Cartan action. Introduce a right invariant metric $d(\cdot, \cdot)$ on $SL(n, \mathbb{K})$, and denote in the same way the induced metric on $SL(n, \mathbb{K})/\Gamma$. Let $1 \leq i, j \leq n, i \neq j$, be two fixed indices, and let exp be the exponential map in $SL(n, \mathbb{K})$. Let $v_{i,j}$ be the elementary $n \times n$ matrix with only one non-zero entry, that in position (i, j). We denote $e_{ij}(s) = \exp(s v_{i,j})$ and define a foliation F_{ij} on $SL(n, \mathbb{K})/\Gamma$ with leaves

$$F_{ij}(x) = \{e_{ij}(s)x | s \in \mathbb{K}\}. \tag{4.4.27}$$

Note that it is immediate from the definition of the foliation F_{ij} that its leaves are invariant under left multiplication by $e_{ij}(s)$. Vice-versa, since the leaves are one $\mathbb{K}$-dimensional, the motion along the leaves can be described in terms of multiplication by $e_{ij}(s)$.

The foliation F_{ij} is invariant under the action α. Indeed, $e_{ij}(s) = \text{Id} + s v_{i,j}$, and a direct calculation shows that

$$\alpha(\mathbf{t})(\text{Id} + s v_{i,j})x = (\text{Id} + s e^{t_i - t_j} v_{i,j})\alpha(\mathbf{t})x. \tag{4.4.28}$$

Formula (4.4.28) also shows that the foliation F_{ij} is contracting under the action of $\alpha(\mathbf{t})$ if $t_i < t_j$, and is expanding if $t_i > t_j$. Consequently, any element in D_n^+ that has the entries pairwise different acts as a partially hyperbolic diffeomorphism on $SL(n, \mathbb{K})/\Gamma$. The dimension of the center distribution is $n-1$ if $\mathbb{K} = \mathbb{R}$ and $2(n-1)$ if $\mathbb{K} = \mathbb{C}$.

We are ready to present the main result of this section, which appeared in [21].

Theorem 4.4.21 *Let $n \geq 3$, $G = SL(n, \mathbb{K})$, $\Gamma \subset G$ a co-compact torsion free lattice, and $M = G/\Gamma$. Let $\alpha : D_n^+ \times M \to M$ be the Cartan action. Let $\beta : D_n^+ \times M \to \mathbb{R}$ be a C^K-cocycle. Then β is cohomologous to a constant cocycle via a $C^{[K/2-\dim(M)/2]}$ transfer function $h : M \to \mathbb{R}$. Moreover, if β is Hölder or smooth, then the transfer function is Hölder, respectively smooth.*

Proof Let $F_{i,j}$, $i \neq j$, be the α-invariant foliations introduced before. These foliations are smooth and their brackets generate the whole tangent space. As shown in [12], this facts imply that the system of foliations is locally transitive. Each F_{ij}-path built using these foliations can be described by a product of elements of type $e_{ij}(t)$. Indeed, each piece of an F_{ij}-leaf can be parameterized by $t \to e_{ij}(t)x$, $t \in I$, for some $x \in G$ and a compact interval I. The path is a cycle if and only if the product of these elements belongs to Γ.

It follows from Proposition 4.3.14 that if the heights $H(\beta, \mathcal{C})$ are equal to zero for all cycles $\mathcal{C}$ determined by a family of locally transitive foliations, then the cocycle β is cohomologous to a constant cocycle. Furthermore, it follows from Proposition 4.3.13 that if the cycle $\mathcal{C}$ is included in a stable or unstable leaf then the height $H(\beta, \mathcal{C})$ is equal to zero.

The height over a cycle is, so far, dependent of the word in $e_{ij}(t)$s describing the cycle. Changing the word, without changing the value of the product, can produce a different height. We show first that, if the product of $e_{ij}(t)$s is equal to identity, then the height over the cycle is trivial.

Using the presentation for $SL(n, \mathbb{K})$ given above, each word in $e_{ij}(t)$s representing the product can be written as a concatenation of conjugates of the basic relations (4.4.22), (4.4.23), and (4.4.24). Each of these relations defines an F_{ij}-cycle.

The relations of type (4.4.22) or (4.4.23) give cycles that are contained in stable leaves for regular elements of the action α. Indeed, in the case of (4.4.23), the motion along the cycle is described by multiplication by $e_{ij}(t)$, for various values of t, so the cycle is included in the stable leaf of an element $\mathbf{t} \in D_n^+$ with $t_i < t_j$. In the case of (4.4.22), we split the proof into three cases:

(i) If $j \neq k$, $i \neq l$ then the cycle is contained in the stable leaf of an element $\mathbf{t} \in D_n^+$ with $t_i < t_j$, $t_k < t_l$.

(ii) If $j = k, i \neq l$ then the cycle is contained in the stable leaf of an element $\mathbf{t} \in D_n^+$ with $t_i < t_j < t_k$.
(iii) If $j \neq k, i = l$ then the cycle is contained in the stable leaf of an element $\mathbf{t} \in D_n^+$ with $t_k < t_l < t_j$.

We consider now, separately, the relations (4.4.24) for the cases $\mathbb{K} = \mathbb{R}$ and $\mathbb{K} = \mathbb{C}$.

Assume $\mathbb{K} = \mathbb{R}$. Let $A = \{(1, 1), (-1, 1), (1, -1), (-1, -1)\}$.

It follows from Section 4.4.5.1 that $(t, s) \to \{t, s\}$ is a Steinberg symbol with values in $K_2(\mathbb{R})$. We show that when (t, s) varies over $\mathbb{R}^* \times \mathbb{R}^*$ the exponential of the height $H(t, s)$ over $C(t, s)$ gives a continuous Steinberg symbol with values in a subgroup of $\mathbb{R}$. For $t_1, t_2, s \in \mathbb{R}^*$ consider the product $\Pi := \{t_1t_2, s\}\{t_2, s\}^{-1}\{t_1, s\}^{-1}$. The Steinberg symbol $\{t, s\}$ is bi-multiplicative in $St_n(\mathbb{R})$, so Π is equal to identity of $St_n(\mathbb{R})$. Using the presentation for $St_n(\mathbb{R})$, any word representing Π can be written as a concatenation of conjugates of the basic relations (4.4.22) and (4.4.23) (which respectively coincide with (4.4.20) and (4.4.21)). So using the discussion above, the height over the cycle determined by Π is trivial.

The height over a concatenation of two cycles is the sum of the heights over the cycles. This implies $H(t_1t_2, s) = H(t_1, s) + H(t_2, s)$. In a similar way one can show $H(s, t)$ is additive in the second variable. To show that the height satisfies skew-symmetry, consider the relation $\{t, 1-t\}$. This relation is equal to identity in the Steinberg group, because the Steinberg symbol $\{t, s\}$ is skew-symmetric, so any word representing $\{t, 1-t\}$ is a product of conjugates of the standard relations (4.4.22) and (4.4.23). Using the discussion above it follows that the height $H(t, 1-t)$ over the cycle determined by $\{t, 1-t\}$ is trivial. We show that $H(s, t)$ takes values in an abelian group. This follows from the fact that any two symbols $\{t_1, s_1\}$ and $\{t_2, s_2\}$ belong to the abelian group $K_2(\mathbb{R})$, so the following equality $\{t_1, s_1\}\{t_2, s_2\} = \{t_2, s_2\}\{t_1, s_1\}$ holds in $St_n(\mathbb{R})$. As before, the identity in $St_n(\mathbb{R})$ implies $H(t_1, s_1) + H(t_2, s_2) = H(t_2, s_2) + H(t_1, s_1)$. The continuity of the symbol $H(t, s)$ follows from its definition and from the fact that the abelian group in which the symbol takes values has a Hausdorff topology induced from $\mathbb{R}$.

So $(t, s) \to H(t, s)$ is a continuous Steinberg symbol. Due to Theorem 4.4.20(b), the only possible values for the height are zero or an element of order 2 in $\mathbb{R}$. Since there are no elements of order 2 in $\mathbb{R}$, the height has to be trivial.

If $\mathbb{K} = \mathbb{C}$ the proof is similar, but simpler, because Theorem 4.4.20(a), implies that the continuous Steinberg symbol is trivial in this case.

After eliminating the contribution to the height that appears due to the relations, the product contains only the elements that conjugate the relations.

Cancelations of type $e_{ij}(t)e_{ij}(-t) = Id_N$ do not change the height because the cycle determined by the product $e_{ij}(t)e_{ij}(-t)$ is contained in a stable leaf of an element of the action α, so the height over it has to be trivial.

We consider now the height over an arbitrary cycle, not necessarily with the product of the e_{ij}s trivial. Any cycle induces an element in the first fundamental group $\pi_1(G/\Gamma)$. One has an exact sequence

$$1 \to \pi_1(G) \to \pi_1(G/\Gamma) \to \pi_1(\Gamma) \to 1.$$

It is well known that for $n \geq 3$ one has $\pi_1(SL(n,\mathbb{R})) = \mathbb{Z}_2$ and $\pi_1(SL(n,\mathbb{C})$ is trivial. If $\mathbb{K} = \mathbb{R}$ then the cycles that induce the nontrivial element in $\pi_1(SL(n,\mathbb{R}))$ are homotopic to the cycle determined by the extra relation. See [115].

Two cycles that induce the same element in the fundamental group have the same height. Indeed, if their products are Π_1 and Π_2, then the concatenated product $\Pi_1\Pi_2^{-1}$ gives a word that is equal to identity, so the height over the cycle determined by $\Pi_1\Pi_2^{-1}$ is trivial, so the heights over Π_1 and Π_2 are equal. Since the height over a concatenation of two cycles is the sum of the heights over the cycles, the height determines a homomorphism ψ from $\pi_1(G)$ into $\mathbb{R}$. The above remarks about the fundamental group of G imply that ψ factors to a homomorphism from Γ into $\mathbb{R}$. Note that any co-compact lattice in $SL(n,\mathbb{K}), n \geq 3$, has property (T) [45]. Since there are no nontrivial homomorphisms from property (T) groups into $\mathbb{R}$ (see [45]), ψ has to be trivial. So all the heights over the cycles are trivial, and the cocycle is cohomologous to a constant cocycle.

So far, the transfer map $h : M \to \mathbb{R}$ is only continuous. To show higher regularity for h we employ standard regularity results. It is standard to show that for any partially hyperbolic element in D_n^+, h is C^K along its stable and unstable directions. See Section 4.2.2. This gives C^K regularity along a finite set of directions, that have the vectors tangent to their distributions, and their length 2 commutators, generating the whole tangent space TM. The commutators needed to consider are of type $[e_{ij}(t), e_{ji}(s)]$. Now Theorem 3.7.2, applied for $r = 2$, implies that h is $C^{[K/2-\dim(M)/2]}$ in the M direction. The statement about smooth cocycles also follows from Theorem 3.7.2. □

4.5 Cocycles over generic Anosov actions

In this section we extend Livshitz' results about the trivialization of cohomology over rank-one Anosov actions to more general Anosov group actions. The material in this section appears in [125].

We need to introduce the following weak condition about the relative position of a generic diffeomorphism with respect to a fixed Anosov diffeomorphism.

Definition 4.5.1 Let $f : M \to M$ be an Anosov C^1 diffeomorphism and let $g \in \mathrm{Diff}^1(M)$. Let $TM = E^s \oplus E^u$ be the invariant splitting induced by Df, the derivative of f. Then g is said to be in *generic position* with respect to f if

$$P_+(Dg)|_{E^u} : E^u \to E^u \text{ and } P_-(Dg^{-1})|_{E^s} : E^s \to E^s$$

can be inverted fiber-wise, where $P_\pm : TM \to E^{u,s}$ is the projection along $E^{s,u}$.

Theorem 4.5.2 (Non-commutative closing lemma) *Let $f : M \to M$ be an Anosov C^1 diffeomorphism and let $g \in \mathrm{Diff}^1(M)$. Let $TM = E^s \oplus E^u$ be the invariant splitting induced by Df, the derivative of f. Assume that g is in generic position with respect to f.*

Then there are $N_0 \in \mathbb{N}$, $\delta_0 > 0$, $K > 0$ such that for any $N \geq N_0$, $0 < \delta < \delta_0$, and for any sequence $x_0, x_1, \ldots, x_N \in M$ such that $d_M(fx_k, x_{k+1}) < \delta$, $0 \leq k \leq N-1$ and $d_M(gx_N, x_0) < \delta$, there is a point $x \in M$ with the following properties:

(i) $gf^n x = x$; and
(ii) the sequence $x_0, x_1, \ldots, x_N$ and the orbit of x are $K\delta$-close, that is

$$d_M(x_k, f^k x) < K\delta, \quad 0 \leq k \leq N.$$

Proof It is convenient to introduce local coordinates on M. For $\delta > 0$ denote by $(TM)_\delta$ the bundle of balls of radius δ centered in the origin. By the compactness of M, there are $\delta_1, \delta_1' > 0$ such that the exponential map $\exp : (TM)_{\delta_1'} \to M$ given in each fiber by

$$\phi_x = \exp_x : (T_x M)_{\delta_1'} \to M$$

has the following properties:

- for any $x \in M$ the image of ϕ_x contains a $B(x, \delta_1)$;
- the dependence of ϕ_x on x is continuous in the C^1 topology;
- $\|\phi_x^\pm\|_{Lip} \leq 2$ for each $x \in M$.

Since f is Anosov and the angle between the distributions E^s, E^u is bounded from below, there is a constant $C_1 > 0$ and linear isomorphisms $\Lambda_x : T_x M \to \mathbb{R}^d$ for all $x \in M$ such that:

- Λ_x carries the decomposition $T_xM = E_x^s \oplus E_x^u$ into $\mathbb{R}^d = \mathbb{R}^{d_+} \oplus \mathbb{R}^{d_-}$;
- the restrictions of Λ_x to $E_x^{s,u}$ are isometries;
- $\|\Lambda_x^{\pm 1}\| \leq C_1$.

Consider a $\delta_1/2$-pseudo-orbit $x_0, x_1, \ldots, x_N$. The maps f, g, viewed in local coordinates along the pseudo-orbit, become

$$f_k^0 = \Lambda_{x_{k+1}} \phi_{x_{k+1}}^{-1} f \phi_{x_k} \Lambda_{x_k}^{-1}, 0 \leq k \leq N-1,$$
$$g_N^0 = \Lambda_{x_0} \phi_{x_0}^{-1} g \phi_{x_N} \Lambda_{x_N}^{-1}.$$

Using the decomposition $\mathbb{R}^d = \mathbb{R}^{d_+} \oplus \mathbb{R}^{d_-}$ one has the following formulas for the derivatives of f_k^0, g_N^0:

$$D_0 f_k^0 = \begin{pmatrix} A_k^+ & * \\ * & A_k^- \end{pmatrix}$$
$$B_N := D_0 g_N^0 = \begin{pmatrix} B_{11} & B_{12} \\ B_{21} & B_{22} \end{pmatrix}$$
$$\bar{B}_N := (D_0 g_N^0)^{-1} = \begin{pmatrix} \bar{B}_{11} & \bar{B}_{12} \\ \bar{B}_{21} & \bar{B}_{22} \end{pmatrix},$$

where we denote by $*$ the entries that are not relevant in the sequel.

By the properties of the local coordinate system and those of f, g, there exist constants $\delta_2 > 0$, $0 < \lambda_1 < 1$ and $C_2, C_3 \geq 1$ such that:

for any $\epsilon > 0$, there exists $0 < \delta_3(\epsilon) < \delta_1/2$ such that if $x_0, \ldots, x_N$ is a $\delta_3(\epsilon)$-pseudo-orbit then:

- the maps f_k^0, $0 \leq k \leq N-1$, g_N^0 are defined on a ball of radius δ_2 centered in the origin;
- on these balls

$$\left\| f_k^0 - \begin{pmatrix} A_k^+ & 0 \\ 0 & A_k^- \end{pmatrix} \right\|_{C^1} < \epsilon,$$
$$\left\| g_N^0 - \begin{pmatrix} B_{11} & B_{12} \\ B_{21} & B_{22} \end{pmatrix} \right\|_{C^1} < \epsilon; \tag{4.5.1}$$

- the linear parts satisfy

$$\|A_k^-\| \leq \lambda_1, \quad \|(A_k^+)^{-1}\| \leq \lambda_1,$$
$$\|(B_{11})^{-1}\| \leq C_2, \quad \|(\bar{B}_{22})^{-1}\| \leq C_2,$$
$$\|B_N\| \leq C_3, \quad \|(B_N)^{-1}\| \leq C_3.$$

One extends now the maps f_k^0 and g_N^0 to the whole space $\mathbb{R}^d$. We recall that if U, V are open bounded sets in $\mathbb{R}^d$ with $\bar{U} \subset V$, then for any $\epsilon > 0$

there is $\delta_4(\epsilon) > 0$ such that if $f \in C^1(V, V)$ and Λ is a linear map on $\mathbb{R}^d$ with $\|f - \Lambda\|_{C^1} < \delta_4(\epsilon)$, then there is a function $\tilde{f} \in C^1(\mathbb{R}^d, \mathbb{R}^d)$ such that $\|\tilde{f} - \Lambda\|_{C^1} < \epsilon$ and $\tilde{f}|_U = f|_U$. Indeed, if $\xi : \mathbb{R}^d \to [0, 1]$ be a C^1 function such that $\xi|_U \equiv 1$ and $\xi|_{\mathbb{R}^d \setminus V} \equiv 0$, and set $\tilde{f} = \xi f + (1 - \xi)\Lambda$.)

Let U, V be $B(0, \delta_1/2)$, respectively $B(0, \delta_1)$, and denote the extended maps by f_k, g_N.

Now proving the theorem becomes a problem involving small C^1 perturbations of linear maps. If the constant ϵ that appears in (4.5.1) is less than $\delta_4(\omega)$, then:

$$\begin{gathered} f_k(x^+, x^-) = \\ (A_k^+ x^+ + \alpha_k^+(x^-, x^+), A_k^- x^+ + \alpha_k^-(x^-, x^+)),\ 0 \le k \le N-1, \\ g_N(x^+, x^-) = \\ (B_{11}x^+ + B_{12}x^- + \beta_N^+(x^-, x^+), B_{21}x^+ + B_{22}x^- + \beta_N^-(x^-, x^+)), \end{gathered}$$

where $A_k^{\pm}$, B_{ij} are linear maps and $\alpha_k^{\pm}$, $\beta_N^{\pm}$ have C^1 norm less than ω.

Let $C_4 = (1 - \lambda_1)^{-1}(2C_3 + 1)^3 \max\{\Lambda_1^{-1}, C_3\}$. Then the constants that appear in the theorem are

$$\begin{gathered} \delta_0 = \min\{(1/16)\delta_2 C_1^{-1} C_4^{-1}, \delta_3(\delta_4((1/2)C_4^{-1}))\}, \\ K = 8C_1^8, \end{gathered}$$

and N_0 satisfies

$$\begin{gathered} \lambda_1^{N_0} C_2 < 1/2, \\ \lambda_1^{N_0} C_3(1 + 2C_2C_3) < 1/2. \end{gathered} \tag{4.5.2}$$

Consider a δ pseudo-orbit $\bar{x}_0, \bar{x}_1, \dots, \bar{x}_N$ with $\delta < \delta_0$ and $N \ge N_0$. The equation $gf^n x = x$ becomes, via the change of coordinates

$$F(\tilde{x}) = \tilde{x}, \tag{4.5.3}$$

where $\tilde{x} = \text{column}(x_0, x_1, \dots, x_N) \in (\mathbb{R}^d)^{N+1}$ and

$$F(x_0, x_1, \dots, x_N) = (f_0 x_1, f_1 x_2, \dots, f_{N-1} x_N, g_N x_0).$$

Let $F = L_N + S_N$, where $L_N, S_N : (\mathbb{R}^d)^{N+1} \to (\mathbb{R}^d)^{N+1}$ are given by:

$$\begin{aligned} L_N(x_0, x_1, \dots, x_N) &= (A_0 x_1, \dots, A_{N-1} x_N, B_N x_0), \\ S_N(x_0, x_1, \dots, x_N) &= (\alpha_0 x_1, \dots, \alpha_{N-1} x_N, \beta_N x_0). \end{aligned}$$

Consider on $(\mathbb{R}^d)^{N+1}$ the norm $\|(x_0, x_1, \dots, x_N)\| = \max_{0 \le i \le N} \|x_i\|$. Note that $\|S_N\|_{C^1} \le 1/2C_4^{-1}$ and

$$\|S(\tilde{0})\| = \|F(\tilde{0})| = \|(\Lambda_{x_1}\phi_{x_1}^{-1} f x_0, \Lambda_{x_2}\phi_{x_2}^{-1} f x_1, \dots, \Lambda_{x_0}\phi_{x_0}^{-1} g x_N| \le 2C_1\delta.$$

We will show that if $N \geq N_0$ then $1 - L_N$ is invertible and

$$\|(1 - L_N)^{-1}\| \leq C_4. \tag{4.5.4}$$

Then the equation (4.5.3) becomes $\tilde{x} = \mathcal{F}(\tilde{x})$, where

$$\mathcal{F} = (I - L_N)^{-1} S_N.$$

Since $\mathcal{F}$ is a contraction, the equation above has a unique solution $\tilde{x}_*$ and

$$\|\tilde{x}_*\| \leq \frac{\|\mathcal{F}(\tilde{0})\|}{1 - \|\mathcal{F}\|_{Lip}} \leq \frac{C_4\|S(\tilde{0})\|}{1 - \|\mathcal{F}\|_{Lip}} \leq 4C_1C_4\delta \leq \frac{\delta_2}{4}.$$

Thus $\tilde{x}_*$ gives a solution of $gf^N x = x$ which satisfies claim (b) as well.

The last claim in the theorem follows from the fact that the M-balls of radius δ_0 centered at the x_ks are mapped inside balls of radius $2C_1\delta_0 \leq 2C_1(1/16)\delta_2 C_1 C_4 < \delta_2/2$ centered at the origin in the corresponding $\mathbb{R}^d$s. Therefore, restricted to this set, the action on the manifold is conjugate to the f-action on $(\mathbb{R}^d)^{N+1}$ and F has a unique fixed point.

To finish the proof of the theorem we need to prove (4.5.4).

After denoting

$$\tilde{\eta} = \text{column}(\eta_0, \eta_1, \ldots, \eta_N), \qquad \tilde{y} = \text{column}(y_0, y_1, \ldots, y_N)$$

one has to solve for $\tilde{y}$ the equation $(I - L_N)\tilde{y} = \tilde{\eta}$, which is equivalent to solving the system of equations:

$$\begin{aligned} y_0 - A_0 y_1 &= \eta_0, \\ y_1 - A_1 y_2 &= \eta_1, \\ &\cdots \\ y_{N-1} - A_{N-1} y_N &= \eta_{N-1}, \\ y_N - B_N y_0 &= \eta_N. \end{aligned} \tag{4.5.5}$$

Denote $\tilde{A} = A_0 A_1 \ldots A_{N-1}$ and

$$\begin{aligned} E = B_N A_0 \cdots A_{N-2}\eta_{N-1} + B_N A_0 \cdots A_{N-3}\eta_{N-2} + \cdots \\ \cdots + B_N A_0 \eta_1 + B_N \eta_0 + \eta_N. \end{aligned} \tag{4.5.6}$$

Decompose both variables and maps with respect to the decomposition $\mathbb{R}^d = \mathbb{R}^{d_+} \oplus \mathbb{R}^{d_-}$:

$$\tilde{A} = \begin{pmatrix} \tilde{A}_+ & 0 \\ 0 & \tilde{A}_- \end{pmatrix}, \begin{pmatrix} E^+ \\ E^- \end{pmatrix}$$

$$y_k = \begin{pmatrix} y_k^+ \\ y_k^- \end{pmatrix}, \; \eta_k = \begin{pmatrix} \eta_k^+ \\ \eta_k^- \end{pmatrix}, \; 0 \leq k \leq N.$$

Note that $\|\tilde{A}_-\| \leq \lambda_1^N$, $\|(\tilde{A}_+)^{-1}\| \leq \lambda_1^N$.

Consider now the equation (4.5.5). One obtains, after eliminating the other variables, that

$$y_N = B_N \tilde{A} y_N + E. \tag{4.5.7}$$

One separates (4.5.7) into two parts, according to the decomposition of $\mathbb{R}^d$:

$$y_N^+ = B_{11}\tilde{A}_+ y_N^+ + B_{12}\tilde{A}_- y_N^- + E^+, \tag{4.5.8}$$

$$y_N^- = B_{21}\tilde{A}_+ y_N^+ + B_{22}\tilde{A}_- y_N^- + E^-. \tag{4.5.9}$$

Rewrite (4.5.8) as $F_0\tilde{A}_+ y_N^+ = B_{12}\tilde{A}_- y_N^- + E^+$, where

$$F_0 = (\tilde{A}_+ B_{11}^{-1} - I)B_{11}.$$

Now $N \geq N_0$ and (4.5.2) implies that $\|\tilde{A}_+ B_{11}^{-1}\| \leq 1/2$. Therefore, F_0 is invertible with

$$\|F_0^{-1}\| \leq 2\|B_{11}^{-1}\| \leq 2C_2, \tag{4.5.10}$$

and

$$y_N^+ = \tilde{A}_+^{-1}(F_0^{-1}B_{12}\tilde{A}_- y_N^- + F_0^{-1}E^+). \tag{4.5.11}$$

Substituting into (4.5.9) one has

$$y_N^- = F_1 y_N^- + B_{21}^{-1}F_0^{-1}E^+ + E^-, \tag{4.5.12}$$

where

$$F_1 = B_{22}\tilde{A}_- + B_{21}F_0^{-1}B_{12}\tilde{A}_-.$$

Compute now using (4.5.10) that

$$\begin{aligned}\|F_1\| &\leq \|B_{22}\|\|\tilde{A}_-\| + \|B_{21}\|\|F_0^{-1}\|\|B_{12}\|\|\tilde{A}_-\| \\ &\leq \|B_N\|\lambda_1^N + \|B_N\|2C_2\|B_N\|\lambda_1^N.\end{aligned}$$

By (4.5.2), if $N \geq N_0$, then $\|F_1\| \leq 1/2$. So $I - F_1$ is invertible, $\|(I - F_1)^{-1}\| \leq 2$, and the equation (4.5.12) has the unique solution

$$y_N^- = (I - F_1)^{-1}(B_{21}^{-1}F_0^{-1}E^+ + E^-),$$

with

$$\|Y_N^-\| \leq 2(\|B_N\|2\|E^+\| + \|E^-\|) \leq (4C_3 + 2)\|E\|. \tag{4.5.13}$$

In conclusion, for $N \geq N_0$ the system (4.5.5) has a unique solution and $1 - L_N$ is invertible. It remains to estimate the norm of $(1 - L_N)^{-1}$. Denote $(\mathbb{R}^{d_+})^{N+1}$ by V^+ and $(\mathbb{R}^{d_-})^{N+1}$ by V^-. In order to make the computation easier we show separately that $\|(1 - L_N)^{-1}|_{V^\pm}\|$ are bounded by C_4.

Let $\tilde{\eta} \in V^-$. For $0 \le k \le N-1$ it follows from (4.5.5) and (4.5.11) that

$$\begin{aligned} y_k^- &= A_k^- \cdots A_{N-1}^- y_N^- + A_k^- \cdots A_{N-2}^- y_{N-1}^- + \cdots + A_k^- \eta_{k+1}^- + \eta_k^-, \\ y_k^+ &= A_k^- \cdots A_{N-1}^- \tilde{A}_+^{-1} (F_0^{-1} B_{12} \tilde{A}_- y_N^- + F_0^{-1} E^+) \\ &= (A_{k-1}^+)^{-1} \cdots (A_0^+)^{-1} (F_0^{-1} B_1 \tilde{A}_- y_N^- + F_0^{-1} E^+). \end{aligned}$$

Hence

$$\begin{aligned} \|y_k^-\| &\le \lambda_1^{N-k} \|y_N^-\| + (\lambda_1^{N-k-1} \|\tilde{\eta}\| + \lambda_1^{N-k-2} \|\tilde{\eta}\| + \cdots \|\tilde{\eta}\|) \\ &\le \lambda_1^{N-k} \|y_N^-\| + (1-\lambda_1)^{-1} \|\tilde{\eta}\|, \\ \|y_k^+\| &\le \lambda_1^k (2C_3 \lambda_1^N \|y_N^-\| + 2\|E\|). \end{aligned}$$

By (4.5.13) it is enough to estimate $\|E\|$. By (4.5.6), for $\tilde{\eta} \in V^-$ one has

$$\|E\| \le C_3 (1-\lambda_1)^{-1} \|\tilde{\eta}\|.$$

Hence

$$\|(1-L_N)^{-1}|_{V^-}\| \le (1-\lambda_1)^{-1} (2C_3+1)^3.$$

In order to estimate $\|(1-L_N)^{-1}|_{V^+}\|$ note that this case can be reduced to the last one. Indeed:

- $(1-L_N)^{-1}|_{V^+} = -L_N^{-1}(1-L_N^{-1})^{-1}|_{V^+}$ and $\|L_N^{-1}\| \le \max\{\lambda_1^{-1}, C_3\}$;
- the matrix of $J(L_N^{-1})J^{-1}$ is of the same form as that of L_N, where $J = (j_{kl})_{0\le k,l\le N}$ is the isometry of $(\mathbb{R}^d)^{N+1}$ given by $j_{kl} = \delta_{k+1,N+1} Id_{\mathbb{R}^d}$;
- the contracting space of L_N^{-1} is V^+, which is invariant under J^{-1}, and the corresponding entry of $\bar{B}$ is $\bar{B}_{22}$.

This completes the proof of the theorem. □

Remark 4.5.3 The proof of Theorem 4.5.2 is an adaptation of the proof of Anosov closing lemma as presented in [67, Theorem 6.4.15].

Definition 4.5.4 Let G be a discrete group, M a compact manifold and $\alpha : G \times M \to M$ a C^1-action of G by diffeomorphisms of M. We say that α is Anosov if there is an element $g \in G$ such that $\alpha(g)$ is an Anosov diffeomorphism. Let Γ be a topological group and $\beta : G \times M \to \Gamma$ be a continuous cocycle. We say that β satisfies the *isolated closing conditions* if for any $g \in G$ and $x \in M$ isolated fixed point of g one has $\beta(g, x) = 1_\Gamma$.

Remark 4.5.5 The next theorem shows that the notion of isolated closing conditions is useful. For an Anosov rank-one action isolated closing conditions are satisfied if and only if closing conditions are satisfied. For general actions this is not the case. As one can see in [125], isolated closing conditions can be easily obtained for natural examples of higher rank lattice group actions on compact manifolds, while general closing conditions might be harder to show.

Theorem 4.5.6 *Let M be a compact manifold. Let $f, g \in Diff^1(M)$, f a topologically mixing Anosov diffeomorphism and g in generic position with respect to f, as in Definition 4.5.1. Let $G =< f, g >$ be the group generated by f and g in $Diff^1(M)$. Let Γ be a topological group and $\beta : G \times M \to \Gamma$ be a continuous cocycle which satisfies the isolated closing conditions. denote by $< f >$ the group generated by f in $Diff^1(M)$ and assume that the cocycle $\beta|_{<f>\times M}$ is cohomologous to a trivial cocycle, that is, there exists a continuous map $P : M \to \Gamma$ such that $\beta(f, x) = P(fx)P(x)^{-1}$ for all $x \in M$. Then β is cohomologous to a trivial cocycle with transfer map P.*

Proof For any $g \in G$ define $\bar{\beta}(g, x) = P(gx)^{-1}\beta(gx)P(x)$. Then $\bar{\beta}(f, x) = 1$ and to finish the proof of the theorem we need to show that $\bar{\beta}(g, x) = 1$ for any fixed $x \in M$.

Pick x_0 close to Bx such that $f^n x_0$ is close to x for some large n. Apply now Theorem 4.5.2 and find x_* such that $gf^n x_* = x_*$, x_* is close to x_0 and $f^n x_*$ is close to $f^n x_0$. Hence $f^n x_*$ is close to x_0. But $gf^n x_* = x_*$ implies that

$$\bar{\beta}(g, f^n x_*) = \bar{\beta}(gf^n, x_*)\bar{\beta}(f^n, x_*)^{-1} = \bar{\beta}(gf^n, x_*) = 1_\Gamma.$$

Using now the fact that $f^n x_*$ approaches x when $f^n x_0$ approaches x and the continuity of $\bar{\beta}$ we have $\bar{\beta}(g, x) = 1_\Gamma$. □

For many Anosov actions of large groups one may apply Theorem 4.5.6 several times for the generators of the action and obtain Livshitz type results. This motivates the following definition.

Definition 4.5.7 Let M be a compact manifold and G a discrete group. A C^1 G-action by diffeomorphisms of M, $\alpha : G \times M \to M$, is said to be *generically Anosov* if G has a set of generators $\{A, B_1, \dots, B_n\}$ such that A is Anosov and each B_k, $1 \le k \le n$, is in generic position with respect to A.

The following theorem is a generalization of Livshitz' cohomological result 4.2.2 and an immediate consequence of Theorem 4.5.6.

Theorem 4.5.8 *Let $\alpha : G \times M \to M$ be a generically Anosov action. Let $\beta : G \times M \to \mathbb{R}$ be a Hölder that satisfies the isolated closing conditions. Then β is cohomologous to a trivial cocycle via a Holder coboundary.*

We show now an explicit example of generically Anosov action.

Proposition 4.5.9 *The standard linear action $SL(n, \mathbb{Z}) \times \mathbb{T}^n \to \mathbb{T}^n$ is generically Anosov.*

Proof We first find an Anosov element A in $SL(n, \mathbb{Z})$. The following matrices are diagonalizable hyperbolic matrices:

$$A_1 = \begin{pmatrix} 2 & 3 \\ 1 & 2 \end{pmatrix}, \quad A_2 = \begin{pmatrix} 2 & 3 & 0 \\ 1 & 2 & -1 \\ -1 & -2 & 2 \end{pmatrix}.$$

If n is even, let $A = \mathrm{diag}(A_1, \ldots, A_1)$, and if n is odd take $A = \mathrm{diag}(A_1, \ldots, A_2)$. Then A is a diagonalizable hyperbolic matrix that induces an Anosov action on $\mathbb{T}^n$. Moreover, there exists an $n \times n$ matrix B, which has the same diagonal block form as A, such that $A = B\mathrm{diag}(\lambda_1, \ldots, \lambda_n)B^{-1}$. Let $E^s \oplus E^u$ be the splitting of $\mathbb{R}^n$ into contracting and expanding parts for $\mathrm{diag}(\lambda_1, \ldots, \lambda_n)$. Let $P_\pm$ be the projection onto $E^{u(s)}$.

Now it is known that $SL(n, \mathbb{Z})$ is generated by $\{E_{ij} | 1 \le i, j \le n, i \ne j\}$, E_{ij} being the matrix with 1 on the diagonal and in the position (i, j), and 0 elsewhere. To finish the proof it is enough to show that $P_+ B^{-1} E_{ij} B|_{E^+}$ and $P_- B^{-1} E_{ij} B|_{E^-}$ are invertible.

One has

$$B^{-1} E_{ij} B = I + \sum_{k,l=1}^{n} \bar{b}_{ki} B_{jl} e_{kl},$$

where $B = (b_{ij})$, $B^{-1} = (\bar{b}_{ij})$ and e_{kl}, $1 \le k, l \le n$ are the matrix units in $\mathrm{Mat}(n, \mathbb{R})$. Because B and B^{-1} are block diagonal, having the same pattern as the matrix A, in the last sum non-zero terms appear only in the block to which the position (i, j) belongs. Hence, unless (i, j) lies in one of the diagonal blocks, the matrix $B^{-1} E_{ij} B$ is triangular and $P_\pm B^{-1} E_{ij} B|_{E^{u(s)}}$ are invertible. The remaining cases can be checked by hand numerically. □

4.6 Twisted cocycles

Twisted cohomology is important because it appears in applications to the differentiable rigidity of Anosov diffeomorphisms [98], to the regularity of the transfer map for non-abelian cocycles over Anosov actions [127], and, more recently, to the local rigidity of higher rank abelian partially hyperbolic actions [22]. In this section we briefly discuss some of these applications as well as extensions via twisted cocycles. The application related to the regularity of the transfer map will be treated in more detail in Chapter 5.

In many problems concerning perturbation stability and rigidity of actions twisted cocycles arise naturally. Given an action $\alpha : G \times M \to M$ in some space of actions X (with some natural topology: C^a, C^∞, C^ω) the question is whether there exists a neighborhood $U = U(\alpha) \subset X$ such that any $\tilde{\alpha} \in U$ is conjugated to α, i.e., $\tilde{\alpha}(g, Hx) = H(\alpha(g, x))$ via a map $H \in Y$, where Y is a space of diffeomorphisms of M (with some topology as strong as or weaker than that of X). Assuming the existence of a linear structure on $X \times Y$ in the neighborhood of (α, Id), the conjugacy problem corresponds to the non-linear equation

$$\mathcal{F}(\tilde{\alpha}, H) = \tilde{\alpha} \circ H - H \circ \alpha = 0.$$

Therefore, the first thing to look at is the derivative of $\mathcal{F}$ at (α, Id) with respect to the second variable. If the derivative is invertible (or invertible in some weaker topology, or at least "almost" invertible) than the inverse function theorem (or the Nash–Moser iterative scheme) may imply the existence of the solution to the non-linear problem in some neighborhood of α.

Inverting the derivative reduces to solving the equation

$$D_H\mathcal{F}(\alpha, Id)H = D\tilde{\alpha}(g)H(x) - H(\alpha(g, x)) = R(g, x).$$

In case α is a $\mathbb{Z}$-action generated by a single diffeomorphism f, the above equation is a *twisted* cohomology equation $D\tilde{f}H - H \circ f = R$ the twist $T = D\tilde{f}$ being a linear map. In case α is a $\mathbb{Z}^2$-action, the right-hand side, i.e., the map R is not necessarily a cocycle over the action. However, if it is at least infinitesimally a twisted cocycle then the twisted equation above may be considered as being approximately a twisted coboundary equation for a twisted cocycle.

For various applications of rigidity for twisted cocycles in the setting described above (rank-one case) we refer to the survey article [102]. For application of the trivialization of twisted cocycles over higher rank $\mathbb{Z}^k$-actions by partially hyperbolic toral automorphisms we refer to [22].

Another application of twisted cocycles are twisted extensions:

$$\tilde{\alpha}(g, (x, h)) = (\alpha(g, x), \beta(g, x)T(g, x)(h)),$$

where α is the given action, β a twisted cocycle, and T is the twist taking values in a Lie group H or in the group of automorphisms of H. However, in order to make sense of cohomology in the space of twisted cocycles, the map T is usually restricted to being a homomorphism from G to the group of automorphisms $\mathrm{Aut}(H)$ of H, constant in x.

Definition 4.6.1 Given an action $\alpha : G \times M \to M$ and a constant cocycle $T : G \to \mathrm{Aut}(H)$, a map $\beta : G \times M \to H$ is called a *T-(twisted) cocycle* if

$$\beta(g_1 g_2, x) = \beta(g_1, \alpha(g_2, x))T(g_1)\beta(g_2, x).$$

Two T-twisted cocycles β, β^* are called *T-cohomologous* if

$$\beta = P(\alpha)\beta^* T P,$$

for some $P : M \to H$.

An action α is $C_H^{a,b}$-T-cocycle rigid if for a given twist T any T-twisted cocycle is T-cohomologous to a constant T-cocycle.

$C_H^{a,b}$-T-cocycle stability is defined in the same way as the (untwisted) cocycle stability.

The following result from [170] is a version of the Livshitz' theorem for twisted cocycles.

Theorem 4.6.2 *Let f be a hyperbolic diffeomorphism of M. Let H be a connected Lie group with a left-right invariant metric and let T be an automorphism of H satisfying the spectral bunching condition which insures that the extension over f by any cocycle with the twist T is partially hyperbolic. Than any two C^a T-twisted cocycles over f which agree on periodic orbits of f are C^a T-cohomologous providing $0 < a < 1$ is sufficiently close to 1.*

In addition, it is shown in [170] that if H is a torus and T an ergodic automorphism of the torus such that T and the map induced by f on the first cohomology group $H^1(M, \mathbb{R})$ have disjoint set of eigenvalues, then there are at most finitely many solutions to the T-twisted cocycle equation over f.

Let A be an ergodic toral automorphism of the torus $\mathbb{T}^N$ and let f be a C^∞ function on $\mathbb{T}^N$ that defines a cocycle over A. Let λ be a non-zero complex number. Then the equation

$$\lambda P - P \circ A = f$$

is a twisted cohomology equation. If it has a C^∞ solution P then it is easy to check this by looking at the dual equation and iterating it so that f must satisfy

$$\mathcal{O}_n^\lambda(f) = \sum_{k=-\infty}^{k=\infty} \lambda^{-k-1} \hat{f}_{(A^*)^k n} = 0,$$

for all non-zero $n \in \mathbb{Z}^N$. This may not be true for solutions of lower regularity, even though the converse always holds, namely if the obstructions above vanish for a C^∞ function f, then there exists a C^∞ solution P to the twisted coboundary equation [22, 169]. There is a *critical regularity* which depends on the twist and the (hyperbolic) properties of the map A such that any solution of regularity higher than the critical regularity has to be C^∞.

This fact was used by de la Llave [99, Theorem 6.3] to show that for a dense set of values $r \in (0, \infty)$ there is a choice of a hyperbolic map $A \in SL(2, \mathbb{Z})$

and a constant $SL(d, \mathbb{Z})$ ($d \geq 2$)-valued cocycle B over A such that there are arbitrarily C^∞-small smooth perturbations β of B with the property that for any $\epsilon > 0$ the cocycles β and B are cohomologous by a $C^{r-\epsilon}$ transfer map but not by a $C^{r+\epsilon}$ transfer map. We prove this result later in Chapter 5.

We note that over a partially hyperbolic genuinely higher rank action by toral automorphisms every smooth twisted cocycle trivializes via a C^∞ map [22, Section 3.2].

In the paper [86] Kononenko studied the actions on homogeneous spaces G/Γ, where G is a semi-simple Lie group and Γ a co-compact lattice in G. The acting group N is a subgroup of G containing the connected component of the maximal split Cartan subgroup A.

Aside from proving $C^\infty_{\mathbb{R}^l}$-cocycle rigidity for such actions for G of $\mathbb{R}$-rank ≥ 2 satisfying the assumption (*) below, he also proved rigidity for twisted cocycles, as in the following theorem.

Theorem 4.6.3 *Let G be as above, of $\mathbb{R}$-rank ≥ 3.*
Assume that the Lie algebra of G has no factors isomorphic to $\mathfrak{so}(m, 1)$ or $\mathfrak{su}(m, 1)$.

(*) *Let π be a non-zero linear form on $\mathcal{A}$, extendable to a representation of $\mathcal{N}$ and not proportional to a root of rank one factor of $\mathcal{G}$. Then every $e^{-\pi}$-twisted cocycle over the N action on G/Γ is C^∞ cohomologous to a constant twisted cocycle. Moreover, if π is not proportional to any root of $\mathcal{G}$ then the cohomology group is trivial.*

In contrast with non-twisted cocycles, which are highly unstable on a continuous level, twisted cocycles show regularity already in continuous category. The following result is proved in [86].

Theorem 4.6.4 *Let G be of $\mathbb{R}$-rank ≥ 3. Then for continuous cocycles there is an open set U in the set of cohomology classes of cocycles over the N action on G/Γ such that for any $T \in U$ all T-twisted continuous cocycles over the N action are T-coboundaries.*

5
First cohomology and rigidity for general cocycles

5.1 Cocycles with values in compact abelian groups

Before proceeding to a general discussion of cocycles with values in non-abelian groups, which constitutes the main theme of this chapter, we present a higher rank cohomological result which demonstrates an important difference between vector-valued cocycles and those with values in other abelian groups, more specifically compact. The result is about classification of extensions of higher rank abelian TNS actions by compact abelian groups. A more complete discussion of this topic, including the case of large cocycles with values in general Lie groups, can be found in [130].

The main observation is that, up to a constant, there are only a finite number of cohomology classes. One also shows the existence of cocycles over higher rank abelian TNS actions that are not cohomologous to constant cocycles. These results are complementary to those for vector-valued cocycles which exhibit vanishing of the cohomology.

Definition 5.1.1 A *horizontal foliation* $\mathcal{F}$ of the direct product bundle $\mathbb{T}^m \times \mathbb{T}^k$ is a continuous foliation for which, in local charts, the leaves are $\mathbb{T}^k$ translates of images of continuous sections $\gamma_x : V_x \to \mathbb{T}^m \times \mathbb{T}^k$, where $V_x \subset \mathbb{T}^m$ is a neighborhood of $x \in \mathbb{T}^m$.

In what follows we denote by A a higher rank abelian group $\mathbb{Z}^l, l \geq 2$. Recall that TNS actions are introduced in Section 4.4.3.

Theorem 5.1.2 *Assume that the smoothly TNS action $\alpha : A \times \mathbb{T}^m \to \mathbb{T}^m$ has a fixed point x_0. Then the continuous cohomology classes of $\mathbb{T}^k$-valued cocycles are in one-to-one correspondence with pairs of homomorphisms*

$$\bar{H} : V \to \mathbb{T}^k, \beta : A \to \mathbb{T}^k,$$

where

$$V := \pi_1(\mathbb{T}^m, x_0)/span\{a_*(\omega) - \omega |\, \omega \in \pi_1(\mathbb{T}^m, x_0), a \in A\}$$

is a finite group determined completely by the action α.

Up to conjugacy, there are only finitely many homomorphisms $\bar{H}$, and each one has a finite image. Therefore, each extension has a finite cover which is a trivial $\mathbb{T}^k$-bundle and on which the lifted A-action is cohomologous to a constant cocycle. This cover can be chosen to be the same for all $\mathbb{T}^k$-extensions of α.

Proof Let $\beta : \mathbb{T}^m \to \mathbb{T}^k$ be a smooth cocycle. As seen in Section 4.2.3, the lifted action $\widetilde{\alpha} : A \times (\mathbb{T}^m \times \mathbb{T}^k) \to (\mathbb{T}^m \times \mathbb{T}^k)$ can be viewed as a partially hyperbolic transformation and possesses strong stable and strong unstable foliations. Moreover, due to the TNS property, the strong stable and unstable foliations commute locally. Thus W^u and W^s are jointly integrable into an $\widetilde{\alpha}$ invariant horizontal foliation. See Section 4.4.3 for details. Denote the holonomy of this foliation by $H : \pi_1(\mathbb{T}^m, x_0) \to \mathbb{T}^k$.

Due to the invariance of the foliation under the action α one has that

$$H(\alpha_*(a)(\omega)) = H(\omega), \text{ for all } a \in A, \omega \in \pi_1(\mathbb{T}^m, x_0). \tag{5.1.1}$$

Formula (5.1.1) allows us to factorize H by

$$\text{span}\{a_*(\omega) - \omega |\, \omega \in \pi_1(\mathbb{T}^m, x_0), a \in A\}.$$

Let $\bar{H}$ be the homomorphism induced by H to V. The homomorphism β is the restriction of $\widetilde{\alpha}$ to the fiber over the fixed point x_0.

One shows that, up to conjugacy, there are only finitely many homomorphisms $\bar{H}$. Note that, for a hyperbolic, the map

$$a_* - \text{Id} : \pi_1(\mathbb{T}^m, x_0)(\cong \mathbb{Z}^m) \to \pi_1(\mathbb{T}^m, x_0)$$

is invertible over the rationals, so the quotient V has to be a finite group. Let $\text{Rep}(V, \mathbb{T}^k)$ be the space of homomorphisms endowed with the compact-open topology. Since $\text{Rep}(V, \mathbb{T}^k)$ is compact, the proposition will follow if we show that any two nearby homomorphisms are conjugate. Notice first that if two homomorphisms are close to each other then they have the same kernel, because there are no small subgroups in $\mathbb{T}^k$. Then the result follows from a theorem of Montgomery and Zippin [116] claiming that if G is a Lie group and G^* a compact subgroup in G, then there exists an open set $\mathcal{O} \subset G$ containing G^* with the property that for any subgroup $G_0 \subset G$ lying in $\mathcal{O}$ there exists an element $g \in G$ such that $g^{-1} G_0 g \subset G^*$.

Consider now the homomorphisms $\bar{H} : V \to \mathbb{T}^k$ and $\beta : A \to \mathbb{T}^k$, and define $H : \pi_1(\mathbb{T}^m, x_0) \to \mathbb{T}^k$ by $H(\gamma) = \bar{H}(\hat{\gamma})$, where $\hat{\gamma}$ is the image of γ in V under the quotient map. In order to finish the proof of Theorem 5.1.2, we construct an extension $\widetilde{\alpha}$ of the action α which correspond to $\bar{H}$ and β. The construction is done in two steps:

Step 1. Construct a horizontal foliation $\mathcal{F}$ of $\mathbb{T}^m \times \mathbb{T}^k$ with holonomy H.
Step 2. Construct an extension $\widetilde{\alpha}$ of the action α that preserves the foliation $\mathcal{F}$ and has the restriction to the fiber over the fixed point x_0 equal to β.

Since an extension of a TNS action can have only one invariant horizontal foliation it follows that the extension we construct corresponds to the homomorphisms $\bar{H}$ and β.

Step 1 is a consequence of the following standard theorem. A proof can be found in [165, Theorem 13.9].

Theorem 5.1.3 *Let X be a compact metric space, and G a compact connected Lie group. Let $H : \pi_1(X, x_1) \to G$ be a homomorphism. Then there exists P a continuous principal G-bundle over X with a horizontal foliation $\mathcal{F}$. Fix $x_1 \in X$. Two continuous principal G-bundles over X with continuous horizontal foliation $(P_1, \mathcal{F}_1)$ and $(P_2, \mathcal{F}_2)$ are isomorphic (i.e., there exists a bundle isomorphism $F : P_1 \to P_2$ covering the identity map on M and such that $F(\mathcal{F}_1) = \mathcal{F}_2$) if and only if the holonomies*

$$H_1, H_2 : \pi_1(X, x_1) \to G$$

are conjugate in G, that is, there exists $g \in G$ such that

$$H_1(\omega) = gH_2(\omega)g^{-1}, \omega \in \pi_1(X, x_1). \tag{5.1.2}$$

We will apply the theorem only for the case when G is a torus, so (5.1.2) becomes $H_1 = H_2$. To finish the first step of the construction we need to show that the principal $\mathbb{T}^k$-bundle given by Theorem 5.1.3 is isomorphic to a trivial bundle $\mathbb{T}^m \times \mathbb{T}^k$. This follows from the fact that any homomorphism H from $\mathbb{Z}^k$ into a torus can be homotopically deformed to the trivial one. One deforms the image of each generator of $\mathbb{Z}^k$ separately.

We start the proof of Step 2. Let P be the $\mathbb{T}^k$ trivial bundle over $\mathbb{T}^m$ given by Theorem 5.1.3 and let $\mathcal{F}$ be the horizontal foliation of P with holonomy H.

We recall the definition of the holonomy. Let Ω be the set of paths in $\mathbb{T}^m$ and let $\Omega_{x,x}$ be the subset of Ω consisting of paths that start and end at $x \in \mathbb{T}^k$. One defines $s, t : \Omega \to \mathbb{T}^k$ to be the *source* and *target* of a path, that is, the initial point and the endpoint. Let P_x be the fiber of P over $x \in \mathbb{T}^m$ and let

$\phi_x : P_x \to \mathbb{T}^k$ be the projection on the $\mathbb{T}^k$ factor. Note that ϕ_x is an equivariant map, that is, $\phi_x(\xi g) = \phi_x(\xi)g$.

Each $\gamma \in \Omega$ determines a parallel transport $\gamma : P_{s(\gamma)} \to P_{t(\gamma)}$ defined by $\gamma(\xi) = \eta$, where η is the endpoint of the horizontal lift of γ starting from $\xi \in P_{s(\gamma)}$. Moreover, for $\gamma \in \Omega$ one can define the *holonomy* $K(\gamma)$ of γ:

$$K(\gamma) = \phi_{t(\gamma)}(\gamma(\xi)) - \phi_{s(t)}(\xi), \tag{5.1.3}$$

which belongs to $\mathbb{T}^k$ and is independent of $\xi \in P_{s(\gamma)}$. Since $\mathcal{F}$ is a foliation, $K(\gamma)$ depends only of the homotopy class with fixed endpoints of γ. In particular, $K : \Omega_{x,x} \to \mathbb{T}^k$ coincides with the holonomy $H : \pi_1(X, x) \to \mathbb{T}^k$ of the foliation. Note that for $\gamma, \omega \in \Omega$ one has

$$K(\omega\gamma) = K(\omega) + K(\gamma), \tag{5.1.4}$$

provided that $t(\gamma) = s(\omega)$. We use the convention that the path $\omega\gamma$ is obtained by covering first γ and then ω.

Recall that $x_0 \in \mathbb{T}^k$ is a fixed point for the action α. Choose a family of curves $\{\omega_x\}_{x\in\mathbb{T}^k}$ with $s(\omega_x) = x_0$ and $t(\omega_x) = x$. As ω_{x_0} choose the constant curve. For $a \in A$ define $\beta(a, \cdot)$ by

$$\beta(a, x) := K(a(\omega_x)) + \beta(a, x_0) - K(\omega_x), \quad x \in \mathbb{T}^k. \tag{5.1.5}$$

Observe first that the right-hand side of (5.1.5) satisfies the cocycle equation. Due to the fact that $\beta(\cdot, x_0) = \beta(\cdot) : A \to \mathbb{T}^k$ is a homomorphism, due to the properties of the holonomy, and by taking

$$\omega_{(a+b)x} = (a + b)(\omega(x)) = a(b(\omega_x)),$$

one has

$$\begin{aligned}
\beta(a + b, x) &= K((\omega_{(a+b)x})) + \beta(a + b, x_0) - K(\omega_x)\\
&= K(a(b(\omega_x))) + \beta(a, x_0) - K(b(\omega_x))\\
&\quad + K(b(\omega_x)) + \beta(b, x_0) - K(\omega_x)\\
&= \beta(a, bx) + \beta(b, x).
\end{aligned}$$

Thus $\beta(\cdot, \cdot) : A \times \mathbb{T}^k \to \mathbb{T}^k$ is a cocycle.

Let $\widetilde{\alpha}$ be the lifted action constructed using the cocycle β. The cocycle β is related to the lift $\widetilde{\alpha}$ via the formula

$$\beta(a, x) = \phi_{a(x)}(\widetilde{a}(\xi)) - \phi_x(\xi). \tag{5.1.6}$$

The lift $\widetilde{\alpha}$ preserves the foliation $\mathcal{F}$ if and only if

$$K(a(\gamma)) + \beta(a, s(\gamma)) = \beta(a, t(\gamma)) + K(\gamma), \qquad \gamma \in \Omega. \tag{5.1.7}$$

Indeed, $\widetilde{a}$ preserves the foliation $\mathcal{F}$ if and only if, given ξ, η in the same leaf of $\mathcal{F}$, $\widetilde{a}(\xi)$ and $\widetilde{a}(\eta)$ are in the same leaf of $\mathcal{F}$. Connect ξ to η by a horizontal curve and denote by γ its projection on $\mathbb{T}^k$. Then $K(\gamma)(\xi) = \eta$ and $K(a(\gamma))(\widetilde{a}(\xi)) = \widetilde{a}(\eta)$. Using formulas (5.1.6) and (5.1.3) now give (5.1.7).

It remains to prove (5.1.7). But this follows from (5.1.5), the definition of the cocycle β. □

We describe now a family of TNS actions that can be used to construct cocycles that are not cohomologous to constants.

Example 5.1.4 Let $\mathbf{T} \subset SL(n, \mathbb{R})$ be a maximal torus such that $\mathbf{T} \cong \mathbb{R}^{n-1}$. Let Γ be a subgroup of finite index in $SL(n, \mathbb{Z})$. Then Γ is a lattice in $SL(n, \mathbb{R})$ and it follows from a theorem of Prasad–Ragunathan [142] that there exists $g \in SL(n, \mathbb{R})$ such that $A := g\mathbf{T}g^{-1} \cap \Gamma$ is a co-compact lattice in $\mathbf{T}$. In particular, the induced action of $A \cong \mathbb{Z}^{n-1}$ on the n-dimensional torus $\mathbb{T}^n$ is a TNS action.

Proposition 5.1.5 *Let $n \geq 3$ and A a higher rank abelian group. Then there are TNS actions $\alpha : A \times \mathbb{T}^n \to \mathbb{T}^n$ that have a fixed point $x_0 \in \mathbb{T}^n$ and non-trivial homomorphisms $\bar{H} : V \to \mathbb{T}^k$, where*

$$V := \pi_1(\mathbb{T}^n, x_0)/span\{a_*(\omega) - \omega | \omega \in \pi_1(\mathbb{T}^n, x_0), a \in A\}.$$

In particular, any cocycle in a cohomology class corresponding to such a homomorphism is not cohomologous to a constant.

Proof Let $k \geq 2$ be an integer. Let $\Gamma_k \subset \mathrm{SL}(n, \mathbb{Z})$ be the congruence group of order k, that is,

$$\Gamma_k = \{m \in \mathrm{SL}(n, \mathbb{Z}) | m \cong \mathrm{Id}(\mathrm{mod}\, k)\}.$$

Then Γ_k has finite index in $\mathrm{SL}(n, \mathbb{Z})$ and we obtain a TNS action of an abelian group $A \cong \mathbb{Z}^{n-1}$ by the method described in the previous example. The subgroup

$$\mathrm{span}\{a_*(\omega) - \omega | \omega \in \pi_1(\mathbb{T}^n, x_0), a \in A\}$$

is included in $k\pi_1(\mathbb{T}^n, x_0) \cong k\mathbb{Z}^n$. Hence V is a non-trivial finite abelian group. Any such group is a direct product of cyclic finite groups. If $\mathbb{Z}_p$ is a non-trivial factor of order p one can map the generator of $\mathbb{Z}_p$ into an element of order p in $\mathbb{T}^k$ and find a non-trivial homomorphism $\bar{H}$. □

5.2 Introduction to rank-one results for non-abelian group-valued cocycles

5.2.1 An overview

While many types of groups were considered as the range for cocycles over hyperbolic dynamical systems we will concentrate on two classes that are arguably of greatest interest: finite-dimensional connected Lie groups and diffeomorphism groups of compact manifolds.

Throughout this chapter we use the terminology from Sections 1.8.1 and 4.1.

When one passes from cocycles with abelian range to those with values in non-commutative groups, a fundamental difference appears: cohomology classes no longer possess group structure. In the abelian case the question of classification of cocycles up to a cohomology reduces to the question of describing coboundaries, i.e., finding conditions for a cocycle to be cohomologous to the trivial identity cocycle. In the general case these two problems are quite different and the latter forms only a particular case of the former.

Another difference appears between cocycles with the values in "small" groups, such as compact, abelian, or nilpotent, and those with values in groups of exponential growth. In the former case the extension of the hyperbolic action via the cocycle can be viewed as a partially hyperbolic transformation, that is, one that has the expansion/contraction in the base greater than the expansion/contraction in the fiber. In the latter case, to ensure that one should make a certain assumption on the cocycle called center bunching, see Definition 5.2.1. Until recently all results in this area involved this condition in some form; this is in particular true for the results presented in this chapter, both for cocycles over rank-one and higher rank systems. Notice that any continuous cocycle with values in a "small group" satisfies any center-bunching conditions so that results for center-bunched cocycles imply corresponding results for cocycles with values in such groups.

The first natural problem in this area is the generalization of the Livshitz theorem (Theorem 4.2.2). Due to the difference between the coboundary characterization problem and the cocycle classification problem, this generalization takes two forms:

(i) Do the closing conditions (5.3.1) for a Hölder cocycle guarantee that the cocycle is a Hölder coboundary?
(ii) Do the coincidence of the periodic data, i.e, the closing conditions (5.6.1) for two Hölder cocycles guarantee that the cocycles are cohomologous with a Hölder transfer function?

Two additional problems are similar to those we already considered for the vector-valued cocycles in the previous chapter:

(iii) Lift the regularity of the transfer map between two cohomologous C^K cocycles from Hölder to C^K, $K = 1, \ldots, \infty, \omega$.
(iv) Lift the regularity of the transfer map between two cohomologous Hölder cocycles from measurable to Hölder.

The first positive result in the direction of problem (i) was found by Livshitz [97], who proved that any "small" H-valued α-Hölder cocycle over an Anosov diffeomorphism, which satisfies the closing conditions (1.4.6), is cohomologous to the trivial cocycle via an α-Hölder transfer map. We prove Livshitz' result and its extension to cocycles with values in diffeomorphism groups in Section 5.3 (Theorems 5.3.1 and 5.3.6 correspondingly). Recently Kalinin [59] solved problem (i) for linear Lie groups, although he formulates his result only for $GL(n, \mathbb{R})$-valued cocycles. An essential part of Kalinin's work is showing that the closing condition (5.3.1) implies arbitrary good center bunching.

In the direction of problem (ii), Parry [135] showed that two α-Hölder cocycles over an Anosov diffeomorphism with values in a compact Lie group with identical periodic data are cohomologous via an α-Hölder map. Later Schmidt [158] extended Parry's result to center-bunched cocycles with values in Lie groups. We present a version of Parry's result in Section 5.6. The method of proof is a combined version of the extension along the orbits and of the extension along stable/unstable foliations. The method of [59] is not directly applicable to problem (ii), which remains open for general cocycles with values in Lie groups.

The rest of this chapter contains constructions and results based on the extension along the stable manifolds method. The main technical tools, which are similar to the case of vector-valued cocycles, will be the lifting of the stable and unstable foliations to the extension associated with the cocycles and the periodic cycle functionals as the obstruction to vanishing of cohomology. Applications include regularity results for the solution of cohomological equations for smooth cocycles in the rank-one case (the solution of problems (iii) and (iv) for cocycles cohomologous to center-bunched cocycles) and the trivialization of cohomology in various settings for the higher rank actions.

5.2.2 The center-bunching condition

Let H be a connected Lie group with Lie algebra LH and identity e. We introduce a convenient metric on H. The tangent space $T_h H$ at $h \in H$ can be identified with LH. Let L_h, R_h denote left, respectively right, multiplication by h. Let $D_g L_h$, $D_g R_h$ denote the derivatives of L_h, R_h at g. The adjoint map $\mathrm{Ad} : H \to \mathrm{Aut}(LH)$ is given by

$$\mathrm{Ad}(g)X = D_g R_{g^{-1}} D_e L_g X, \text{ for all } X \in T_e H,$$

that is, $\mathrm{Ad}(g)$ is the derivative of conjugation by g.

Let $\| \cdot \|_e$ be a norm on $T_e H$. Define a Finsler metric on TH by $\|X\|_h = \|D_h R_{h^{-1}} X\|_e$ for $X \in T_h H$. The Finsler defines a metric d_H on H, which is right invariant. Indeed, if $X \in T_h H$, then

$$\begin{aligned}\|D_h R_g X\|_{hg} &= \|D_{hg} R_{(hg)^{-1}} D_h R_g X\|_e \\ &= \|D_{hg} \left(R_{h^{-1}} \circ R_{g^{-1}}\right) D_h R_g X\|_e \\ &= \|D_h R_{h^{-1}} X\|_e = \|X\|_h.\end{aligned}$$

The metric is not left invariant, but has a useful sub-multiplicative property. Let $X \in T_h H$. Then

$$\begin{aligned}\|D_h L_g X\| &= \|D_{gh} R_{(gh)^{-1}} D_H L_g X\|_e \\ &= \|D_{gh} \left(R_{g^{-1}} \circ R_{h^{-1}}\right) D_h L_g X\|_e \\ &= \|D_g R_{g^{-1}} D_e L_g D_h R_{h^{-1}} X\|_e \\ &= \|\mathrm{Ad}(g) D_h R_{h^{-1}} X\|_e \leq \|\mathrm{Ad}(g)\| \|X\|_h.\end{aligned}$$

In conclusion, there exists a metric d_H on H with the following properties:

$$\begin{aligned}d(h_1 g, h_2 g) &= d(h_1, h_2), \\ d(g h_1, g h_2) &\leq \|\mathrm{Ad}(g)\| d(h_1, h_2),\end{aligned} \tag{5.2.1}$$

for any $h_1, h_2, g \in H$.

The estimates we will need in the future are related to the fact that certain skew-extensions can be viewed as a partially hyperbolic transformation.

Throughout this section $\mathcal{A}$ denotes $\mathbb{Z}$ or $\mathbb{R}$.

Definition 5.2.1 Let M be a compact manifold, $d \geq 1$ be an integer, and $\alpha : \mathcal{A}^d \times M \to M$ be an action. Let $\beta : \mathcal{A}^d \times M \to H$ be a cocycle. Let S be a compact set of generators for $\mathcal{A}^d$.

Define μ_-, μ_+ to be

$$\begin{aligned}\mu_- &= \lim_{n\to\infty} \sup_{x\in M, a\in S} \|\mathrm{Ad}(\beta(na, x))^{-1}\|^{-1/n}, \\ \mu_+ &= \lim_{n\to\infty} \sup_{x\in M, a\in S} \|\mathrm{Ad}(\beta(na, x))\|^{1/n}.\end{aligned} \tag{5.2.2}$$

Note that $\mu_- \leq \mu_+$. Throughout this chapter we will use only μ_-^{-1}, which for the simplicity of the notation we denote by μ.

For $\theta \in (0, 1)$, and $0 < \lambda < 1$, we say that a C^θ cocycle β is *λ-center bunched* with respect to S if $\mu\lambda^\theta < 1$.

For $K \geq 1$, and $0 < \lambda < 1$, we say that a C^K cocycle β is *λ-center bunched* with respect to S if $\mu\lambda < 1$.

We assume in the future, without loss of generality, that if $d = 1$ then the set of generators S appearing in the Definition 5.2.1 is either 1, if $\mathcal{A} = \mathbb{Z}$, or the interval $(0.1, 1]$, if $\mathcal{A} = \mathbb{R}$. We will skip the reference to S in these cases.

The notion of λ-center bunching is sufficient for several of the results in this chapter. Nevertheless, some of them will require stronger center bunching or additional smallness assumptions. We will present these requirements when they become necessary.

If H is compact or nilpotent, then $\mu = 1$, so that Hölder cocycles are automatically λ-center bunched. The same is true for semidirect products of type $K \ltimes N$, with K compact Lie group and N nilpotent Lie group. Thus, for such groups some of the results presented here can be stated without the assumption of λ-center bunching.

5.3 Calculation of cohomology for non-abelian cocycles over rank-one Anosov actions

In this section we discuss the non-abelian version of Livshitz's cohomological result. We use notations and results from Section 1.8.1 and Section 4.1. The method of proof is the non-abelian version of the extension along the orbits.

5.3.1 Livshitz' theorem for cocycles with values in Lie groups

Note that for cocycles with range a Lie group H, the closing conditions (4.2.2) become

$$\beta(f^{n-1}x) \cdots \beta(fx)\beta(x) = Id_H, \qquad \text{whenever} \qquad f^n x = x. \tag{5.3.1}$$

Theorem 5.3.1 (Livshitz [97]) *Let M be a compact Riemannian manifold, $f : M \to M$ a topologically transitive C^1 Anosov diffeomorphism. Let H be a connected Lie group endowed with a metric d_H that satisfies* (5.2.1). *Let $\beta : M \to H$ be a center-bunched α-Hölder cocycle that satisfies* (5.3.1).

Then there exists an α-Hölder function P on M such that

$$\beta(x) = P(fx)P(x)^{-1}.$$

Proof Let x be the transitive point of f and set

$$\phi_n(x) = \beta(f^{n-1}x) \cdots \beta(fx)\beta(x).$$

We will prove first that there is a constant C' such that

$$d_H(\phi_n(x), e_H) < C' d_M(f^n x, x)^\alpha.$$

This needs to be proved only for small $d_M(f^n x, x)$, say for

$$d_M(f^n x, x) < \eta < \Delta/K.$$

Let $y, z \in M$ be provided by Lemma 1.8.7. Since y is a periodic point for f, and by the closing conditions,

$$\beta(f^{n-1}y)\cdots\beta(fy)\beta(y) = Id_H.$$

Using now the properties of the metric d_H, and again the closing conditions, one has

$$\begin{aligned}
d_H(\phi_n(x), Id_H) &= d_H(\phi_n(x), \phi_n(y)) \\
&= d_H(\beta(f^{n-1}x)\cdots\beta(fx)\beta(x), \beta(f^{n-1}y)\cdots\beta(fy)\beta(y)) \\
&\le \sum_{k=0}^{n-1} d_H(\beta(f^{n-1}y)\cdots\beta(f^{k+1}y)\beta(f^k y)\beta(f^{k-1}x)\cdots \\
&\qquad \beta(fx)\beta(x), \beta(f^{n-1}y)\cdots\beta(f^{k+1}y)\beta(f^k x) \\
&\qquad \beta(f^{k-1}x)\cdots\beta(fx)\beta(x)) \\
&\le \sum_{k=0}^{n-1} d_H(\beta(f^{n-1}y)\cdots\beta(f^{k+1}y)\beta(f^k y), \beta(f^{n-1}y)\cdots \\
&\qquad \beta(f^{k+1}y)\beta(f^k x)) \\
&= \sum_{k=0}^{n-1} d_H(\beta(y)^{-1}\beta(fy)^{-1}\cdots\beta(f^k y)^{-1}\beta(f^k y), \\
&\qquad \beta(z)^{-1}\beta(fy)^{-1}\cdots\beta(f^k y)^{-1}\beta(f^k x)) \\
&\le \sum_{k=0}^{n-1} \|\mathrm{Ad}(\beta(k+1, y))^{-1}\| d_H(\beta(f^k y), \beta(f^k x)) \\
&\le \sum_{k=0}^{n-1} \mu'^k \left(d_H(\beta(f^k y), \beta(f^k z)) + d_H(\beta(f^k z), \beta(f^k x))\right) \\
&\le 2\sum_{k=0}^{n-1} c^\alpha (\mu'\lambda^\alpha)^k \|\beta\|_{\text{Hölder}} d_M(f^n x, x)^\alpha \le C d_M(f^n x, x)^\alpha,
\end{aligned} \tag{5.3.2}$$

where $\mu < \mu'$, $\mu'\lambda^\alpha < 1$, and the constant $C > 0$ is independent of n, x.

Consider now $f^{n+k}x$ and $f^k x$ such that

$$d_M(f^{n+k}x, f^k x) < \eta < \Delta/K.$$

The invariance property of the metric d_H and the above argument applied to $f^k x$ in place of x shows that:

$$\begin{aligned} d_H(\phi_{n+k}(x), \phi_k(x)) & \\ &= d_H(\phi_n(f^k x)\phi_k(x), \phi_k(x)) \\ &= d_H(\phi_n(f^k x), Id_H) \\ &\leq C d_M(f^{n+k}x, f^k x)^\alpha. \end{aligned} \tag{5.3.3}$$

Since the orbit of x is dense in M we obtain an α-Hölder function $P : M \to H$ such that $P(f^n x) = \phi_n(x)$ for $n = 0, 1, 2, \ldots$. Finally we have $\beta(x) = P(fx)P(x)^{-1}$ on a dense set, and hence everywhere. □

5.3.2 Livshitz theorem for cocycles with values in certain diffeomorphism groups

The results in this subsection summarize part of [125], in which one can find the first Livshitz's type result for cocycles with values in diffeomorphism groups.

One starts by introducing the usual C^k metrics on the set of diffeomorphisms of an open connected set U in an Euclidean space $\mathbb{R}^d$.

Definition 5.3.2 Let U be an open connected set in $\mathbb{R}^d$. For any path γ in U denote by $l(\gamma)$ the length of γ, and for any $x, y \in U$ define the distance $d_U(x, y) := \inf l(\gamma)$, where the infimum is taken over all paths γ in U connecting x and y. Let $U \subset \mathbb{R}^d$ be an open connected set. For any integer $K \geq 0$ denote by $C^K(U, U)$ the set of C^K-functions from U to U. For $a \in C^K(U, U)$ and for $1 \leq k \leq K$ denote by $D_x^k a$ the kth derivative of a at $x \in U$, which is a symmetric k-linear map $D_x^k a : (\mathbb{R}^{dimU})^k \to \mathbb{R}^{dimU}$. Denote by $\|D_x^k a\|$ the operator norm of D_x^k.

For $a, b \in C^K(U, U)$ and $1 \leq k \leq K$ define:

$$\begin{aligned} d_0(a) &:= \sup_{x \in U} d_U(a(x), 0), \text{ if } 0 \in U, \\ d_0(a, b) &:= \sup_{x \in U} d_U(a(x), b(x)), \\ d_{(k)} &:= \sup_{x \in U} \|D_x^k a\|, \end{aligned}$$

$$d'_K(a) := \sum_{k=1}^{K} \frac{d_{(k)}(a)}{k!}, \tag{5.3.4}$$

$$d''_K(a) := \sum_{k=2}^{K} \frac{d_{(k)}(a)}{k!},$$

$$d_K(a) := d_0(a) + d'_K(a), \text{ if } 0 \in V,$$

$$d_K(a, b) := d_0(a, b) + d'_K(a - b).$$

If $U \subset \mathbb{R}^d$ is an open connected subset denote by $\mathrm{Diff}^K(U)$ the set of C^K-diffeomorphisms of U. Then

$$\mathrm{Diff}^K_{bdd}(U) := \{a \in \mathrm{Diff}^K(U) | d_K(a^{\pm 1}, Id_N) < \infty\}$$

is a group with the usual operation of composition. Moreover, it is a complete metric space with respect to the metric

$$\bar{d}_K(a, b) := \max\{d_K(a, b), d_{(1)}(a^{-1}, b^{-1}\}.$$

The following proposition describes the quasi-invariance properties of quantities introduced above. Complete proofs for these statements can be found in [125].

Proposition 5.3.3 *Let $K \geq 1$ be an integer. Let $U \subset \mathbb{R}^d$ be an open set. Then for $a, b, u, v \in C^K(U, U)$, the following hold:*

(D1) $d'_K(a) \leq 1 + d_K(a, Id_N)$;
(D2) $d''_K(a) = d''_K(a, Id_N)$;
(R1) $d'_K(au) \leq d'_K(a) \cdot [\max\{1, d'_K(u)\}]^K$;
(R2) $d'_K(au, bu) \leq d'_K(a, b) \cdot [\max\{1, d'_K(u)\}]^K$;
(R3) $d_K(au, bu) \leq d_K(a, b) \cdot [\max\{1, d'_K(u)\}]^K$;
(L1) $d_K(va, vb) \leq d_K(a, b)(K + 1)d'_{K+1}(v) \cdot [1 + 2d'_K(a, Id) + 2d'_K(b, Id)]^K$.

The following definition describes a class of manifolds for which the diffeomorphism groups possess convenient quasi-invariant metrics.

Definition 5.3.4 Let N be a closed C^∞ manifold. N is called a *good manifold* if there is an embedding of N into an Euclidean space $\mathbb{R}^d$ such that N has a tubular open neighborhood U which is C^∞ diffeomorphic to $N \times D$, where D is an open disk in $\mathbb{R}^{d-dim(N)}$.

Let $K \geq 1$ be an integer, and N be a good manifold. Let U be the open neighborhood defined above such that $U \cong N \times D$. Any diffeomorphism $a \in \mathrm{Diff}^K(N)$ can be extended to a C^K-diffeomorphism $\tilde{a}$ of U by

$$\tilde{a}(x, y) = (a(x), y), \text{ for } x \in N, y \in D. \tag{5.3.5}$$

All quantities introduced in (5.3.4) can be defined for $a, b \in \mathrm{Diff}^K(N)$ using (5.3.5). $\mathrm{Diff}^K(N)$ becomes a complete metric space with the induced metric $\bar{d}_K$. Moreover, formulas (D1), (D2), (R1), (R2), (R3), and (L1) still hold.

Remark 5.3.5 Any good manifold has stably trivial tangent bundle. The converse is also true. Examples of good manifolds are the n-torus, the n-sphere, and any closed hypersurface in $\mathbb{R}^d$.

The following theorem is the main result in this subsection.

Theorem 5.3.6 *Let M be a closed manifold, and $f : M \to M$ be a C^1 topologically transitive Anosov diffeomorphism. Let N be a good manifold. Fix $0 \leq \alpha \leq 1$, $K \in \mathbb{N}$, $K \geq 4$. Assume that the diffeomorphism group* $\mathrm{Diff}^K(N)$ *is endowed with the metric d_K. Then there exists $\delta > 0$ with the following property: if $\beta : M \to$* $\mathrm{Diff}^K(N)$ *satisfies:*

(i) $d_K(\beta(x)^{\pm 1}, Id_N) < \delta$;
(ii) *β fulfills the closing conditions:*

$$\beta(f^{n-1}x) \cdots \beta(fx)\beta(x) = Id_N, \qquad \textit{whenever } f^n x = x;$$

(iii) β is α-Hölder,

then there is an α-Hölder map $P : M \to$ $\mathrm{Diff}^{k-2}(N)$ *such that*

$$\beta(x) = P(fx)P(x)^{-1}.$$

The following lemma isolates the main computational part from the proof of Theorem 5.3.6.

Lemma 5.3.7 *Let $0 < \mu < 1$ and $\delta > 0$ such that $\mu(1 + 4\delta)^{2K} < 1$. Let $n \geq 1$ integer and $a_k, b_k, c_k \in$* $\mathrm{Diff}^K(N)$, $1 \leq k \leq n$, *be such that:*

(i) $d_{K-1}(a_k, b_k) \leq C \cdot \mu^k$;
(ii) $d_{K-2}(b_k, c_k) \leq C \cdot \mu^{n-k}$;
(iii) $d_{K-1}(a^{\pm 1}, Id_N) < \delta$;
(iv) $d_K(b^{\pm 1}, Id_N) < \delta$;
(v) $d_{K-2}(c^{\pm 1}, Id_N) < \delta$.

Denote:

$$a = a_1a_2 \cdots a_n,$$
$$b = b_1b_2 \cdots b_n,$$
$$c = c_1c_2 \cdots c_n.$$

Then:

(a) $d_{K-1}(a,b) \leq C\frac{K}{1-\mu(1+4\delta)^{2K}} \max\{1, d'_{K-1}(a)\}^{K-1}$;
(b) $d_{K-2}(b,c) \leq C\frac{K-1}{1-\mu(1+4\delta)^{2K}} d'_{K-1}(b)$.

Proof Start by observing that

$$d_{K-1}(a,b) = d_{K-1}(a_1a_2 \cdots a_N, b_1b_2 \cdots b_N) \leq \Lambda_1 + \Lambda_2 + \cdots + \Lambda_N,$$

where

$$\Lambda_k = d_{K-1}(b_1 \cdots b_{k-1}a_ka_{k+1} \cdots a_N, b_1 \ldots b_{k-1}b_ka_{k+1} \cdots a_N).$$

In order to obtain good estimates for the terms Λ_k, one replaces in Λ_k the product $a_{k+1} \cdots a_n$ by the product $a_k^{-1}a_{k-1}^{-1} \cdots a_1^{-1}a$, then applies formula (R3) successively for the final a factors, and then applies formula (L1) once for the product of the leading b factors:

$$\begin{aligned}\Lambda_k &\leq d_{K-1}(b_1 \cdots b_{k-1}a_k, b_1 \cdots b_k) \\ &\quad \cdot \left[\prod_{j=1}^{k} \left(\max\{1, d'_{K-1}(a_j^{-1})\}\right)^{K-1}\right] \left(\max\{1, d'_{K-1}(a)\}\right)^{K-1} \\ &\leq d_{K-1}(a_k, b_k)K \left[\prod_{j=1}^{k} \left(\max\{1, d'_{K-1}(a_j^{-1})\}\right)^{K-1}\right] \left(\max\{1, d'_{K-1}(a)\}\right) \\ &\quad \cdot d'_K(b_1 \cdots b_{K-1}) \cdot [1 + 2d'_{K-1}(a_k, Id_N) + 2d'_{K-1}(b_k, Id_N)]^{K-1}.\end{aligned}$$

Applying successively (R1) for each factor, one gets (for $k \geq 2$) that

$$d'_K(b_1 \cdots b_{k-1}) \leq d'_K(b_1) \cdot \prod_{l=2}^{k-1} \left(\max\{1, d'_K(b_l)\}\right)^{K-1}.$$

By (D1) and (iii) one gets:

$$\begin{aligned}\Lambda_k &\leq d_{K-1}(a_k, b_k) \cdot K \cdot \left[(1+\delta)^{K-1}\right]^k \cdot \left(\max\{1, d'_{K-1}(a)\}\right)^{K-1} \\ &\quad \cdot (1+\delta)^{1+K(k-2)} \cdot (1+4\delta)^{K-1} \\ &\leq d_{K-1}(a_k, b_k) \cdot K \cdot (1+4\delta)^{K-1} \left(\max\{1, d'_{K-1}(a)\}\right)^{K-1}.\end{aligned}$$

Together with (i) this yields (a).

(b) Start by observing that

$$d_{K-2}(b, c) = d_{K-2}(b_1 b_2 \cdots b_n, c_1 c_2 \cdots c_n) \le \Lambda_1 + \Lambda_2 + \cdots + \Lambda_n,$$

with

$$\Lambda_k \le d_{K-2}(b_1 \cdots b_{n-k} b_{n-k+1} c_{n-k+2} \cdots c_n, b_1 \cdots b_{n-k} c_{n-k+2} \cdots c_n).$$

Note that here one starts interchanging factors from the right-hand side of the products.

Replace in Λ_k the product $b_1 \cdots b_{n-k}$ by the product $bb_n^{-1} \cdots b_{n-k+1}^{-1}$, then apply formula (R1) successively for the last c-factors and formula (L1) once for the leading b-factors in order to get

$$\begin{aligned}\Lambda_k \le\ & d_{K-2}(b_{n-k-1}, c_{n-k+1}) \cdot (K-1) \\ & \cdot \left[\prod_{j=n-k+2}^{n} \left(\max\{1, d_{K-2}(c_j)\}\right)^{K-2} \right] \cdot d'_{K-1}(bb_n^{-1} \cdots b_{n-k+1}^{-1}) \\ & \cdot [1 + d'_{K-2}(b_{n-k+1}, Id_N) + d'_{K-2}(c_{n-k+1}, Id_N)]^{K-2}.\end{aligned}$$

Now by repeatedly applying (R1) one gets

$$\begin{aligned}& d'_{K-1}(bb_n^{-1} \cdots\ b_{n-k+1}^{-1}) \\ & \qquad \le d'_{K-1}(b) \cdot \prod_{j=n-k+1}^{n} \left(\max\{1, d'_{K-1}(b_{n-k+1})\}\right)^{K-1},\end{aligned}$$

hence by (D1) and (iii) it follows that

$$\begin{aligned}\Lambda_k & \le d_{K-2}(b_{n-k+1}, c_{n-k+1}) \cdot (K-1) \cdot (1+\delta)^{(K-2)(k-1)} \cdot d'_{K-1}(b) \\ & \qquad \cdot (1+\delta)^{(K-1)k} \cdot (1+4\delta)^{K-2} \\ & \le d_{K-2}(b_{n-k+1}, c_{n-k+1})(K-1)(1+4\delta)^{(2kK-3k)} \cdot d'_{K-1}(b) \\ & \le d_{K-2}(b_{n-k+1}, c_{n-k+1})(K-1)(1+4\delta)^{2kK} \cdot d'_{K-1}(b)\end{aligned}$$

which together with (ii) implies (b). □

Proof of Theorem 5.3.6 Let x be a transitive point of f and for n positive integer set

$$\phi_n = \beta(f^{n-1}x) \cdots \beta(fx)\beta(x).$$

One starts by proving that there is a constant $C_1 > 1$ such that for n positive integer and for

$$d_M(f^n x, x) < \eta < \Delta/K,$$

one has

$$d_{K-2}(\phi_n, Id_N) < C_1 d_M(f^n x, x)^\alpha. \tag{5.3.6}$$

Let $y, z \in M$ be provided by Lemma 1.8.7. Since y is a periodic point for f, and by the closing conditions

$$\beta(f^{n-1}y)\cdots\beta(fy)\beta(y) = Id_N.$$

Moreover, the following estimates from Lemma 1.8.7 hold:

$$\begin{aligned} d_M(f^k x, f^k z) &\le c\mu^k d_M(f^N x, x), \quad k = 0, 1, \ldots, n-1, \\ d_M(f^k z, f^k y) &\le c\mu^{N-k} d_M(f^N x, x), \quad k = 0, 1, \ldots, n-1. \end{aligned} \tag{5.3.7}$$

For $1 \le k \le n$ denote:

$$\begin{aligned} a_k &= \beta(f^{n-k}y), \\ b_k &= \beta(f^{n-k}z), \\ c_k &= \beta(f^{n-k}x), \\ a &= a_1 a_2 \cdots a_n, \\ b &= b_1 b_2 \cdots b_n, \\ c &= c_1 c_2 \cdots c_n. \end{aligned}$$

Then $\phi_n = c$, while $a = Id_N$ due to the closing conditions. We have to show that c is close to a.

Since β is α-Hölder, one has

$$\begin{aligned} d_K(a_k, b_k) &\le C'\mu^{k\alpha} d_M(f^n x, x)^\alpha, \\ d_K(b_k, c_k) &\le C'\mu^{(n-k)\alpha} d_M(f^n x, x)^\alpha, \end{aligned}$$

where $C' = c^\alpha \|\beta\|_{\text{Hölder}}$.

Apply now Lemma 5.3.7 to get

$$\begin{aligned} d_{K-2}(\phi_n, Id_N) &= d_{K-2}(c, a) \\ &\le d_{K-2}(b, a) + d_{K-2}(c, b) \le C_1 d_M(f^n x, x)^\alpha, \end{aligned}$$

where C_1 is a constant independent of x and n that, in addition, can be chosen larger than 1. So (5.3.6) holds.

Observe now that for $a \in \mathrm{Diff}^2(N)$ one has

$$d_2(a, Id_N) < 1/2 \quad \text{implies} \quad d_2'(a^{-1}, Id_N) < 4. \tag{5.3.8}$$

One shows now that for m positive integer, the quantities $d_{K-2}(\phi_m, Id_N)$ and $d_2'(\phi_m^{-1})$ can be bounded independent of m, that is,

$$M_1 := \sup_m d_{K-2}(\phi_m, Id_N) < \infty,$$

$$M_2 := \sup_m d_2'(\phi_m^{-1}) < \infty.$$

Define $\eta' = 0.5\max\{\eta, (2A)^{-1/\alpha}\}$. Since M is compact, there is a finite set $I \subset \mathbb{N}$ such that the d_M-balls $U_i, i \in I$, of radius η' centered in $\{f^i x\}_{i\in I}$ cover M.

Let $m \in \mathbb{N}$ be such that $x_m \in U_i, m \geq i$. Then, using first formula (R1) and then formula (5.3.6) applied for the transitive point $f^i(x)$, one has

$$\begin{aligned} d_{K-2}(\phi_m, \phi_i) &= d_{K-2}((\phi_m\phi_i^{-1})\phi_i, \phi_i) \\ &\leq d_{K-2}(\phi_m\phi_i^{-1}, Id_N) \cdot [1 + d_{K-2}(\phi_i, Id_N)]^{K-2} \\ &\leq C_1\eta'^{\alpha}[1 + d_{K-2}(\phi_i, Id_N)]^{K-2}. \end{aligned}$$

The values of $m \in \mathbb{N}$ not covered by the argument above belong to a finite set $J = \{m \in \mathbb{N} | x_m \in U_i, m < i\}$. Therefore, M_1 can be bounded by

$$\begin{aligned} &\sup_{m\in\mathbb{N}} d_{K-2}(\phi_m, Id_N) \\ &\leq \max_{i\in I, m\in J} \{(1 + C_1\eta'^{\alpha})[1 + d_{K-2}(\phi_i, Id_N)]^{K-2}, d_{K-2}(\phi_m, Id_N)\} \\ &< \infty. \end{aligned}$$

To estimate M_2, observe that using formula (L1) one has

$$\begin{aligned} d_2'(\phi_m^{-1}, \phi_i^{-1}) &\leq d_2(\phi_i^{-1}(\phi_n\phi_i^{-1})^{-1}, \phi_i^{-1}) \\ &\leq 3d_2((\phi_n\phi_i^{-1})^{-1}, Id_N)d_3'(\phi_i^{-1})[1 + 2d_2'((\phi_n\phi_i^{-1})^{-1}, Id_N) \\ &\quad + 2d_2'(Id_n, Id_n)]^2 \leq d_3'(\phi_i^{-1})[1 + 3d_2((\phi_n\phi_i^{-1})^{-1}, Id_N)]^3. \end{aligned}$$

Using now (5.3.6) applied for the transitive point $f^i(x)$ and (5.3.8) it follows that

$$d_2(\phi_n\phi_i^{-1}, Id_N) \leq C_1\eta'^{\alpha} < 1/2, \text{ for } 2 \leq K - 2,$$

and

$$d_2((\phi_n\phi_i^{-1})^{-1}, Id_N) \leq 4,$$

and from here one can proceed as in the proof of the finiteness of M_1.

Consider now m, n positive integers, $n > m$, such that $f^n x$ and $f^m x$ satisfy

$$d_M(f^{n+k}x, f^k x) < \eta < \Delta/K.$$

We estimate now the distance $\bar{d}_{K-2}(\phi_m, \phi_n)$. We estimate separately the two components of $\bar{d}_{K-2}$.

Using (R3) one has

$$\begin{aligned} d_{K-2}(\phi_n, \phi_m) &= d_{K-2}((\phi_n\phi_m^{-1})\phi_m, \phi_m) \\ &\leq d_{K-2}(\phi_n\phi_m^{-1}, Id_N)[1 + d_{K-2}(\phi_m, Id_N)]^{K-2} \\ &\leq M_1^{K-2} C_1 d_M(f^m x, f^n x)^\alpha. \end{aligned}$$

Using (L1) one has

$$\begin{aligned} d_{(1)}(\phi_n^{-1}, \phi_m^{-1}) &\leq d_1(\phi_n^{-1}, \phi_n^{-1}(\phi_n, \phi_m^{-1})) \\ &\leq d_1(\phi_n\phi_m^{-1}, Id_N) d_2'(\phi_n^{-1})[1 + 2d_1(\phi_n\phi_m^{-1}, Id_N)] \\ &\leq M_2(1 + 2C_1 d_M(f^m x, f^n x)) C_1 d_M(f^m x, f^n x)^\alpha. \end{aligned}$$

Since the orbit of x is dense in M we obtain an α-Hölder function $P : M \to \mathrm{Diff}^{K-2}(N)$ such that $P(f^n x) = \phi_n$ for $n = 0, 1, 2, \ldots$. Finally, we have $\beta(x) = P(fx)P(x)^{-1}$ on a dense set in the complete metric space $\mathrm{Diff}^{K-2}(N)$, and hence everywhere. □

Remark 5.3.8 Let us note that one can pursue these techniques for other infinite dimensional groups in the fiber. For example, [5] contains a version of Livshitz' theorem for cocycles with values in Banach algebras.

5.4 Invariant foliations for Lie group and diffeomorphism group-valued extensions

Throughout this section $\mathcal{A}$ denotes $\mathbb{Z}$ or $\mathbb{R}$. We will describe in detail a construction which directly generalizes that of Section 4.2.3. We use the method of extension along the stable manifolds in the non-abelian setting.

5.4.1 Cocycles with values in Lie groups: the C^1 case

Let M be a compact manifold, H a connected Lie group, $\alpha : \mathcal{A} \times M \to M$ a smooth action, and $\beta : \mathcal{A} \times M \to H$ a λ-center-bunched Hölder cocycle. Assume that M is foliated by a continuous, α-invariant, and contracting foliation W, with C^1 leaves and contraction constant λ. A basic result proved in this section, which appears for the first time in [127], is the existence of a lifted contracting invariant foliation for the extension of α by β. The proof is constructive and gives explicit formulas for the leaves of the lifted foliation as graphs of invariant functions over the leaves of W. If the action is higher rank abelian, we also show that the lifted foliation is independent of the particular map used to build it.

A related result presented here is the existence of the lifted foliation if the cocycle takes values in a diffeomorphism group.

These results apply immediately to Anosov diffeomorphisms and flows, as well as to higher rank partially hyperbolic actions, and will be used extensively throughout this chapter.

Definition 5.4.1 Let M be a compact manifold, and $\alpha : \mathcal{A} \times M \to M$ be a C^1-action on M. Let W be a continuous foliation of M with C^1 leaves $W(x)$, $x \in M$. The foliation W is called *α-invariant* if

$$\alpha(n, W(x)) \subset W(\alpha(n, x)), \quad x \in M, \quad n \in \mathcal{A}.$$

Definition 5.4.2 Let M be a compact manifold and $\alpha : \mathcal{A} \times M \to M$ is a C^1-action on M. Let $0 < \lambda < 1$. An α-invariant foliation W is called *contracting with contraction constant* λ, if there exists a constant $C > 0$, such that

$$\mathrm{dist}_{W(\alpha(n,x))}(\alpha(n, x), \alpha(n, y)) \leq C\lambda^n \mathrm{dist}_{W(x)}(x, y), \tag{5.4.1}$$

for all $x, y \in M$, $n \geq 0$.

An α-invariant foliation W is called *expanding* if there exists a constant $C > 0$, such that

$$\mathrm{dist}_{W(\alpha(-n,x))}(\alpha(-n, x), \alpha(-n, y)) \leq C_2\lambda^n \mathrm{dist}_{W(x)}(x, y), \tag{5.4.2}$$

for all $x, y \in M$, $n \geq 0$.

Remark 5.4.3 The contracting/expanding foliations we will use in the future are intersections of stable or unstable foliations of a partially hyperbolic element of an $\mathbb{R}^k$- or $\mathbb{Z}^k$-action, and the $\mathcal{A}$-action α is the restriction of that action to the one-parameter subgroup generated by the element. These foliations have the property that the distance between pairs of points in the same local leaf is equivalent to the distance between points on the manifold. This will allow to replace the induced metric on the leaves $d_{W(x)}(x, y)$ by $d_M(x, y)$ in future arguments.

Recall that the extension of an action $\alpha : \mathcal{A} \times M \to M$ via a cocycle $\beta : \mathcal{A} \times M \to H$ is the action $\alpha_\beta : \mathcal{A} \times M \times H \to M \times H$ defined by

$$\alpha_\beta(n, x, g) = (\alpha(n, x), \beta(n, x)g). \tag{5.4.3}$$

If n is a positive integer and $f(\cdot) := \alpha(1, \cdot)$, then

$$\beta(n, x) = \beta(f^{n-1}x) \ldots \beta(fx)\beta(x). \tag{5.4.4}$$

Let W be an α-invariant contracting foliation. Our goal is to find a contracting foliation $\{\mathbf{W}(x, h)\}_{(x,h) \in M \times H}$ of the product space $M \times H$ that is invariant under the extended action α_β. In addition, we want the lifted foliation $\mathbf{W}$ to have the property that the projections of the its leaves into M coincide with leaves of W.

The last condition follows if there exists a family of continuous functions $\{\gamma_x | \gamma_x : W(x) \to H\}_x$ such that each leaf $\mathbf{W}(x, h)$ is given as the graph of a function γ_x:

$$\mathbf{W}(x, h) := \{(t, \gamma_x(t)h) | t \in W(x)\}, x \in M, g \in H.$$

The α_β-invariance of the foliation $\{\mathbf{W}(x, h)\}_{(x,h)}$ amounts to the relation

$$\beta(n, t)\gamma_x(t) = \gamma_{\alpha(n)(x)}(\alpha(n)(t))\beta(n, x),$$

or

$$\gamma_x(t) = \beta(n, t)^{-1}\gamma_{\alpha(n)(x)}(\alpha(n)(t))\beta(n, x), \tag{5.4.5}$$

for $n \in \mathcal{A},\ t \in W(x)$.

Since we want the functions γ_x to be continuous, (5.4.5) suggests that we should define γ_x by $\gamma_x(t) = \lim_{n\to\infty} \beta(n, t)^{-1}\beta(n, x)$.

Proposition 5.4.4 *Let M be a compact manifold. Let H be a connected Lie group endowed with a metric d_H that satisfies* (5.2.1). *Let $\alpha : \mathcal{A} \times M \to M$ be a smooth action. Let W be a contracting α-invariant foliation with contraction constant λ. Let $\beta : M \to H$ be a λ-center-bunched θ-Hölder cocycle.*

For any $x \in M$ and $n \in \mathcal{A}$ define the family of functions $\gamma_{x,n} : W(x) \to H$ by

$$\gamma_{x,n}(t) = \beta(n, t)^{-1}\beta(n, x) \tag{5.4.6}$$

Then the following statements are true:

(i) The family of functions $\{\gamma_{x,n}\}_n$ converges pointwise in H, as $n \to \infty$.

(ii) The map $\gamma_x : (W(x), d_{W(x)}) \to (H, d_H)$ given by

$$\gamma_x(t) = \lim_{n\to\infty} \gamma_{x,n}(t)$$

is uniformly θ-Hölder.

(iii) $\gamma_x(x) = e$, e the identity in H.

(iv) The family of graphs

$$\mathbf{W}(x, h) := \{(t, \gamma_x(t)h) | t \in W(x)\}, x \in M, h \in H,$$

gives an α_β-invariant foliation of $M \times H$. This is equivalent to

$$\beta(n,t)\gamma_x(t) = \gamma_{\alpha(n)(x)}(\alpha(n)(t))\beta(n,x), \ n \in \mathcal{A}, \ t \in W(x).$$

(v) *If $y \in W(x)$ and $\nu > \lambda^\theta$ then*

$$\lim_{n\to\infty} \nu^{-n} d_H(\beta(n,x), \beta(n,t)\gamma_x(t)) = 0.$$

In particular, the foliation **W** *is α_β contracting.*

(vi) *The family of functions $\{\gamma_x\}_x$ is uniquely determined by the properties (ii), (iii), and (iv).*

(vii) *The family of functions $\{\gamma_x\}_x$ is uniquely determined by the property (v), satisfied for a value $0 < \nu < \mu^{-1}$. Moreover, if*

$$\lim_{n\to\infty} \sup_{x\in M} \|\mathrm{Ad}(\beta(n,x))^{-1}\| = 0,$$

then $\{\gamma_x\}_x$ is characterized by (v) with $\mu = 1$.

(viii) *The foliation* **W** *depends continuously on the cocycle β, that is, the application*

$$\{\beta(1,\cdot) : M \to H\} \to \{\gamma_x : W(x) \to H\}, \ x \in M,$$

is continuous from the topology of uniform convergence of maps from M into H, to the topology of uniform convergence on compact sets of maps from the leafs of W to H.

Proof In order to simplify the exposition we consider only the case $\mathcal{A} = \mathbb{Z}$. The proof of the case $\mathcal{A} = \mathbb{R}$ is similar. We use the notation $f(\cdot) := \beta(1,\cdot)$.

(i) We claim that the sequence $\{\gamma_{x,n}(t)\}_n$ is uniformly Cauchy, so the limit $\gamma_x(t) = \lim_{n\to\infty} \gamma_{x,n}(t)$ exists uniformly for $t \in W_{\mathrm{loc}}(x)$ and defines a continuous function $\gamma_x : W_{\mathrm{loc}}(x) \to H$.

Let $m > n$ be large positive integers and $t \in W_{\mathrm{loc}}(x)$. Then, using the properties of the metric d_H, one has

$$\begin{aligned} d_H(\gamma_{x,m}(t), \gamma_{x,n}(t)) &\le \sum_{k=n}^{m-1} d_H(\gamma_{x,k+1}(t), \gamma_{x,k}(t)) \\ &= \sum_{k=n}^{m-1} d_H(\beta(k,t)^{-1}\beta(f^{k+1}t)^{-1}\beta(f^{k+1}x)\beta(k,x), \\ &\qquad \beta(k,t)^{-1}\beta(f^{k+1}t)^{-1}\beta(f^{k+1}t)\beta(k,x)) \\ &\le \sum_{k=n}^{m-1} d_H(\beta(f^{k+1}x), \beta(f^{k+1}t))\mu'^{k+1} \end{aligned}$$

$$\leq \sum_{k=n}^{m-1} \|\beta\|_{\text{Hölder}} \mu'^k \lambda^{(k+1)\theta} d_M(x,t)^\alpha$$
$$\leq C(\mu'\lambda^\theta)^n d_M(x,t)^\alpha,$$

where $C > 0$ is a constant independent of n, x, t and, due to the center-bunching condition, we have chosen $\mu < \mu'$ and $\mu'\lambda^\theta < 1$. So the claim follows.

Notice now that the identity $\gamma_{x,n}(t) = \gamma_{x',n}(t)\gamma_{x,n}(x')$ implies $\gamma_x(t) = \gamma_{x'}(t)\gamma_x(x')$, which allows to extend the function γ_x to a whole leaf $W(x)$.

(ii) We show that the functions $\gamma_x : W(x) \to H$ are θ-Hölder and their Hölder norm is bounded by a constant $C > 0$ independent of $x \in M$.

Let $t, t' \in W(x)$ and $n > 0$. Then using the properties of the metric d_H one has

$$d_H(\gamma_{x,n+1}(t), \gamma_{x,n+1}(t'))$$
$$= d_H(\beta^{-1}(t)\cdots\beta^{-1}(f^n t), \beta^{-1}(t')\cdots\beta^{-1}(f^n t'))$$
$$\leq \sum_{k=0}^{n} d_H(\beta^{-1}(t')\cdots\beta^{-1}(f^{k-1}t')\beta^{-1}(f^k t)\beta^{-1}(f^{k+1}t)\cdots\beta^{-1}(f^n t),$$
$$\beta^{-1}(t')\cdots\beta^{-1}(f^{k-1}t')\beta^{-1}(f^k t')\beta^{-1}(f^{k+1}t)\cdots\beta^{-1}(f^n t))$$
$$\leq \sum_{k=0}^{n} \mu'^k d_H(\beta^{-1}(f^k t), \beta^{-1}(f^k t'))$$
$$\leq \sum_{k=0}^{n} (\mu'\lambda^\theta)^k \|\beta^{-1}\|_{\text{Hölder}} d_M(t,t')^\theta \leq C d_M(t,t')^\theta,$$

where $C > 0$ is a constant independent of n, t, t' and, due to the center-bunching condition, we have chosen $\mu < \mu'$ and $\mu'\lambda^\theta < 1$. The claim follows now if we take limit as $n \to \infty$.

(iii) This follows from the definition of γ_x.

(iv) This follows from the identity $\gamma_{x,n+1}(t) = \beta(t)^{-1}\gamma_{fx,n}(ft)\beta(x)$, and then taking limit as $n \to \infty$.

(v) From (5.4.3) it follows that

$$f_\beta^n(t, \gamma_x(t)g) = (f^n t, \beta(n,x)\gamma_x(t)g). \tag{5.4.7}$$

Since f is already contracting along W, with contracting constant λ, it remains to show

$$\lim_{n\to\infty} \nu^{-n} d_H(\beta(n,x)g, \beta(n,t)\gamma_x(t)g) = 0. \tag{5.4.8}$$

By the invariance of the metric d_H and statements (iv), (iii), and (ii), one has

$$\begin{aligned} d_H(\beta(n,x)g, \beta(n,t)\gamma_x(t)g) &= d_H(\beta(n,x), \beta(n,t)\gamma_x(t)) \\ &= d_H(\beta(n,x), \gamma_{f^n x}(f^n t)\beta(n,x)) = d_H(e, \gamma_{f^n x}(f^n t)) \\ &= d_H(\gamma_{f^n x}(f^n x), \gamma_{f^n x}(f^n t)) \leq C d_{W(f^n x)}(f^n x, f^n t)^\theta \\ &\leq C\lambda^{n\alpha} d_M(x,t)^\alpha, \end{aligned} \tag{5.4.9}$$

and (5.4.8) follows.

(vi) Let $\{\omega_x : W(x) \to H\}_{x\in M}$ be a family of functions that satisfies statements (ii), (iii), and (iv).

From statement (iv) we have

$$\omega_x(t) = \beta(n,t)^{-1}\omega_{f^n x}(f^n t)\beta(n,x), \quad t \in W(x). \tag{5.4.10}$$

Then, using statements (iii) and (ii) it follows that

$$\begin{aligned} &d_H(\omega_x(t), \gamma_{x,n}(t)) \\ &\quad = d_H(\beta(n,t)^{-1}\omega_{f^n x}(f^n t)\beta(n,x), \beta(n,t)^{-1}\beta(n,x)) \\ &\quad \leq \|\mathrm{Ad}(\beta(n,t)^{-1}\| d_H(\omega_{f^n x}(f^n t), e) \\ &\quad = \|\mathrm{Ad}(\beta(n,t)^{-1}\| d_H(\omega_{f^n x}(f^n t), \omega_{f^n x}(f^n x)) \\ &\quad \leq C(\mu'\lambda^\alpha)^n d_M(t,x)^\alpha, \end{aligned} \tag{5.4.11}$$

where $C > 0$ is a constant independent of x, n, t, and, due to the center-bunching condition, we have chosen $\mu < \mu'$ and $\mu'\lambda^\theta < 1$. Now $n \to \infty$ implies that $\omega_x(t) = \gamma_x(t)$.

(vii) Let $\{\omega_x : W(x) \to H\}_{x\in M}$ be a family of functions that satisfies statement (v). Then

$$\begin{aligned} &d_H(\gamma_{x,n}(t), \ \omega_x(t)) \\ &\qquad = d_H(\beta(n,t)^{-1}\beta(n,x), \beta(n,t)^{-1}\beta(n,t)\omega_x(t)) \\ &\qquad \leq \|\mathrm{Ad}(\beta(n,t)^{-1})\| d_H(\beta(n,x), \beta(n,t)\omega_x(t)) \\ &\qquad \leq (\mu'\nu)^n \nu^{-n} d_H(\beta(n,x), \beta(n,t)\omega_x(t)), \end{aligned} \tag{5.4.12}$$

where $\mu < \mu' < \nu^{-1}$. Now $n \to \infty$ and (v) satisfied by ω_x imply that $\omega_x(t) = \gamma_x(t)$.

The proof for the case when $\lim_{n\to\infty} \sup_{x\in M} \|\mathrm{Ad}(\beta(n,x))^{-1}\| = 0$ follows also from (5.4.12).

(viii) This statement is a consequence of the fact that the application $\beta \to \gamma_x$ is the "uniform limit of a sequence of continuous functions".

Let $\epsilon > 0$ and $x \in M$ fixed. Let $t \in K$, $K \subset W(x)$ compact. Let $\beta, \beta' : M \to H$ be center-bunched cocycles, with corresponding families of functions

γ_x, γ'_x. If n is large enough and fixed, then statement (ii) in the proposition implies that

$$d_H(\gamma_{x,n}(t), \gamma_x(t)) < \frac{\epsilon}{4} \text{ and } d_H(\gamma'_{x,n}(t), \gamma'_x(t)) < \frac{\epsilon}{4} \qquad (5.4.13)$$

for all $t \in K$. Moreover, if the cocycles β, β' are close in uniform topology, then, for the same n, one can show using the properties of the metric d_H that

$$d_H(\gamma_{x,n}(t), \gamma'_{x,n}(t)) < \frac{\epsilon}{2}. \qquad (5.4.14)$$

Combining (5.4.13) and (5.4.14) gives

$$d_H(\gamma_x(t), \gamma'_x(t)) < \epsilon, \qquad (5.4.15)$$

and the statement follows. □

Remark 5.4.5 One can think of the quantity $\gamma_x(y)$ as the "height" of a point on the leaf of a lifted foliation in the space $M \times N$. The lifted foliation projects on the contracting foliation W.

Note that a similar invariant structure can be introduced over an expanding foliation. In this case $\gamma_x(y)$ is defined by

$$\gamma_x(y) = \lim_{t \to -\infty} \beta(t, y)^{-1} \beta(t, x). \qquad (5.4.16)$$

If the cocycle β exhibits more regularity, the leaves of the lifted foliation introduced in Proposition 5.4.4 exhibit more regularity as well. In general, the leaves are as differentiable as the cocycle. This follows from the regularity of the functions γ_x showed in the following proposition.

Proposition 5.4.6 *Let $K = 1, 2, \dots, \infty, \omega$. Let M be a C^K compact manifold. Let H be a connected Lie group endowed with a metric d_H that satisfies* (5.2.1). *Let $\alpha : \mathcal{A} \times M \to M$ be a smooth action. Let W be a contracting α-invariant foliation with C^K-leaves and contraction constant λ. Let $\beta : M \to H$ be a λ-center-bunched C^K cocycle.*

Then the functions γ_x are C^K. In particular, the leaves of the foliation **W** *are as differentiable as the cocycle β.*

Proof In order to simplify the exposition we consider only the case $\mathcal{A} = \mathbb{Z}$. The proof of the case $\mathcal{A} = \mathbb{R}$ is similar. We use the notation $f(\cdot) := \beta(1, \cdot)$. Moreover, we assume during the proof that H is a closed subgroup in $\mathrm{GL}(n, \mathbb{R})$. This reduces the action of the derivatives to ordinary matrix multiplications. In particular, one can use operator norm on $\mathrm{GL}(n, \mathbb{R})$, which is submultiplicative, in order to estimate the derivatives.

We show first that if the cocycle β is C^1, then the functions γ_x are C^1.

We denote by D_W the derivative along the leaves of W. Let $\gamma_{x,n}$ defined by (5.4.6). Then, for $t \in W_{\text{loc}}(x)$, one has

$$
\begin{aligned}
&D_W\gamma_{x,n+1}|_t \\
&= D_W\left[\beta^{-1}(t)\beta^{-1}(ft)\cdots\beta^{-1}(f^nt)\beta(n+1,x)\right] \\
&= \sum_{k=0}^{n}\beta^{-1}(t)\cdots\beta^{-1}(f^{k-1}t)\cdot D_W\beta^{-1}|_{f^kt}\cdots\beta^{-1}(f^nt)\beta(n+1,x)D_Wf^k|_t \\
&= \sum_{k=0}^{n}\beta(k,t)^{-1}\cdot D_W\beta^{-1}|_{f^kt}\cdot\beta(n-k,f^{k+1}t)^{-1}\beta(n+1,x)D_Wf^k|_t.
\end{aligned}
$$

We use now (5.4.5) and the cocycle equation to obtain

$$
\begin{aligned}
&D_W\gamma_{x,n+1}|_t \\
&= \sum_{k=0}^{n}\beta(k,t)^{-1}D_W\beta^{-1}|_{f^kt}\cdot\gamma_{f^{k+1}x}(f^{k+1}t)\beta(n-k,f^{k+1}x)^{-1} \\
&\qquad\cdot\gamma^{-1}_{f^{n+1}x}(f^{n+1}t)\beta(n-k,f^{k+1}x)\beta(k+1,x)D_Wf^k|_t \\
&= \sum_{k=0}^{n}\beta(k,t)^{-1}D_W\beta^{-1}|_{f^kt}\cdot\gamma_{f^{k+1}x}(f^{k+1}t)\beta(k+1,x)D_Wf^k|_t \\
&\quad+\sum_{k=0}^{n}\beta(k,t)^{-1}D_W\beta^{-1}|_{f^kt}\cdot\gamma_{f^{k+1}x}(f^{k+1}t)\cdot\beta(n-k,f^{k+1}x)^{-1} \\
&\qquad\cdot[\gamma_{f^{n+1}x}(f^{n+1}t)-e]\beta(n-k,f^{k+1}x)\beta(k+1,x)D_Wf^k|_t.
\end{aligned}
\tag{5.4.17}
$$

From the proof of Proposition 5.4.4 (b), it follows that the family of functions $\{\gamma_x\}_x$ is uniform Lipschitz. Let $\|\gamma\|_{\text{Lip}}$ be a common upper bound for the Lipschitz norms. For $t \in W_{\text{loc}}(x)$ and $k \le n$ one has that

$$
\begin{aligned}
&\|\beta(n-k,f^{k+1}x)^{-1}\cdot[\gamma_{f^{n+1}x}(f^{n+1}t)-e]\cdot\beta(n-k,f^{k+1}x)\| \\
&\le \sup_{x,n}\|\mathrm{Ad}(\beta(n-k,x)^{\pm1})\|\|\gamma_{f^{n+1}x}(f^{n+1}t)-\gamma_{f^{n+1}x}(f^{n+1}e)\| \\
&\le \|\gamma\|_{\text{Lip}}\mu'^{n-k}\lambda^n \le C(\max\{\lambda,\mu'\lambda\})^n,
\end{aligned}
\tag{5.4.18}
$$

where $\mu < \mu' < \lambda^{-1}$ and $C > 0$ is a constant independent of x, t, k, n. Equation (5.4.18) implies that the second sum from the right-hand side (5.4.17) converges to zero.

Using the estimate (4.2.15) in formula (5.4.17) we obtain that the series for $D_W\gamma_{x,n}|_t$ is absolutely convergent and, moreover, γ_x is differentiable in the stable direction with derivative

$$D_W\gamma_x = \sum_{k=0}^{\infty} \gamma_x(t)\beta(k,x)^{-1}\gamma_{f^kx}(f^kt)^{-1}D_W\beta^{-1}|_{f^kt} \cdot \gamma_{f^{k+1}x}(f^{k+1}t)\cdot\beta(k+1,x)D_Wf^k|_t. \tag{5.4.19}$$

The derivative $D_W\gamma_x$ is a continuous function with respect to $x \in M$ and $t \in W(x)$.

5.4.2 Higher regularity cases

A direct computation, which takes into account (4.2.15) and the fact that only a finite number of derivatives of β and γ_x appear in (5.4.19), shows that the higher derivatives of γ_x can be found by term-wise differentiation of formula (5.4.19). The resulting series is a continuous function with respect to $x \in M$ and $t \in W(x)$. This solves the case $K \leq \infty$.

The case $K = \omega$ follows from the observation that the uniform convergence proved in Proposition 5.4.4(b), also holds on the complexification of W_{loc} (W is a C^ω-foliation by Theorem 3.1 in [127]).

Indeed, since M, f, and β are real analytic, there is a complex analytic manifold $\widetilde{M}$ containing M and analytic extensions of f and β to $\widetilde{f} : \widetilde{M}_0 \to \widetilde{M}$ and respectively $\widetilde{\beta}_0 : \widetilde{M}_0 \to H$, where $\widetilde{M}_0$ is an open neighborhood of M in $\widetilde{M}$. Each leaf $W_{\text{loc}}(x)$, $x \in M$, has a uniform extension to a complex analytic sub-manifold $\widetilde{W}_{\text{loc}}(x)$ of $\widetilde{M}_0$, $\widetilde{f}(\widetilde{W}_{\text{loc}}(x)) \subset \widetilde{W}_{\text{loc}}(f(x))$, and $\widetilde{f}$ acts as a contraction on these complexified leaves. On $\widetilde{M} := \bigcup_{x\in M}\widetilde{W}_{\text{loc}}(x)$ one can define a $\mathbb{Z}_+$-cocycle $\widetilde{\beta}_+ : \mathbb{Z}_+ \times \widetilde{W} \to H$ over $\widetilde{f}$. The corresponding constants $\widetilde{\lambda}_-$ and $\widetilde{\widetilde{\mu}}_\pm$ can be made as close to λ_-, respectively $\widetilde{\mu}_\pm$, as desired by shrinking $\widetilde{M}_0$ (note that the limits defining these numbers are actually infimums).

Then $\widetilde{\gamma}_{x,n}(t) := \widetilde{\beta}_+(n,t)^{-1}\cdot\widetilde{\beta}_+(n,x)$ is defined on $\widetilde{W}_{\text{loc}}(x)$ for $x \in M$ and is an analytic extension of $\gamma_{x,n}$. The computation of Proposition 5.4.4(b), carries over to $\widetilde{\gamma}_{x,n}$ and we obtain the existence of a uniform limit as $n \to \infty$, bounded on $\widetilde{W}_{\text{loc}}(x)$ by some constant which is independent of x. We conclude that γ_x is analytic and depends continuously in C^ω on $x \in M$. □

5.4.3 Cocycles with values in diffeomorphism groups

A result similar to Proposition 5.4.4 can be obtained for cocycles with values in diffeomorphism groups. We start by presenting the relevant setup.

Let M, N be smooth compact Riemannian manifolds endowed with metrics d_M, d_N. Let $\text{Homeo}(N)$ be the set of homeomorphisms of N, endowed with the operation $\circ$ of composition. Let K be an integer or $K = \infty$ and $\text{Diff}^K(N) \subset$

Homeo(N) be the set of C^K-diffeomorphisms of N. We introduce the standard C^0 metric on Homeo(N):

$$\mathbf{d}_N(u, v) := \sup_{y \in N} d_N(u(y), v(y)), \quad u, v \in \text{Homeo}(N). \tag{5.4.20}$$

Notice that Homeo(N) is not complete in that metric. A complete metric on Homeo(N) is given by

$$\max\{\mathbf{d}_N(u, v), \mathbf{d}_N(u^{-1}, v^{-1})\}, u, v \in \text{Homeo}(N). \tag{5.4.21}$$

For $0 < \theta \leq 1$ and $u \in$ Homeo(N), define the θ-Hölder norm

$$\|u\|_\theta = \sup_{x,y \in N} \frac{d_N(u(x), u(y))}{d_N(x, y)^\theta}, \tag{5.4.22}$$

which may be finite or infinite. The 1-Hölder norm is called the *Lipschitz norm*. Note that for any $u \in \text{Diff}^K(N)$, the Lipschitz norm $\|u\|_1$ is finite and $\|u\|_1 \leq \|Du\|$, where Du is the derivative. For $u, v, w, z \in$ Homeo(N) the metric $\mathbf{d}_N$ has the following properties:

$$\begin{aligned} \mathbf{d}_N(vu, wu) &= \mathbf{d}_N(v, w), \\ \mathbf{d}_N(uv, uw) &\leq \|u\|_1 \mathbf{d}_N(v, w). \end{aligned} \tag{5.4.23}$$

Using (5.4.23) we deduce:

$$\mathbf{d}_N(wz, uv) \leq \mathbf{d}_N(wz, uz) + \mathbf{d}_N(uz, uv) \leq \mathbf{d}_N(w, u) + \|u\|_1 \mathbf{d}_N(z, v), \tag{5.4.24}$$

and

$$\begin{aligned} \mathbf{d}_N(u^{-1}, v^{-1}) &= \mathbf{d}_N(u^{-1}v, v^{-1}v) \\ &= \mathbf{d}_N(u^{-1}v, u^{-1}u) \leq \|u^{-1}\|_1 \mathbf{d}_N(u, v). \end{aligned} \tag{5.4.25}$$

If K is a finite integer, one can introduce on $\text{Diff}^K(N)$ a structure of Banach manifold, and in particular a structure of complete metric space. We briefly review this standard construction. Let $n = \dim(N)$. If $\Omega \subset \mathbb{R}^n$ is a compact set, then a function $f : \Omega \to \mathbb{R}$ is C^K if it has continuous derivatives of order K. The space $C^K(\Omega)$ of all C^K functions on Ω, endowed with the norm defined as the supremum of the derivatives of order up to K, has a structure of Banach space. Using coordinate charts that cover the manifold, and taking maximum over the C^K norms of the coordinate expressions, one can define a norm on the space of C^K vector fields on N that makes it a Banach space. After that the exponential map in Riemannian geometry can be used to construct charts from $\text{Diff}^K(N)$ into the set of vector fields. Note that it is standard to endow $\text{Diff}^\infty(N)$ with a structure of a Frechet manifold [44]. The C^∞ topology is defined as the inverse limit of the C^K topologies. We recall that two C^∞ diffeomorphisms are close if they are C^K close for some large K.

Let $\alpha : \mathcal{A} \times M \to M$ be a smooth action on M. For simplicity, in what follows we denote $\alpha(a, x)$ by ax. Let $\beta : \mathcal{A} \times M \to \mathrm{Diff}^K(N)$ be a *cocycle* over α that is

$$\beta(a + b, x) = \beta(a, bx) \circ \beta(b, x), \quad a, b \in \mathcal{A}, x \in M. \tag{5.4.26}$$

The *extended action* $\widetilde{\alpha} : \mathcal{A} \times (M \times N) \to M \times N$ is defined by

$$\widetilde{\alpha}(a, x, y) = (\alpha(a, x), \beta(a, x)(y)).$$

Any cocycle $\beta : \mathcal{A} \times M \to \mathrm{Diff}^K(N)$ can be viewed as a map $\beta : \mathcal{A} \times M \times N \to N$. In order to state our results, we need to assume certain regularity for cocycles. We assume that the regularity in $\mathcal{A}$ and N variables is C^K. For the M variable we need at least Hölder regularity. One way to obtain this is to require that the cocycle, viewed as a map $\beta : \mathcal{A} \times M \times N \to N$, to be C^K, or C^∞. If this is the case, we will call the cocycle a C^K*-cocycle*, respectively *smooth cocycle*. A broader class of cocycles which we will consider later is that of θ-Hölder cocycles.

Definition 5.4.7 Let $\theta \in (0, 1]$ and $d \geq 1$ integer. The cocycle $\beta : \mathcal{A} \times M \to \mathrm{Diff}^K(N)$ is said to be θ*-Hölder* if there exists a $C > 0$ constant and a compact set of generators $S \subset \mathcal{A}^k$ such that for any $a \in S$

$$\mathbf{d}_N(\beta(a, x), \beta(a, y)) \leq C\mathrm{dist}_M(x, y)^\theta, \quad x, y \in M. \tag{5.4.27}$$

The smallest value C for which (5.4.27) holds is denoted by $\|\beta\|_\theta$ and is called the θ*-norm of* β.

Any C^K-cocycle is θ-Hölder for any $0 < \theta \leq 1$, as immediately follows from the mean value theorem.

Definition 5.4.8 Let $0 < \lambda < 1$ and $d \geq 1$ integer. A θ-Hölder cocycle $\beta : \mathcal{A}^d \times M \to \mathrm{Diff}^K(N)$ is said to be λ*-center bunched* if there is a compact set of hyperbolic generators $S \subset \mathcal{A}^d$ such that

$$\Lambda := \sup_{a \in S, x \in M} \|D_N\beta(a, x)^{\pm 1}\|\lambda^\theta < 1, \tag{5.4.28}$$

where D_N is the derivative in the N direction.

A C^K-cocycle is said to be λ*-center bunched* if it satisfies (5.4.28) with $\theta = 1$.

If β is θ-Hölder and close to the identity, then (5.4.25), (5.4.27), and (5.4.28) imply that β^{-1} is θ-Hölder as well, and

$$\|\beta^{-1}\|_\theta \leq \lambda^{-\theta}\|\beta\|_\theta. \tag{5.4.29}$$

We assume in the future, without loss of generality, that if $d = 1$ then the set of generators S appearing in Definitions 5.4.7 and 5.4.8 is either 1, if $\mathcal{A} = \mathbb{Z}$, or the interval $(0.1, 1]$, if $\mathcal{A} = \mathbb{R}$. We will skip the reference to S in these cases.

Proposition 5.4.9 *Let M, N be compact manifolds, $\alpha : \mathcal{A} \times M \to M$ a smooth action, and W an α-invariant contracting foliation of M with contraction constant $0 < \lambda < 1$. Let $\beta : \mathcal{A} \times M \to$ Diff$^K(N)$ be a θ-Holder cocycle over α that is λ-center bunched. For $x \in M, t \geq 0$, define $\gamma_{x,t} : W(x) \to$ Diff$^K(N) \subset$ Homeo(N) by*

$$\gamma_{x,t}(y) = \beta(t, y)^{-1} \circ \beta(t, x). \tag{5.4.30}$$

Then the following statements hold:

(i) The family of homeomorphisms $\{\gamma_{x,t}(y)\}_{t\geq 0}$ converges in Homeo(N) *as $t \to \infty$ to a homeomorphism:*

$$\gamma_x(y) := \lim_{t\to\infty} \gamma_{x,t}(y). \tag{5.4.31}$$

(ii) The map $\gamma_x : (W(x), d_{W(x)}) \to$ (Homeo$(N), \mathbf{d}_N$) is uniformly θ-Holder.

(iii) $\gamma_x(x) = Id_N$.

(iv) The family of functions $\{\gamma_x\}_{x\in M}$ is invariant under the extended action on $M \times N$, that is,

$$\beta(t, y) \circ \gamma_x(y) = \gamma_{tx}(ty) \circ \beta(t, x),\ y \in W(x), t \geq 0. \tag{5.4.32}$$

In particular, the family of graphs

$$\mathbf{W}(x, y) := \{(t, \gamma_x(t)(y) | t \in W(x)\}, x \in M, y \in N, \tag{5.4.33}$$

gives an α_β-invariant foliation of $M \times N$.

(v) If $y \in W(x)$ and $\nu > \lambda^\theta$ then

$$\lim_{t\to\infty} \nu^{-t}\mathbf{d}_N(\beta(t, x), \beta(t, y) \circ \gamma_x(y)) = 0. \tag{5.4.34}$$

(vi) The family of functions $\{\gamma_x\}_{x\in M}$ is uniquely determined by (ii), (iii), and (iv).

(vii) The family of functions $\{\gamma_x\}_x$ is uniquely determined by the property (v), satisfied for a value $0 < \nu < \mu^{-1}$. Moreover, if

$$\lim_{n\to\infty} \sup_{x\in M} \|D_N\beta(n, x)^{-1}\| < \infty,$$

then $\{\gamma_x\}_x$ *is characterized by* (5.4.34) *with* $\nu = 1$.

(viii) The foliation $\mathbf{W}$ *depends continuously on the cocycle* β*, that is, the application*

$$\{\beta(1,\cdot) : M \to \mathrm{Diff}^K(N)\} \to \{\gamma_x : W(x) \to H\},\ x \in M,$$

is continuous from the topology of uniform convergence of maps from M *into* $\mathrm{Homeo}(N)$*, to the topology of uniform convergence on compact sets of maps from the leafs of* W *to* H.

Proof In order to simplify the exposition we consider only the case $\mathcal{A} = \mathbb{Z}$. The proof of the case $\mathcal{A} = \mathbb{R}$ is similar. We use the notation $f(\cdot) := \beta(1,\cdot)$.

(i) Let $x \in M$, $y \in W(x)$. We show that the family $\{\gamma_{x,n}(y)\}_n$ is uniformly convergent in the complete metric (5.4.21) as $n \to \infty$. Let $m \geq n > 0$. To simplify the notation, during the proof we denote $\beta(1,x)$ by $\beta(x)$. We estimate only $d := \mathbf{d}_N(\gamma_{x,m}(y), \gamma_{x,n}(y))$. The estimation for $\mathbf{d}_N(\gamma_{x,m}^{-1}(y), \gamma_{x,n}^{-1}(y))$ is similar.

Using (5.4.26) for any $z \in M$ we have

$$\beta(m,z) = \beta(m-n+n,z) = \beta(m-n,tz)\circ\beta(n,z). \tag{5.4.35}$$

By (5.4.35), applied for $z = x$, by (5.4.23), and again by (5.4.26), we have

$$\begin{aligned} d &= \mathbf{d}_N(\beta(m,y)^{-1}\circ\beta(m,x), \beta(n,y)^{-1}\circ\beta(n,x)) \\ &= \mathbf{d}_N(\beta(m,y)^{-1}\circ\beta(m-n,nx)\circ\beta(n,x), \\ &\quad \beta(n,y)^{-1}\circ\beta(m-n,ny)^{-1}\circ\beta(m-n,ny)\circ\beta(n,x)) \\ &= \mathbf{d}_N(\beta(m,y)^{-1}\circ\beta(m-n,nx), \beta(m,y)^{-1}\circ\beta(m-n,ny)). \end{aligned} \tag{5.4.36}$$

Using now repeatedly (5.4.26), (5.4.24), and the triangle inequality for the metric $\mathbf{d}_N$, (5.4.36) becomes

$$\begin{aligned} d &= \mathbf{d}_N(\beta(m,y)^{-1}\circ\beta((m-1)x)\circ\beta((m-2)x)\circ\cdots\circ\beta(nx), \\ &\quad \beta(m,y)^{-1}\circ\beta((m-1)y)\circ\beta((m-2)y)\circ\cdots\circ\beta(ny)) \\ &\leq \sum_{k=1}^{m-n} \mathbf{d}_N(\beta(m,y)^{-1}\circ\beta((m-1)y)\circ\cdots\circ\beta((m-k+1)y)\circ\beta((m-k)x), \\ &\quad \beta(m,y)^{-1}\circ\beta((m-1)y)\circ\cdots\circ\beta((m-k+1)y)\circ\beta((m-k)y)). \end{aligned} \tag{5.4.37}$$

Note that by (5.4.26) we have

$$\beta(m,y)^{-1}\circ\beta((m-1)y)\circ\cdots\circ\beta((m-k+1)y) = \beta(m-k+1,y)^{-1}. \tag{5.4.38}$$

Using now first (5.4.38) and (5.4.23), and later the chain rule, (5.4.27), (5.4.28), and (5.4.1), one can bound the right-hand side in (5.4.37) by

$$\begin{aligned} d &\leq \sum_{k=1}^{m-n} \|\beta(m-k+1, y)^{-1}\|_1 \mathbf{d}_N(\beta((m-k)x), \beta((m-k)y)) \\ &\leq C_1 \sum_{k=1}^{m-n} \sup_{x \in M} \|D_N \beta(1, x)^{-1}\|^{m-k} \lambda^{(m-k)\theta} \mathrm{dist}_M(x, y)^\theta \\ &\leq C_2 \Lambda^n \mathrm{dist}_M(x, y)^\theta, \end{aligned} \tag{5.4.39}$$

where C_1, C_2 are constants independent of x, y, t, t', and $\Lambda < 1$ is defined in (5.4.28). Formula (5.4.39) implies (i).

(ii) We show that the map $y \to \gamma_x(y)$ is uniformly θ-Holder. Let $y', y'' \in W(x)$ and $n > 0$. Then

$$\begin{aligned} &\mathbf{d}_N(\gamma_{x,n}(y'), \gamma_{x,n}(y'')) \\ &\quad = \mathbf{d}_N(\beta(n, y')^{-1} \circ \beta(n, x), \beta(n, y'')^{-1} \circ \beta(n, x)) \\ &\quad = \mathbf{d}_N(\beta(n, y')^{-1}, \beta(n, y'')^{-1}) \\ &\quad \leq C_1 \sum_{k=0}^{n} \sup_{x \in M} \|D\beta(1, x)^{-1}\|^k \lambda^{k\theta} \mathrm{dist}_M(x, y)^\theta \leq C_2 \mathrm{dist}_M(x, y)^\theta, \end{aligned}$$

where C_1, C_2 are independent of x, y.

(iii) This is obvious.

(iv) This follows from the identity $\beta(n, y) \circ \gamma_{x,m}(y) = \gamma_{nx,m-n}(ny) \circ \beta(n, x)$, which is a consequence of (5.4.30), by taking limit as $m \to \infty$.

(v) By (ii), (iii), and (iv) it follows that

$$\begin{aligned} \mathbf{d}_N(\beta(n, x), \beta(n, y) \circ \gamma_x(y)) &= \mathbf{d}_N(\beta(n, x), \gamma_{nx}(ny) \circ \beta(n, x)) \\ &= \mathbf{d}_N(Id_N, \gamma_{nx}(ny)) \leq \mathbf{d}_N(\gamma_{nx}(nx), \gamma_{nx}(ny)) \\ &\leq C\lambda^{n\theta} \mathrm{dist}_M(x, y)^\theta. \end{aligned} \tag{5.4.40}$$

Then (5.4.34) follows by taking the limit as $n \to \infty$.

(vi) Let $\Gamma_x : W(x) \to \mathrm{Homeo}(N)$ be another family of functions that satisfies (ii)–(iv). From (iv) it follows that

$$\Gamma_x(y) = \beta(n, y)^{-1} \circ \Gamma_{nx}(ny) \circ \beta(n, x). \tag{5.4.41}$$

Then (5.4.41), (ii), (iii), and (5.4.24) imply

$$\begin{aligned} &\mathbf{d}_N(\Gamma_x(y), \gamma_{x,n}(y)) \\ &\quad = \mathbf{d}_N(\beta(n, y)^{-1} \circ \Gamma_{nx}(ny) \circ \beta(n, x), \beta(n, y)^{-1} \circ \beta(n, x))) \end{aligned}$$

$$
\begin{aligned}
&\le \|D_N\beta(n,y)^{-1}\|\mathbf{d}_N(\Gamma_{nx}(ny), Id_N)\\
&\le \|D_N\beta(n,y)^{-1}\|\mathbf{d}_N(\Gamma_{nx}(ny), \Gamma_{nx}(nx))\\
&\le \|D_N\beta(n,y)^{-1}\|\lambda^{n\theta}\|\Gamma\|_\theta d_M(x,y)^\theta\\
&\le C_4\Lambda^t \mathrm{dist}_M(x,y)^\theta,
\end{aligned}
\tag{5.4.42}
$$

where C_4 is a constant independent of n and $\Lambda < 1$ is defined in (5.4.28).

From (5.4.42) and (5.4.31) it follows that $\Gamma_x = \gamma_x$.

(vii) Let $\{\omega_x : W(x) \to \mathrm{Diff}^K(N)\}_{x\in M}$ be a family of functions that satisfies statement (v). Then

$$
\begin{aligned}
&\mathbf{d}_N(\gamma_{x,n}(y), \omega_x(y))\\
&\quad = \mathbf{d}_N(\beta(n,y)^{-1}\beta(n,x), \beta(n,y)^{-1}\beta(n,y)\omega_x(y))\\
&\quad \le \|D_N\beta(n,y)^{-1}\|\mathbf{d}_N(\beta(n,x), \beta(n,y)\omega_x(y))\\
&\quad \le (\mu'\nu)^n\nu^{-n}\mathbf{d}_N(\beta(n,x), \beta(n,y)\omega_x(y)),
\end{aligned}
\tag{5.4.43}
$$

where $\mu < \mu' < \nu^{-1}$. Now $n \to \infty$ and (v) satisfied by ω_x imply that $\omega_x(y) = \gamma_x(y)$.

The proof for the case when $\lim_{n\to\infty}\sup_{x\in M}\|D_N\beta(n,x)^{-1}\| = 0$ follows also from (5.4.43).

(viii) This statement is a consequence of the fact that the application $\beta \to \gamma_x$ is the "uniform limit of a sequence of continuous functions."

Let $\epsilon > 0$ and $x \in M$ be fixed. Let $t \in K$, $K \subset W(x)$ be compact. Let $\beta, \beta' : M \to \mathrm{Diff}^K(N)$ be center-bunched cocycles, with corresponding families of functions γ_x, γ'_x. If n is large enough and fixed, then statement (ii) in the proposition implies that

$$
\mathbf{d}_N(\gamma_{x,n}(y), \gamma_x(y)) < \frac{\epsilon}{4} \text{ and } \mathbf{d}_N(\gamma'_{x,n}(y), \gamma'_x(y)) < \frac{\epsilon}{4}
\tag{5.4.44}
$$

for all $t \in K$. Moreover, if the cocycles β, β' are close in uniform topology, then, for the same n, one can show using the properties of the metric $\mathbf{d}_N$ that

$$
\mathbf{d}_N(\gamma_{x,n}(y), \gamma'_{x,n}(y)) < \frac{\epsilon}{2}.
\tag{5.4.45}
$$

Combining (5.4.44) and (5.4.45) gives

$$
d_H(\gamma_x(t), \gamma'_x(t)) < \epsilon,
\tag{5.4.46}
$$

and the statement follows. □

Remark 5.4.10 If H is a connected Lie group that has a co-compact lattice Δ, then Proposition 5.4.4 follows from Proposition 5.4.9. Indeed, one observes that H acts on the compact manifold $N := H/\Delta$ by left multiplications, which are diffeomorphisms of H, so β can be seen as taking values in $\mathrm{Diff}^\infty(N)$. Full details can be found in [127].

Similar to the situation for Lie group valued cocycles, if β exhibits more regularity, the leaves of the lifted foliation introduced in Proposition 5.4.9 exhibit more regularity as well. In general, the leaves are as differentiable as the cocycle. This follows from the regularity of the functions γ_x showed in the following proposition.

Proposition 5.4.11 *Let $K = 1, 2, \ldots, \infty, \omega$. Let M, N be C^K compact manifolds. Let $\alpha : \mathcal{A} \times M \to \mathit{Diff}^K(N)$ be a C^K-action. Let W be a contracting α-invariant foliation with C^K-leaves and contraction constant λ. Let $\beta : M \to$* Diff $^K(N)$ *be a λ-center-bunched C^K cocycle.*

Then the functions γ_x takes values in Diff $^K(N)$, *and moreover, as functions $W(x) \to$* Diff $^K(N)$ *are C^K. In particular, the leaves of the foliation* **W** *given by* (5.4.33) *are as differentiable as the cocycle β.*

Moreover, if K is finite, if W is the strong stable foliation of a partially hyperbolic diffeomorphism f, and if γ_x, γ'_x correspond to two cocycles $\beta, \beta' : M \to$ Diff $^K(N)$, *respectively, over the $\mathbb{Z}$-action induced by f, then γ_x as a function $W(x) \to$* Diff $^K(N)$ *converges C^K to γ'_x as β converges to β' in C^K-topology.*

Proof The proposition is a consequence of Theorem 1.8.14.

Let $\alpha : \mathbb{R} \times M \to M$ be a smooth action, for which the time one map $\alpha(1)$ is an r-normally hyperbolic to a C^K foliation $W^c_{\alpha(1)}$ of M. Let $W^s_{\alpha(1)}$ be an α-invariant contracting foliation of M with contraction constant $0 < \lambda < 1$, which also is the stable foliation of $\alpha(1)$. Let $\beta : \mathbb{R} \times M \to \mathrm{Diff}^K(N)$ be a C^K-cocycle that is C^K-close to Id_N on the set of generators S. Then the map $y \to \gamma_x(y)$ takes values in $\mathrm{Diff}^K(N)$ and is C^K as a function from $W^s_{\alpha(1)}(x) \to \mathrm{Diff}^K(N)$. Moreover, if γ_x and γ'_x correspond to β and β' respectively, then γ_x converges C^K to γ'_x as β converges to β' in C^K-topology.

We start by observing that formula (5.4.31) gives the same limit γ_x if we take the limit using discrete time $n \in \mathbb{N}$. Note also that β C^K-close to Id_N on the set of generators S implies that β is λ-center bunched. The idea of the proof is to construct a partially hyperbolic diffeomorphism f for which the stable foliation is given by γ_x, and then read the regularity properties of γ_x from the regularity properties of the stable foliation.

Let $X = M \times N, f : X \to X, f(x, z) = (\alpha(1)(x), z)$ and $g : X \to X, g(x, z) = (\alpha(1)(x), \beta(1, x)(z))$. The map f is a partially hyperbolic diffeomorphism with neutral foliation with leaves $W^c_f(x, z) = W^c_{\alpha(1)}(x) \times N$. It is immediate that f is K-normally hyperbolic to the foliation W^c_f. Since β

is C^K-close to Id_N on the set of generators S, the map g is a C^K-perturbation of f. Observe now that the graph of the function $W^s_{\alpha(1)}(x) \ni y \to (y, \gamma_x(y)(z)) \in X$ coincides with the stable leaf $W^s_g(x, z)$ of g. This follows from the characterization of the stable leaf given in formula (1.8.4) and from the contraction properties of the functions γ_x shown in Proposition 5.4.9,(5). □

Assume now that the foliation W is contracting/expanding under the several $\mathcal{A}$-actions that are parts of a higher rank abelian $\mathcal{A}^k$-action. The following proposition shows that the "height" $\gamma_x(y)$ introduced in Propositions 5.4.4 and 5.4.9 does not depend on the particular $\mathcal{A}$-subflow used to build it. In order to mark the dependence of γ_x on α, we sometimes write γ^a_x if $a \in \mathcal{A}^k$ is a generator of the $\mathcal{A}$-subaction α used to construct the flow.

Proposition 5.4.12 *Let M, N be compact manifolds, H a compact connected Lie group, $\alpha : \mathcal{A}^k \times M \to M$ a smooth action, and W an α-invariant foliation of M. Let $0 < \lambda < 1$ and $\beta : \mathcal{A}^k \times M \to \mathrm{Diff}^K(N)$, or $\beta : \mathcal{A}^k \times M \to H$, be a θ-Holder cocycle over α. Let $a, b \in \mathcal{A}^k$. Assume that W is contracting (with contraction constants λ_a and λ_b) for the subactions induced by $\mathcal{A}a$ and $\mathcal{A}b$, and that β is* $\max(\lambda_a, \lambda_b)$*-center bunched. Then:*

(i) $\beta(b, y)\gamma^a_x(y) = \gamma^a_{bx}(by)\beta(b, x)$, for $y \in W(x)$; and
(ii) $\gamma^a_x = \gamma^b_x$, for all $x \in M$.

Proof We start by proving (i). For $x \in M$, define the function $\Gamma_x : W(x) \to \mathrm{Diff}^K(N)$ by

$$\Gamma_x(y) = \beta(b, y)^{-1}\gamma^a_{bx}(by)\beta(b, y). \tag{5.4.47}$$

Clearly $\Gamma_x(x) = Id_N$, and since the family $\{\gamma_x\}_x$ is uniformly θ-Hölder, the family $\{\Gamma_x\}_x$ is also uniformly θ-Hölder. We will show that Γ_x satisfies condition (iv) in Proposition 5.4.9, and then Proposition 5.4.9 (v), implies that $\Gamma_x = \gamma_x$.

Since $a+b = b+a$, the cocycle equation (5.4.26) gives $\beta(b, ax)\beta(a, x) = \beta(a, bx)\beta(b, x)$. Together with (5.4.32) this gives

$$\begin{aligned} \beta(a, y)\Gamma_x(y) &= [\beta(a, y)\beta(b, y)^{-1}]\gamma^a_{bx}(by)\beta(b, x) \\ &= \beta(b, ay)^{-1}\gamma^a_{(a+b)x}((a+b)y)[\beta(a, bx)\beta(b, x)] \\ &= [\beta(b, ay)^{-1}\gamma^a_{(a+b)x}((a+b)y)\beta(b, ax)]\beta(a, x) \\ &= \Gamma_{ax}(ay)\beta(a, x). \end{aligned} \tag{5.4.48}$$

To prove (ii), observe that γ_x^a satisfies Proposition 5.4.9(iv), as applies to γ_x^b; this is exactly (i) in this proposition. So, using Proposition 5.4.9(v) again, it follows that $\gamma_x^a = \gamma_x^b$. □

5.4.4 Invariant foliations for bundle extensions

In this subsection we show how to carry over the construction of the invariant foliations for the more general case of principal bundle extensions. Throughout we use the terminology introduced in Section 1.3. We follow [130], where the construction is done under the assumption that the extensions are with compact Lie group fiber. This is replaced here by a bunching assumption.

Definition 5.4.13 Let H be a Lie group. Let P_1, P_2 be principal H-bundles and $F : P_1 \to P_2$ an H-map. Then F is said to be θ-Hölder (with respect to a specified metric on the bundles) respectively C^K, if it is given by θ-Hölder, respectively C^K, maps in a family of local trivializations of the bundles.

Definition 5.4.14 Consider an action of a (discrete) group A on the manifold $M, \alpha : A \times M \to M$. An H-extension of α is a principal H-bundle $p : P \to M$ endowed with a lift of α to an action $\tilde{\alpha} : A \times P \to P$ through H-maps.

If $A = \mathbb{Z}$ and α is defined by a diffeomorphism $f := \alpha(1, \cdot) : M \to M$, then $F := \tilde{\alpha}(1, \cdot) : P \to P$ is called an H-extension of f.

Moreover, if f is an Anosov diffeomorphism and if the stable foliation of f has contraction constant $0 < \lambda < 1$, then F is called a λ-center-bunched extension if it satisfies $\mu\lambda^\theta < 1$, where μ is defined using (5.2.2), a specific metric on the bundles, and a family of local trivializations of the bundles.

Remark 5.4.15 If the group H is compact, any H-extension F of an Anosov diffeomorphism f with contraction constant λ is λ-center bunched.

We now describe the natural equivalence relation between extensions, which corresponds to the cohomology of cocycles.

Definition 5.4.16 Two H-extensions $\tilde{\alpha}_i : A \times P_i \to P_i, i = 1, 2$, of α are called *cohomologous* if there is an H-bundle isomorphism between $\tilde{\alpha}_1$ and $\tilde{\alpha}_2$, that is, an H-bundle map $F : P_1 \to P_2$ covering the identity map of M and such that

$$\tilde{\alpha}_2(a) = F \circ \tilde{\alpha}_1(a) \circ F^{-1} \text{ for each } a \in A. \tag{5.4.49}$$

We call F the *transfer function*.

Note that if two H-extensions are cohomologous through a continuous transfer map, then the corresponding principal bundles are isomorphic in the C^K-category as well, $1 \leq K \leq \infty$ [50, §4.3].

Proposition 5.4.17 *Let H be a Lie group, $f : M \to M$ be an Anosov diffeomorphism, and $p : P \to M$ be a principal H-bundle. Let $W = \{W(x)\}_{x \in M}$ be an f-invariant foliation of M whose leaves are included in the stable foliation of f. Assume that the stable foliation of f has contraction constant $0 < \lambda < 1$ and that F is a λ-center-bunched θ-Hölder H-extension of f to P. Then there exists an F- and H-invariant continuous foliation $\tilde{W} = \{\tilde{W}(x)\}_{x \in P}$ of P whose leaves are lifts of the leaves of W, that is, locally, the leaves of $\tilde{W}$ are obtained by H-translates of graphs of continuous sections*

$$\tilde{\gamma}_x^W : W_{loc}(x) \to P,$$

which properly normalized, vary in the uniform metric continuously with respect to x.

Moreover, this is the unique foliation with these properties, and the sections $\tilde{\gamma}_x^W$ are actually uniformly θ-Hölder on $W_{loc}(x)$. If W and F are smooth, then $\tilde{\gamma}_x^W$ are uniformly smooth on $W_{loc}(x)$, that is, $\tilde{W}$ has smooth leaves).

Proof Assuming that W, F, and the stable foliation of f are smooth, the last statement of the proposition follows from the general theory of stable foliations of partially hyperbolic diffeomorphisms. If the stable foliation of f is not smooth, we can use the same techniques, or the formula of $\tilde{\gamma}_x^W$ obtained below for the Hölder case and the explicit computations from Section 5.5, where we show the lift of regularity for the transfer map, to reach the desired conclusion. Thus, we only need to prove the Hölder case.

If W is a subfoliation of the stable foliation W_f^s of f, then $\tilde{\gamma}_x^W$ are the restrictions of $\tilde{\gamma}_x^{W_f^s}$ to the leaves of W_f^s. Therefore, it is enough to prove the existence and regularity of $\tilde{\gamma}_x^W$ for $W = W_f^s$, and then check their uniqueness for a general foliation W. For the simplicity of the notation, $\tilde{\gamma}_x^{W_f^s}$ are denoted $\tilde{\gamma}_x^F$.

Consider a family of H-bundle charts

$$\{\phi_i : p^{-1}(U_i) \subset P \to U_i \times H\}_{i \in I},$$

where $\{U_i\}_{i \in I}$ is a finite open cover of M. Hence $\phi_i(\xi h) = \phi_i(\xi)h$ for $\xi \in p^{-1}(U_i)$ and $h \in H$.

Let $r' > 0$ be a Lebesgue constant of the covering $\{U_i\}_{i \in I}$ (for each x, the open ball $B_{r'}(x)$ of diameter r_0 is contained in one of the U_i's).

Choose a *domain function* $d : M \to I$ such that $B_{r'}(x) \subset U_d(x)$.

We also use the notation ϕ_x for $\phi_{d(x)}$, respectively U_x for $U_{d(x)}$.

For $x \in M$ let us denote $x_n := f^n(x)$.

Since f is uniformly continuous, there is a constant $r > 0$ such that $f(B_r(x)) \subset B_{r'}(f(x))$ for each $x \in M$.

Given $x \in M, n \in \mathbb{Z}$, and $i, j \in I$ such that $x \in U_i$, $f^n(x) \in U_j$, denote by $\beta_{j,i}(n, x) \in H$ the *fiber component* of $F^n : p^{-1}(x) \to p^{-1}(f^n(x))$ in the charts ϕ_i and ϕ_j, that is,

$$F^n(\phi_i^{-1}(x, h)) = \phi_j^{-1}(f^n(x), \beta_{j,i}(n, x)h), h \in H. \tag{5.4.50}$$

Therefore,

$$\beta_{k,j}(n, f^m(x))\beta_{j,i}(m, x) = \beta_{k,i}(n + m, x).$$

If $i = d(x)$ and $j = d(f^n(x))$ then write $\beta(n, x)$ for $\beta_{j,i}(n, x)$. We will also use the notation $\beta_{x,y}$ for $\beta_{d(x),d(y)}$.

With these notations, one claims that the stable manifolds of F are given by the images of

$$t \in W^s(x; f) \subset M \to \tilde{\gamma}_x^F(t) := \phi_t^{-1}(t, \gamma_x^F(t)h) \in P, \tag{5.4.51}$$

where $h \in H$ and

$$\gamma_x^F(t) = \lim_{n\to\infty} [\beta_{x_n,t}(n, t)]^{-1}\beta(n, x). \tag{5.4.52}$$

The expression in (5.4.52) is considered from the domain of the chart ϕ_x to the domain of the chart ϕ_t. Note that for n large, $f^n(t) \in U_d(f^n(x))$.

Indeed, once the convergence in (5.4.52) is proven, one can check that if $t \in W^s(x; f)$, n is large enough, $\xi = \phi_x^{-1}(x, h)$ and $\eta = \phi_t^{-1}(t, x(t)h)$, then

$$\phi_{x_n}(F^n(\xi)) = (x_n, \beta(n, x)h), \phi_{x_n}(F^n(\eta)) = (t_n, \beta_{x_n,t}(n, t)\gamma_x(t)h),$$

and, using the invariance of the metric on H and λ-center bunching it follows

$$\lim_{n\to\infty} d_H(\beta(n, x)h, \beta_{x_n,t}(n, t)\gamma_x(t)h) = 0,$$

which shows that $\eta \in W^s(\xi; F)$.

Next we prove that the limit in (5.4.52) exists, and that the functions $\tilde{\gamma}_x^F$ are Hölder and describe a continuous F-invariant foliation.

Define, for $t \in W^s(x; f)$ and n so large that $t_n \in U_{x_n}$,

$$\gamma_{n,x}(t) = \beta_{x_n,t}(n, t)^{-1}\beta(n, x),$$

(we ignore from now on the superscript F of γ). Then

$$\gamma_{n,x_k}(tk) = \beta(k, t)\gamma_{n+k,x}(t)\beta(k, x)^{-1},$$

which shows that it is enough to prove (5.4.52) for t close to x (i.e., k large), and that the limit satisfies the desired F-invariance.

But then $t_n \in B_r(x_n)$ for all $n \geq 0$ and therefore one can write both $\beta(n, x)$ and $\beta_{x_n,x}(t)$ as products of the cocycles $\beta_k := \beta_{x_{k+1},x_k}(1, \cdot), k \geq 0$, defined on $B_r(x_k)$. Since there are only finitely many charts, these cocycles have a uniformly bounded Hölder norm, independently of x. Also, the constant $\mu := \mu_-^{-1}$ is well defined and can be computed using formula (5.2.2).

Therefore, one can use λ-center bunching and follow the computations of Proposition 5.4.9 for the products of β_k (instead of using the same β) which shows that

$$\gamma_x'(t) := \lim_{n\to\infty} \beta_{x_n,x}(n, t)^{-1}\beta(n, x)$$

has the desired properties: the limit exists and is uniformly Hölder on $W^s_{\text{loc}}(x; f)$. Since for t close to x,

$$\tilde{\gamma}_x^F(t) = \phi_x^{-1}(t, \gamma_x'(t)),$$

we obtain the desired result.

We now prove the uniqueness part of the proposition.

Assume that $\tilde{\omega}_x : W_{\text{loc}}(x) \to P$ is a family of continuous sections whose H-translates give an F-invariant continuous foliation. One can normalize these sections by a translation, so that $\phi_x(\tilde{\omega}_x(x)) = (x, Id)$.

The F-invariance implies that

$$F^n(\tilde{\omega}_x(t)) = \tilde{\omega}_{f^n(x)}(f^n(t))c_{n,x}, \tag{5.4.53}$$

where $c_{n,x} \in H$ does not depend on $t \in W(x)$. Computing at $t = x$ in the charts ϕ_x and $\phi_{f^n(x)}$, one finds that $c_{n,x} = \beta(n, x)$. Note that the same conclusion applies for $\tilde{\gamma}_x$ as well because formulas (5.4.51) and (5.4.52) show that $\phi_x(\tilde{\gamma}_x(x)) = (x, Id)$ too.

Since both $\tilde{\omega}_x$ and $\tilde{\gamma}_x$ are sections defined on $W(x)$, one can write

$$\tilde{\omega}_x(t) = \tilde{\gamma}_x(t)h_x(t), h_x : W_{\text{loc}}(x) \to H,$$

where the functions h_x are continuous, vary continuously with $x \in M$, and $h_x(x) = Id$. Using this in (5.4.53), together with the the same formula for e x and the H-equivariance of F, we obtain

$$h_x(t) = \beta(n, x)^{-1}h_{x_n}(t_n)\beta(n, x).$$

Since t_n, x_n belong to the same stable leaf, one has

$$\lim_{n\to\infty} d_M(t_n, x_n) = 0.$$

Moreover, $h_x(t)$ is continuous in both x and t, so one concludes from the invariance of the distance on H and from λ-bunching that

$$\lim_{n\to\infty} d_H(\beta(n,x)^{-1}h_{x_n}(t_n)\beta(n,x), Id) = \lim_{n\to\infty} d_H(h_{x_n}(t_n), Id) = 0,$$

hence $h_x = Id$ on $W_{\text{loc}}(x)$, which shows that $\tilde{\gamma}_x = \tilde{\omega}_x$, as claimed. □

5.5 Regularity results for non-abelian cocycles over rank-one Anosov actions

Let G be a connected Lie group or a diffeomorphism group. It is shown in [126] that for a C^K G-valued cocycle cohomologous to identity, over an Anosov diffeomorphism, a continuous transfer function is actually $C^{K-\epsilon}$ for any $\epsilon > 0$. More generally, in [127] it is proved that a continuous transfer map between two cohomologous center-bunched C^K cocycles is also $C^{K-\epsilon}$ for any $\epsilon > 0$ if it reaches a level of Hölder regularity that depends only on the action and the cocycles. This result is proved in the case G is a connected Lie group or the group of diffeomorphisms of a compact manifold. Note that a cocycle cohomologous to identity is actually center bunched.

5.5.1 Cocycles with values in Lie groups

We discuss first the case of Lie group valued cocycles.

Theorem 5.5.1 *Let M be a compact Riemannian manifold, $\alpha : \mathcal{A} \times M \to M$ a C^K Anosov action. Let H be a connected Lie group endowed with a metric d_H that satisfies* (5.2.1). *Let $\beta, \widetilde{\beta} : \mathcal{A} \times M \to H$ be two C^K cocycles that are cohomologous through a continuous transfer map $P : M \to H$, that is,*

$$\beta(n,x) = P(\alpha(n,x))\widetilde{\beta}(n,x)P(x)^{-1}. \tag{5.5.1}$$

Define:

$$\begin{aligned} \lambda_- &= \lim_{n\to\infty} \|D\alpha(n)|_{E^s}\|^{1/n}, \\ \lambda_+ &= \lim_{n\to\infty} \|D\alpha(-n)|_{E^s}\|^{-1/n}, \end{aligned} \tag{5.5.2}$$

and

$$\begin{aligned} \mu_- &= \lim_{n\to\infty} \sup_{x\in M} \|\text{Ad}(\beta(n,x))^{-1}\|^{-1/n}, \\ \mu_+ &= \lim_{n\to\infty} \sup_{x\in M} \|\text{Ad}(\beta(n,x))\|^{1/n}. \end{aligned} \tag{5.5.3}$$

Assume that

$$\lambda_- < \mu_- \leq \mu_+ < \lambda_+, \tag{5.5.4}$$

and let

$$\theta_0 = \max\left\{\frac{\ln \mu_+}{\ln \lambda_+}, \frac{\ln \mu_-}{\ln \lambda_-}\right\}. \tag{5.5.5}$$

If $\alpha_0 = 0$, possible only if $\mu_- = \mu_+ = 1$, and if

$$\begin{aligned} \limsup_{n\to\infty} \sup_{\in M} \|\mathrm{Ad}(\beta(n,x))\| < \infty, \\ \limsup_{n\to\infty} \sup_{\in M} \|\mathrm{Ad}(\beta(n,x)^{-1})\| < \infty, \end{aligned} \tag{5.5.6}$$

then P is $C^{K-\epsilon}$ for any $\epsilon > 0$.

In general, if P is θ-Hölder for some $\theta > \theta_0$, then P is $C^{K-\epsilon}$ for any $\epsilon > 0$.

Proof The inequalities (5.5.2) imply that the stable, respectively unstable, foliations of the Anosov action are contracting, respectively expanding, for any $\lambda > \lambda_-$, respectively $\lambda > \lambda_+^{-1}$. Note that the quantities μ_-, μ_+ can be defined for $\widetilde{\beta}$ as well. Nevertheless, due to (5.5.1) and the continuity of P, they have to coincide with the quantities μ_-, μ_+ defined for β.

Consider the extensions $\alpha_\beta, \alpha_{\widetilde{\beta}}$. The inequalities (5.5.4) are exactly the bunching conditions necessary in Proposition 5.4.4 for the existence of the lifted stable/unstable foliations for $\alpha_\beta, \alpha_{\widetilde{\beta}}$. Note that, since $\beta, \widetilde{\beta}$ are C^K, the Hölder exponent θ that appears in Proposition 5.4.4 is equal to 1.

Define $\mathbf{P} : M \times H \to M \times H$ by $\mathbf{P}(x,h) = (x, P(x)h)$. It follows from (5.5.1) that

$$\alpha_\beta(n) \circ \mathbf{P} = \mathbf{P} \circ \alpha_{\widetilde{\beta}}(n), \quad n \in \mathcal{A}. \tag{5.5.7}$$

We show now that $\theta > \theta_0$ implies that the images of the stable/unstable leaves of $\widetilde{\alpha}$ under $\mathbf{P}$ are stable/unstable leaves of α. We show the proof only for the stable leaves. Denote by $\mathbf{W}^s$, $\widetilde{\mathbf{W}}^s$, the stable foliation of α_β, respectively $\alpha_{\widetilde{\beta}}$. Denote by $\mathbf{W}^u$, $\widetilde{\mathbf{W}}^u$, the unstable foliation. The leaves of $\mathbf{W}^s$, $\widetilde{\mathbf{W}}^s$, are graphs of functions $\{\gamma_x\}_x$, respectively $\{\widetilde{\gamma}_x\}_x$. Since $\beta, \widetilde{\beta}$ are C^K, it follows from Proposition 5.4.6 that γ_x and $\widetilde{\gamma}_x$ are C^K and depend continuously on the parameter $x \in M$.

The inequality $\theta > \theta_0$ implies that there exists a constant ν such that

$$\lambda_-^\theta < \nu^\theta < \mu_-. \tag{5.5.8}$$

Since $\lambda_- < \nu$, it follows from the contracting property of the stable leaves of $\alpha_{\widetilde{\beta}}$, given in Proposition 5.4.4 (v), that for $n \in \mathcal{A}$, $u = (x,h) \in M \times H$ and $z \in \widetilde{\mathbf{W}}^s(u)$ one has

$$\lim_{n\to\infty} \nu^{-n} d_{M\times H}\left(\alpha_{\widetilde{\beta}}(n)(z), \alpha_{\widetilde{\beta}}(n)(u)\right) = 0. \tag{5.5.9}$$

Using now (5.5.7), that $\mathbf{P}$ is θ-Hölder, and (5.5.9) gives

$$\begin{aligned}
&\nu^{-n\theta} d_{M\times H}\left(\alpha_\beta(n)(\mathbf{P}(z)), \alpha_\beta(n)(\mathbf{P}(u))\right)\\
&= \nu^{-n\theta} d_{M\times H}\left(\mathbf{P}(\alpha_{\widetilde{\beta}}(n)(z)), \mathbf{P}(\alpha_{\widetilde{\beta}}(n)(u))\right)\\
&\le C\nu^{-n\theta} d_{M\times H}\left(\alpha_{\widetilde{\beta}}(n)(z), \alpha_{\widetilde{\beta}}(n)(u)\right)^\theta\\
&\le C\left[\nu^{-n} d_{M\times H}\left(\alpha_{\widetilde{\beta}}(n)(z), \alpha_{\widetilde{\beta}}(n)(u)\right)\right]^\theta \to 0 \text{ as } n\to\infty.
\end{aligned} \tag{5.5.10}$$

Now (5.5.10) and (5.5.8), and the characterization of the lifted stable foliation given in Proposition 5.4.4(vii), imply that $\mathbf{P}(z) \in \mathbf{W}^s(\mathbf{P}(u))$. Since $z \in \widetilde{\mathbf{W}}^s(u)$ is arbitrary, and due to the special form of $\mathbf{P}$, one has $\mathbf{P}(\widetilde{\mathbf{W}}^s(u)) = \mathbf{W}^s(\mathbf{P}(u))$, or, equivalently,

$$\mathbf{P}(t, \widetilde{\gamma}_x(t)h) = \gamma_x(t)P(x)h, \text{ for all } t \in W^s(x). \tag{5.5.11}$$

One can apply now the inverse function theorem to the function

$$t \in W^s(x) \to (t, \widetilde{\gamma}_x(t)h) \in \widetilde{\mathbf{W}}^s(u),$$

which is C^K in t and depends continuously on the other parameters due to the regularity properties of γ_x and $\widetilde{\gamma}_x$. Since the inverse function is C^K and depends continuously on the parameters, (5.5.11) implies that $\mathbf{P}$ is C^K along the stable leaves of $\widetilde{\mathbf{W}}^s$ and the derivatives are continuous on M.

A similar reasoning done using the unstable foliation shows that $\mathbf{P}$ is C^K along the stable leaves of $\widetilde{\mathbf{W}}^u$ and the derivatives are continuous on M.

The extensions α_β and $\widetilde{\alpha}_\beta$ have an invariant "vertical" foliation with fibers $\mathbf{W}^c(x, h) := \{x\} \times H, x \in M$. The action of $\mathbf{P}$ restricted to one of these leaves is given by the multiplication by a fixed element of H, so $\mathbf{P}$ is C^K along these leaves, and the derivatives are continuous. If $\mathcal{A} = \mathbb{R}$, α_β and $\widetilde{\alpha}_\beta$ also have an invariant foliation given by the orbits of the flow. It follows from (5.5.1) that $\mathbf{P}$ is C^K along the orbits, and the derivatives are continuous.

Since the vertical foliation and the stable foliation are jointly integrable into a foliation $\mathbf{W}^{cs}$, it follows from Journé's theorem (Theorem 3.3.1) that $\mathbf{P}$ is $C^{K-\epsilon}$ along the leaves of $\mathbf{W}^{cs}$, the with continuous derivatives. Since $\mathbf{W}^u$ and $\mathbf{W}^{cs}$ are transverse, one can apply Journé's theorem again to conclude that $\mathbf{P}$ is $C^{K-\epsilon}$. If $\mathcal{A} = \mathbb{R}$, one more application of Journé's theorem is necessary.

The special form of $\mathbf{P}$ implies then that P is $C^{K-\epsilon}$.

Assume now that $\theta_0 = 0$ and P is only continuous, without any Hölder assumption. Then instead of (5.5.10) we obtain only

$$d_{M\times H}\left(\alpha_{\widetilde{\beta}}(n)(z), \alpha_{\widetilde{\beta}}(n)(u)\right) \to 0 \text{ as } n \to \infty.$$

However, due to our additional assumption (5.5.6), Proposition 5.4.4(vii), still implies that $\mathbf{P}(\widetilde{\mathbf{W}}^s(u)) = \mathbf{W}^s(\mathbf{P}(u))$. □

5.5.2 Cocycles with values in diffeomorphism groups

Theorem 5.5.2 *Let M, N be compact Riemannian manifolds, $\alpha : \mathcal{A} \times M \to M$ a C^K Anosov action. Let $\beta, \widetilde{\beta} : \mathcal{A} \times M \to \mathrm{Diff}^K(N)$ be two C^K cocycles that are cohomologous through a continuous transfer map $P : M \to \mathrm{Diff}^K(N)$, that is,*

$$\beta(n, x) = P(\alpha(n, x)) \circ \widetilde{\beta}(n, x) \circ P(x)^{-1}. \tag{5.5.12}$$

Define similarly to the Lie group case:

$$\begin{aligned} \lambda_- &= \lim_{n\to\infty} \|D\alpha(n)|_{E^s}\|^{1/n}, \\ \lambda_+ &= \lim_{n\to\infty} \|D\alpha(-n)|_{E^s}\|^{-1/n}, \end{aligned} \tag{5.5.13}$$

and

$$\begin{aligned} \mu_- &= \lim_{n\to\infty} \sup_{x\in M} \|D_N\beta(n, x)^{-1}\|^{-1/n}, \\ \mu_+ &= \lim_{n\to\infty} \sup_{x\in M} \|D_N\beta(n, x)\|^{1/n}. \end{aligned} \tag{5.5.14}$$

Assume that

$$\lambda_- < \mu_- \le \mu_+ < \lambda_+, \tag{5.5.15}$$

and let

$$\theta_0 = \max\left\{\frac{\ln \mu_+}{\ln \lambda_+}, \frac{\ln \mu_-}{\ln \lambda_-}\right\}. \tag{5.5.16}$$

If $\alpha_0 = 0$, possible only if $\mu_- = \mu_+ = 1$, and if

$$\begin{aligned} &\limsup_{n\to\infty} \sup_{\in M} \|D_N\beta(n, x)\| < \infty, \\ &\limsup_{n\to\infty} \sup_{\in M} \|D_N\beta(n, x)^{-1}\| < \infty, \end{aligned} \tag{5.5.17}$$

then P is $C^{K-\epsilon}$ for any $\epsilon > 0$.

In general, if P is θ-Hölder for some $\theta > \theta_0$, then P is $C^{K-\epsilon}$ for any $\epsilon > 0$.

Proof The proof is similar to that of Theorem 5.5.1. One only needs to replace Proposition 5.4.4 by Proposition 5.4.9, and Proposition 5.4.6 by Proposition 5.4.11. □

5.5.3 Lack of regularity in the absence of bunching

We show now several modifications of a well-known counter-example due to de la Llave [99] that explains why the regularity results presented above in this section, given in terms of a Hölder assumption on the transfer map, are sharp. The counter-example is used in [99] in order to show that, for $n \geq 4$, the values of the Lyapunov exponents in the periodic points are not a complete set of invariants for differentiable conjugacy of two topologically conjugate Anosov systems. Note that for $n = 2$ the Lyapunov exponents are a complete set of invariants, and if $n = 3$ the problem is still open.

We start with the case of Lie group-valued cocycles.

Theorem 5.5.3 *Let $A \in SL(2,\mathbb{Z})$ be a hyperbolic matrix with eigenvalues $\lambda^{\pm 1}, 0 < \lambda < 1$, which acts on $\mathbb{T}^2$, and let $\beta : \mathbb{T}^2 \to GL(2,\mathbb{R})$ be a constant cocycle over the action induced by A and defined by*

$$B := \beta(x) = \begin{pmatrix} \mu & 0 \\ 0 & 1, \end{pmatrix}$$

where $0 < 1 < \mu$. Let $r := \ln\mu / \ln\lambda^{-1}$.

There are arbitrarily C^∞-small perturbations $\widetilde{\beta} : \mathbb{T}^2 \to GL(2,\mathbb{R})$ of β with the property that for any $\epsilon > 0$ the cocycles β and $\widetilde{\beta}$ are cohomologous to a $C^{r-\epsilon}$ transfer map, but not by a $C^{r+\epsilon}$ transfer map.

Proof Consider the cocycle $\widetilde{\beta}$ and the transfer map P to have the form

$$\widetilde{\beta}(x) := \begin{pmatrix} \mu & \phi(x) \\ 0 & 1 \end{pmatrix}, \quad P(x) := \begin{pmatrix} 1 & \psi(x) \\ 0 & 1 \end{pmatrix}.$$

The cohomological equation $\beta(x) = P(Ax)\widetilde{\beta}(x)P(x)^{-1}$ is equivalent to

$$\mu\psi(x) - \psi(Ax) = \phi(x). \tag{5.5.18}$$

The equation (5.5.18) admits a unique bounded solution which can be found, via a straight forward telescopic sum argument, as a sum of a uniformly convergent series:

$$\psi(x) = \mu^{-1}\sum_{k=0}^{\infty}\mu^{-k}\phi(A^k x). \tag{5.5.19}$$

As long as $\alpha < r$, one can differentiate further in (5.5.19) and still obtain a uniformly convergent series. So ψ is of class C^α as long as $\alpha < r$.

By choosing ϕ to be a trigonometric polynomial, one can find the Fourier series expansion for ψ. Let

$$\phi(x) = \sum_{k\in\mathbb{Z}^2} \hat{\phi}_k e^{2\pi ikx} \text{ and } \psi(x) = \sum_{k\in\mathbb{Z}^2} \hat{\psi}_k e^{2\pi ikx}. \tag{5.5.20}$$

Note that the Fourier coefficients symmetric about the origin have to be equal.

Formula (5.5.18) becomes

$$\mu\hat{\psi}_k - \psi_{Ck} = \hat{\phi}_k, \tag{5.5.21}$$

$k \in \mathbb{Z}^2$ and $C = (A^{-1})^t$. If we choose $\phi(x) = \sin(2\pi x_1)$, a solution of (5.5.21) is given by

$$\begin{aligned} \hat{\psi}_k &= 0, \text{ for all } k \notin \{\pm C^n k_0\}, \\ \hat{\psi}_{\pm C^n k_0} &= 0, \text{ for all } n \geq 1, \\ \psi_{\pm C^n k_0} &= \mu^{-n-1}\hat{\phi}_{\pm k_0}, \text{ for all } n \leq 0, \end{aligned} \tag{5.5.22}$$

where $k_0 = \begin{pmatrix} 0 \\ 1 \end{pmatrix}$. The Fourier series given by (5.5.22) is uniformly convergent, so one has a continuous solution ψ. Nevertheless, if $r < \alpha$, one can show, using estimates for the growth of the Fourier coefficients and results about approximation in the Lipschitz spaces Λ_α, that $\psi \notin \Lambda_\alpha$. See [88] for approximation results in Λ_α. We recall that for r positive integer, Λ_r coincides with the space of C^r differentiable functions.

Since multiplying ϕ by a constant factor changes ψ by the same factor, the perturbation of β can be made as small as desired, while preserving the loss of regularity. □

Remark 5.5.4 Note that the constant r in Theorem 5.5.3 can take a dense set of values in $(0, \infty)$ if we replace the matrices A and B by their integer powers.

We discuss now the case of diffeomorphism groups-valued cocycles.

Theorem 5.5.5 *Let $A \in SL(2,\mathbb{Z})$ be a hyperbolic matrix with eigenvalues $\lambda^{\pm 1}, 0 < \lambda < 1$, which acts on $\mathbb{T}^2$, and let $\beta : \mathbb{T}^2 \to \mathrm{Diff}^{\infty}(\mathbb{T})$ be a constant cocycle over the action induced by A defined by another hyperbolic matrix $\beta(x) := B \in SL(2,\mathbb{Z})$. Let $\mu, 0 < 1 < \mu$, be a real eigenvalue for B. Let $r := \ln\mu/\ln\lambda^{-1}$.*

There are arbitrarily C^∞-small perturbations $\bar{\beta} : \mathbb{T}^2 \to \mathrm{Diff}^{\infty}(\mathbb{T})$ of β with the property that for any $\epsilon > 0$ the cocycles β and $\widetilde{\beta}$ are cohomologous to a $C^{r-\epsilon}$ transfer map, but not by a $C^{r+\epsilon}$ transfer map.

Proof Consider the normalized eigenvectors

$$A\mathbf{v}_- = \lambda \mathbf{v}_-,$$
$$A\mathbf{v}_+ = \lambda^{-1}\mathbf{v}_+,$$
$$B\mathbf{e}_\mu = \mu \mathbf{e}_\mu.$$

Consider a function $\phi : \mathbb{T}^2 \to \mathbb{R}$ and the following actions on $\mathbb{T}^2 \times \mathbb{T}^d$:

$$f(x, y) = (Ax, By),$$
$$\bar{f}(x, y) = (Ax, By + \phi(x)\mathbf{e}_\mu).$$

Note that f is a hyperbolic diffeomorphism, hence, for C^1-small ϕ, $\bar{f}$ is also hyperbolic and from structural stability there is $h \in \mathrm{Homeo}(\mathbb{T}^2 \times \mathbb{T}^d)$ close to identity such that

$$h\bar{f} = fh. \tag{5.5.23}$$

The homeomorphism h is unique among the homeomorphisms which are homotopic to identity. The unique solution of (5.5.23) is given by

$$h(x, y) = (x, y + \psi(x)\mathbf{e}_\mu),$$

where, as in the case of Lie group-valued cocycles, μ is given by the twisted cocycle equation

$$\mu\psi(x) - \psi(Ax) = \phi(x).$$

Moreover, one can show as in the proof of Theorem 5.5.3 that ψ is of class C^α as long as $\alpha < r$. Also, by choosing ϕ to be a convenient trigonometric polynomial, one can arrange that ψ is not $C^{\alpha+\epsilon}$ for any $\epsilon > 0$.

Defining the cocycle $\bar{\beta}$ by

$$\bar{\beta}(x)(y) = By + \phi(x)\mathbf{e}_\mu.$$

Since the conjugacy h is a bundle map, it induces a transfer map $P : M \to \mathrm{Diff}(\mathbb{T}^d)$ between β and $\bar{\beta}$ given by

$$P(x)(y) = y - \psi(x)\mathbf{e}_\mu, \quad x \in \mathbb{T}^2, y \in \mathbb{T}^d.$$

The regularity of P is equal to the regularity of ψ. □

Remark 5.5.6 A geometric way to understand the break in the regularity of P is to arrange for the maps f and $\bar{f}$ that appear in the proof of Theorem 5.5.5 to be partially hyperbolic diffeomorphisms, and then to look at the images of the leaves of the stable and unstable foliations of f under the conjugacy h. In what follows we use the notations introduced in Theorem 5.5.5 and its proof.

Note first that if $\|B^{\pm 1}\| < \lambda^{-1}$, then f and $\bar{f}$ are partially hyperbolic diffeomorphisms. This condition also implies that P is not C^1.

Given a partially hyperbolic diffeomorphism of a compact manifold, one calls *horizontal foliation* a C^1-foliation which is transverse and complementary to the neutral distribution. Any C^1 invariant horizontal foliation has to contain, and hence is spanned by, the stable and unstable foliations. This is because any invariant distribution which is not tangent to the one has to be contained in $E^s \oplus E^u$. Moreover, any C^0 foliation spanned by the stable and unstable foliations of a C^r partially hyperbolic diffeomorphism is actually $C^{r-\epsilon}$, because of Journé's theorem (Theorem 3.3.1). Indeed, if L is a leaf of a foliation F spanned by two C^r foliations F_1 and F_2, then one can choose C^r coordinate charts such that a small open set in L can be seen as the graph of a function $\phi : U \to V$, where U is a small domain close to the span of the tangent spaces of F_1 and F_2. Since ϕ is uniformly C^r along each of the transverse foliations obtained by projecting $F_1 \cap L$ and $F_2 \cap L$ to U, one concludes that ϕ is $C^{r-\epsilon}$.

Since the cocycles β and $\bar{\beta}$ are cohomologous, one obtains an C^0-invariant foliation for $\bar{\beta}$, or more precisely for $\bar{f}$, by taking the image under h of the horizontal invariant foliation of f. The horizontal invariant foliation of f has leaves

$$L(x, y) = \mathbb{T}^2 \times \{y\}, (x, y) \in \mathbb{T}^2 \times \mathbb{T}^d.$$

Its image has as leaves

$$\bar{L}(x, y) = \{(z, (P(z) \circ P^{-1}(x))(y)) | z \in \mathbb{T}^2\}, (x, y) \in \mathbb{T}^2 \times \mathbb{T}^d.$$

Since P is not C^1, there exists $\bar{L}(x, y)$ which is not C^1 either, hence it cannot contain the leaves of $\bar{f}$.

Indeed, this can be viewed explicitly if one computes the $D\bar{f}$ invariant splitting $E^s \oplus E^c \oplus E^u$ of the tangent space $T(\mathbb{T}^2 \times \mathbb{T}^d) \cong T\mathbb{T}^2 \times T\mathbb{T}^d$:

$$E^s(x, y) = \mathbb{R}(\mathbf{v}_-, \rho^s(x)\mathbf{e}_\mu),$$
$$E^u(x, y) = \mathbb{R}(\mathbf{v}_+, \rho^u(x)\mathbf{e}_\mu),$$

where

$$\rho^s(x) = -\mu^{-1}\left[\sum_{k=0}^{\infty}\left(\frac{\lambda}{\mu}\right)^k D\phi_{A^k x}(\mathbf{v}_-)\right],$$

$$\rho^u(x) = \lambda\left[\sum_{k=0}^{\infty}(\lambda\mu)^k D\phi_{A^{-k-1}x}(\mathbf{v}_+)\right].$$

The distribution E^s is transversally C^1. The stable leaves of $\bar{f}$ are

$$W^s(x, y) = \{(x + t\mathbf{v}_-, y + \omega^s_{x,y}(t)\mathbf{e}_\mu | t \in \mathbb{R}\},$$

where

$$\omega^s_{x,y}(t) = -\mu^{-1} \sum_{k=0}^{\infty} \mu^{-k}[\phi(A^k(x + t\mathbf{v}_-)) - \phi(A^k(x))],$$

and the unstable leaves of $\bar{f}$ are

$$W^u(x, y) = \{(x + t\mathbf{v}_+, y + \omega^u_{x,y}(t)\mathbf{e}_\mu)| t \in \mathbb{R}\},$$

where

$$\omega^u_{x,y}(t) = \sum_{k=0}^{\infty} \mu^k[\phi(A^{-k-1}(x + t\mathbf{v}_+)) - \phi(A^{-k-1}(x))].$$

Hence

$$\mathbf{P}(W^u(x, y)) \not\subseteq W^u(\mathbf{P}(x, y)),$$

but

$$\mathbf{P}(W^s(x, y)) \subset W^s(\mathbf{P}(x, y)).$$

Remark 5.5.7 For $H = SL(2, \mathbb{R})$ valued cocycles over an irrational rotation of the circle Krikorian [89] proved $\mathcal{C}^\infty_H$-cocycle rigidity for cocycles homotopic to identity and such that the corresponding extension over the rotation has fibered rotation number Diophantine with respect to the rotation in the base (assuming that the fibered products are uniformly bounded in C^0 topology).

5.6 Parry's general cohomological result for cocycles with compact non-abelian range

In this section we present a version of Parry's cohomological result. We restrict to the case of compact non-abelian range and follow closely the presentation in [137].

Related results can be found in [135] and [158].

Theorem 5.6.1 *Let M be a compact Riemannian manifold, and $f : M \to M$ be a topologically transitive C^1 Anosov diffeomorphism. Let H be a compact connected Lie group. Let $\beta_1, \beta_2 : M \to H$ be α-Hölder cocycles that satisfy the closing conditions:*

$$f^n x = x \text{ for } x \in M \text{ implies } \beta_1(n, x) = \beta_2(n, x). \tag{5.6.1}$$

Then there exists an α-Hölder function P on M such that

$$\beta_1(x) = P(fx)\beta_2(x)P(x)^{-1},$$

that is, the cocycles β_1, β_2 are cohomologous.

Proof Note that H can be endowed with a bi-invariant metric d_H.

Let $z \in M$ be a fixed point for f, which we do not change for the rest of the proof. If there is no fixed point for f, the proof can be easily adjusted for z periodic point, which always exists for f topologically transitive by the closing lemma (Theorem 1.8.20). Let $W^s(z)$, $W^u(z)$ be the stable, respectively unstable, leaf of z, and let $W(z) = W^s(z) \cap W^u(z)$ be the set of homoclinic points associated to z. It is well known that, due to the semiconjugacy between f and a shift of finite type defined by a Markov partition, $W(z)$ is dense in M [67].

Let $i \in \{1, 2\}$. If $\beta_i(z)$ is not equal to Id_H, due to compactness of H we can choose and fix a sequence n_k such that

$$\beta_i(n_k, z) = \beta_i(z)^{n_k} \to Id_H. \tag{5.6.2}$$

For $x \in W^s(z)$ it follows from Proposition 5.4.4 and (5.6.2) that

$$\gamma_i(x) := \lim_{n_k\to\infty} \beta_i(n_k, x)^{-1}\beta_i(n_k, z) = \lim_{n_k\to\infty} \beta_i(n_k, x)^{-1}. \tag{5.6.3}$$

Due to (5.6.3) one can define for $x \in W(z)$

$$\beta_{i,+}(x) := \lim_{n_k\to\infty} \beta_i(f^{n_k}x)\cdots\beta_i(fx). \tag{5.6.4}$$

Similarly one has for $x \in W(z)$

$$\beta_{i,-}(x) := \lim_{n_k\to\infty} \beta_i(f^{-1}x)\cdots\beta_i(f^{-n_k}x). \tag{5.6.5}$$

It is immediate from (5.6.3) that

$$\beta_i(x)\beta_{i,-}(x) = \beta_{i,-}(fx)\beta_i(z)^{-1}, \quad i = 1, 2. \tag{5.6.6}$$

Using now $\beta_1(z) = \beta_2(z)$, which is a consequence of the closing conditions (5.6.1), and (5.6.6), it follows that

$$\beta_1(x)\beta_{1,-}(x)\beta_{2,-}(x)^{-1}\beta_2(x)^{-1} = \beta_{1,-}(fx)\beta_{2,-}(fx)^{-1}. \tag{5.6.7}$$

The following is a candidate for the transfer map:

$$h(x) = \beta_{1,-}(x)\beta_{2,-}(x)^{-1}. \tag{5.6.8}$$

It follows from (5.6.7) and (5.6.8) that

$$\beta_1(x) = h(x)^{-1}\beta_2(x)h(fx), \quad x \in W(z). \tag{5.6.9}$$

As $W(z)$ is dense in M, in order to finish the proof it is enough to show that h is an α-Hölder function on $W(z)$. Let $x, y \in W(z)$, such that $d_M(x, y) < \epsilon, \epsilon > 0$, and let w be the homoclinic point associated with $\{f^{-n}(x)\}$ and $\{f^n(y)\}, n \geq 0$. Then

$$d_H(h(x), h(y)) \leq d_H(h(x), h(w)) + d_H(h(w), h(y)). \tag{5.6.10}$$

One has, using the bi-invariance of d_H:

$$\begin{aligned} &d_H(h(x), h(w)) \\ &\quad = d_H(\beta_{1,-}(x)\beta_{2,-}(x)^{-1}, \beta_{1,-}(w)\beta_{2,-}(w)^{-1}) \\ &\quad \leq d_H(\beta_{1,-}(x)\beta_{2,-}(x)^{-1}, \beta_{1,-}(w)\beta_{2,-}(x)^{-1}) \\ &\quad + d_H(\beta_{1,-}(w)\beta_{2,-}(x)^{-1}, \beta_{1,-}(w)\beta_{2,-}(w)^{-1}) \\ &\quad = d_H(\beta_{1,-}(x), \beta_{1,-}(w)) + d_H(\beta_{2,-}(x), \beta_{2,-}(w)). \end{aligned}$$

Moreover, approximating the homoclinic orbits corresponding to the trajectories below by periodic orbits (Theorem 1.8.20) and using the closing conditions (5.6.1) one has

$$\begin{aligned} \beta_{1,+}(w)\beta_1(w)\beta_{1,-}(w) &= \beta_{2,+}(w)\beta_2(w)\beta_{2,-}(w), \\ \beta_{1,+}(y)\beta_1(y)\beta_{1,-}(y) &= \beta_{2,+}(y)\beta_2(y)\beta_{2,-}(y), \end{aligned}$$

which gives

$$\begin{aligned} &d_H(h(w), h(y)) \\ &\quad = d_H(\beta_{1,-}(w)\beta_{2,-}(w)^{-1}, \beta_{1,-}(y)\beta_{2,-}(y)^{-1}) \\ &\quad = d_H(\beta_1(w)^{-1}\beta_{1,+}^{-1}(w)\beta_{2,+}(w)\beta_2(w), \beta_1(y)^{-1}\beta_{1,+}^{-1}(y)\beta_{2,+}(y)\beta_2(y)) \\ &\quad = d_H(\beta_{2,+}(w)\beta_2(w), \beta_{2,+}(y)\beta_2(y)) + d_H(\beta_{1,+}(w)\beta_1(w), \beta_{1,+}(y)\beta_1(y)). \end{aligned}$$

Using the estimations above, one has

$$\begin{aligned} &d_H(h(x), h(y)) \\ &\quad = \sum_{n=-\infty}^{-1} \big(d_H(\beta_2(f^n x), \beta_2(f^n w)) + d_H(\beta_1(f^n x), \beta_1(f^n w))\big) \\ &\quad + \sum_{n=0}^{\infty} \big(d_H(\beta_2(f^n w), \beta_2(f^n y)) + d_H(\beta_1(f^n w), \beta_1(f^n y))\big) \\ &\quad \leq \sum_{n=-\infty}^{-1} (\|\beta_2\|_\alpha + \|\beta_1\|_\alpha)\, d_M(f^n x, f^n w)^\alpha \end{aligned}$$

$$+\sum_{n=0}^{\infty}(\|\beta_2\|_\alpha+\|\beta_1\|_\alpha)\,d_M(f^n y, f^n w)^\alpha$$
$$\leq C(\alpha)d_M(x,y),$$

where for the last inequality follows by using $x, w \in W(z)$ and $y, w \in W(z)$ and bounding the infinite sums by converging geometric series. Thus h is α-Hölder on $W(z)$. □

5.7 Lift of regularity for the transfer map from measurable to Hölder

In this section we present a measurable version of Livshitz's cohomological result for Lie group-valued cocycles. We follow closely [122]. A function defined on a measure space with values in a Lie group is called *essentially bounded* if it is bounded on a set of full measure.

Theorem 5.7.1 *Let M be a compact Riemannian manifold, and $f : M \to M$ be a C^2 Anosov diffeomorphism. Let μ be an ergodic invariant volume. Let H be a connected Lie group endowed with a metric d_H that satisfies* (5.2.1). *Assume that $\beta : M \to H$ is α-Hölder, and there is a μ measurable function $P : M \to H$ such that P and P^{-1} are essentially bounded and a.e.*

$$\beta(x) = P(fx)P(x)^{-1}. \tag{5.7.1}$$

Then there is an α-Hölder function $P' : M \to H$ such that $P' = P$ μ-a.e.

Proof Let $S \subset M$ be set of full measure on which both P and P^{-1} are bounded by a constant $K > 0$. Replacing S by $\cap_{n\in\mathbb{Z}} f^{-n}(S)$ and using (5.7.1), one can assume that S is f-invariant.

From (5.7.1) follows that $\beta(n, x)$ is bounded by K^2 on S. In particular, there is a constant $B > 0$ such that

$$\begin{aligned} d_H(\beta(n,x)h, \beta(n,x)h') &\leq \|\mathrm{Ad}(\beta(n,x))\| d_H(h,h') \\ &\leq B d_H(h,h'), \end{aligned} \tag{5.7.2}$$

for all $x \in S, h, h' \in H$.

The ergodic measure μ is a product measure, that is, for μ a.e. $x \in M$, in a small neighborhood of x, the measure μ is locally equivalent to a product measure $\mu_x^s \times \mu_x^u$, where μ_x^s and μ_x^u are the conditional measures of μ along the local stable and unstable leaves $W^s_{\mathrm{loc}}(x)$, respectively $W^u_{\mathrm{loc}}(x)$.

By Luzin's theorem one can choose a set $U \subset S, \mu(U) > 1/2$, such that P and P^{-1} are uniformly continuous on U. Since f is ergodic with respect to μ, Birkhoff ergodic theorem applied for the characteristic function of U implies that

$$\lim_{n\to\infty} \frac{1}{n}\mathrm{Card}\{i \mid f^i(x) \in U, 0 \le i \le n-1\} = \mu(U) > \frac{1}{2}. \tag{5.7.3}$$

Let $x \in S$ be a point for which (5.7.1) is true. Now from Fubini's theorem it follows that for μ_x^s a.e. $y \in W^s_{\mathrm{loc}}(x)$ equation (5.7.1) holds for x and y. For such a point y one has:

$$d_M(f^n(x), f^n(y)) \le C\lambda^n d_M(x, y). \tag{5.7.4}$$

Then

$$\begin{aligned} d_H(P(x), P(y)) &= d_H(\beta(n,x)P(f^n x), \beta(n,y)P(f^n y)) \\ &\le d_H(\beta(n,x)P(f^n x), \beta(n,x)P(f^n y)) \\ &\quad + d_H(\beta(n,x)P(f^n y), \beta(n,y)P(f^n y)) \\ &\le B d_H(P(f^n x), P(f^n y)) + d_H(\beta(n,x), \beta(n,y)). \end{aligned} \tag{5.7.5}$$

Now:

$$\begin{aligned} & d_H(\beta(n,x), \beta(n,y)) \\ &= d_H(\beta(f^{n-1}x)\cdots\beta(fx)\beta(x), \beta(f^{n-1}y)\cdots\beta(fy)\beta(y)) \\ &\le \sum_{k=0}^{n-1} d_H(\beta(f^{n-1}x)\cdots\beta(f^{k+1}x)\beta(f^k x)\beta(f^{k-1}y)\cdots\beta(x), \\ &\quad \beta(f^{n-1}x)\cdots\beta(f^{k+1}x)\beta(f^k y)\beta(f^{k-1}y)\cdots\beta(x)) \\ &\le D\sum_{k=0}^{n-1} d_H(\beta(f^k x), \beta(f^k y)) \le D\sum_{k=0}^{n-1}\lambda^k d_M(x,y)^\theta \\ &\le C_1 d_M(x,y)^\theta, \end{aligned} \tag{5.7.6}$$

where C_1 is independent of x, y.

Now from Fubini's theorem follows that for μ a.e. $x \in M$ and for μ_x^s a.e. $y \in W^s_{\mathrm{loc}}(x)$ equation (5.7.3) holds for x and y. Hence for μ a.e. $x \in M$ and for μ_x^s a.e. $y \in W^s_{\mathrm{loc}}(x)$ one can repeatedly apply the pigeon-hole principle and choose a subsequence n_i such that $f^{n_i}(x)$, $f^{n_i}(y)$ belong to the set U. Due to the uniform continuity of $P|_U$, this implies that for μ a.e. $x \in M$ and for μ_x^s a.e. $y \in W^s_{\mathrm{loc}}(x)$ one has $\lim_{n_i\to\infty} d_H(P(f^{n_i}x), P(f^{n_i}y)) = 0$. Taking limit through such a subsequence in (5.7.5), and using (5.7.6), one has $d_H(P(x), P(y)) \le C_1 d_M(x,y)^\theta$. This implies that P is equal a.e. $W^s(x)$ to a Hölder function.

Similar considerations can be done along the unstable foliation. The local product structure for μ implies now that P coincides μ a.e. to a Hölder function (see [67, Proposition 19.1.1]). □

More general, one can study the lift of regularity for the transfer map between two general cohomologous equations. The following example shows that Theorem 5.7.1 cannot be extended to cocycles with general Lie group fiber.

Example 5.7.2 Let M be a compact manifold, $f : M \to M$ C^2 Anosov that has a fixed point $x_0 \in M$, and G the connected solvable Lie group given by

$$G = \left\{ \begin{pmatrix} x_1 & x_2 \\ 0 & 1 \end{pmatrix} \middle| x_1 > 0, x_2 \in \mathbb{R} \right\}.$$

We show there are C^∞ cocycles $\beta, \widetilde{\beta} : M \to G$ such that the cohomological equation

$$\beta(x) = P(fx)\widetilde{\beta}(x)P(x)^{-1}, \ x \in M \tag{5.7.7}$$

has a measurable solution $P : M \to G$, but no continuous solution.

Let

$$\beta = \begin{pmatrix} \beta_1 & \beta_2 \\ 0 & 1 \end{pmatrix}, \widetilde{\beta} = \begin{pmatrix} \widetilde{\beta}_1 & \widetilde{\beta}_2 \\ 0 & 1 \end{pmatrix}, P = \begin{pmatrix} p_1 & p_2 \\ 0 & 1 \end{pmatrix}, \tag{5.7.8}$$

where $\beta_i, \widetilde{\beta}_i, p_i : M \to \mathbb{R}, i = 1, 2$, are functions. Choose now $\beta_i, \widetilde{\beta}_i$ such that:

(i) $\beta_1 = \widetilde{\beta}_1$;
(ii) $\beta_1(x_0) = 1, \beta_2(x_0) \neq \widetilde{\beta}_2(x_0)$;
(iii) $\widetilde{\beta}_2(x) = 0$; and
(iv) for a.e. $x \in M$

$$\lim_{n\to\infty} \left[\beta_1(x)\beta_1(fx)\cdots\beta_1(f^{n-1}x)\right]^{1/n} \to \alpha < 1. \tag{5.7.9}$$

Note that since f is ergodic, the Birkhoff ergodic theorem implies that (iv) is equivalent to $\int_M \ln(\beta_1(x))dx < 0$, which can be easily achieved for a C^∞ map β_1 satisfying also (i).

Condition (i) allows us to take $p_1 = 1$. Then the cohomological equation (5.7.7) becomes

$$\beta_2(x) = P_2(fx) - \beta_1(x)P_2(x). \tag{5.7.10}$$

If (5.7.10) has a continuous solution P_2, then it holds everywhere. By setting $x = x_0$ one has a contradiction to assumption (ii).

It remains to show that (5.7.10) has a measurable solution P_2. For $\epsilon > 0$ define

$$S = \{x \in M | \beta_1(x)\beta_1(fx)\cdots\beta_1(f^{n-1}x) < c(x)(\alpha+\epsilon)^n\}, \qquad (5.7.11)$$

where $c(x)$ is a constant depending only on x. Due to condition (iv), S has full measure. Replacing S by $\cup_{n\in\mathbb{Z}} f^{-n}(S)$, one can assume that S is f-invariant. For $n > 0$ and $x \in S$ define:

$$\delta_n(x) = \beta_2(x)+\beta_1(x)\beta_2(fx)+\cdots+\beta_1(x)\cdots\beta_1(f^{n-1}x)\beta_2(f^n x). \quad (5.7.12)$$

Then

$$\delta_n(x) = \delta_{n-1}(x) + \beta_1(x)\cdots\beta_1(f^{n-1}x)\beta_2(f^n x) \qquad (5.7.13)$$

and

$$\delta_n(x) = \beta_2(x) + \beta_1(x)\delta_{n-1}(fx). \qquad (5.7.14)$$

From (5.7.13) it follows that

$$|\delta_{n+k}(x) - \delta_{n+k-1}(x)| \le \|\beta_2\|_\infty c(x)(\alpha+\epsilon)^{n+k},$$

and

$$\begin{aligned} &|\delta_{n+j}(x) - \delta_{n+j-1}(x)| \\ &\le \sum_{k=1}^{m} |\delta_{n+k}(x) - \delta_{n+k-1}(x)| \\ &\le c(x)(\alpha+\epsilon)^n \|\beta_2\|_\infty \sum_{k=0}^{\infty} (\alpha+\epsilon)^k \\ &\le C(X)(\alpha+\epsilon)^n. \end{aligned}$$

So the sequence $\{\delta_n(x)\}_n$ is Cauchy for any x. Denote the limit $u(x)$. Note that $x \to u(x)$ is a measurable function. Taking the limit in (5.7.14) gives $\beta_2(x) = u(x) - \beta_1(x)u(fx)$, so we obtain a measurable solution of (5.7.10).

5.8 Periodic cycle functionals

In this section we review the theory developed by Katok and Kononenko in [71], which allows us to study cocycles over partially hyperbolic actions that have accessibility property. Our setup is that of cocycles over abelian actions with values in diffeomorphism groups. The statements here have analogs for cocycles with values in Lie groups and the proofs are similar.

In what follows $\mathcal{A} = \mathbb{R}$ or $\mathbb{Z}$, $k \geq 1$ is an integer, M and N are compact manifolds, $\alpha : \mathcal{A}^k \times M \to M, k \geq 1$, is a smooth action on M, and $\beta : \mathcal{A}^k \times M \to \mathrm{Diff}^K(N)$ is a θ-Hölder cocycle over α which is λ-center bunched with respect to S, a compact set of generators for $\mathcal{A}$. All the foliations that appear in this section are assumed to be α-invariant, continuous, and with smooth leaves, and contracting or expanding under the action of certain subflows $\mathcal{A}a, a \in S$. Therefore the construction of the functions γ_x introduced in Section 5.4 can be carried over.

In order to emphasize the dependence of the function γ_x on a certain contracting/expanding foliation W, we introduce the notation γ_x^W.

Definition 5.8.1 Let $\mathcal{F}_1, \ldots, \mathcal{F}_r$ be a family of foliations of M. An ordered set of points

$$(x_1, \ldots, x_l, x_{l+1}), x_i \in M,\ 1 \leq i \leq l+1,$$

is called an $\mathcal{F}_{1,\ldots,r}$-*path* of length l if for every $i = 1, \ldots, l$ there exists $j(i) \in \{1, \ldots, r\}$ such that $x_{i+1} \in \mathcal{F}_{j(i)}(x_i)$. If $x_{l+1} = x_1$, the path is called on $\mathcal{F}_{1,\ldots,r}$-*cycle*.

Definition 5.8.2 Let $\mathcal{F}_1, \ldots, \mathcal{F}_r$ be a family of foliations of M, each $\mathcal{F}_i$ either contracting or expanding under the action of a subflow $\mathcal{A}a_i, a_i \in S$, and $\mathcal{P} = (x_1, \ldots, x_l, x_{l+1})$ an $\mathcal{F}_{1,\ldots,r}$-*path*. We define the *height* of β *over the path* $\mathcal{P}$ to be

$$H(\beta, \mathcal{P}) = \gamma_{x_l}^{\mathcal{F}_{j(l)}}(x_{l+1}) \cdots \gamma_{x_2}^{\mathcal{F}_{j(2)}}(x_3)\gamma_{x_1}^{\mathcal{F}_{j(1)}}(x_2). \tag{5.8.1}$$

It follows from Proposition 5.4.12 that the height $H(\beta, \mathcal{P})$ does not depend on the particular subflows $\mathcal{A}a_i$. A different choice of the flows for which the foliations are still contracting/expanding gives the same height.

The following proposition shows a necessary condition for a cocycle to be cohomologous to a constant cocycle.

Proposition 5.8.3 *Let $\mathcal{F}_1, \ldots, \mathcal{F}_r$ be a family of foliations of M, each $\mathcal{F}_i$ either contracting or expanding under the action of a subflow $\mathcal{A}a_i, a_i \in S$. Assume that the cocycle β is cohomologous to a constant cocycle. Then all the heights of β over $\mathcal{F}_{1,\ldots,r}$-cycles are trivial, that is, equal to Id_N.*

Proof Let $\pi : \mathcal{A}^k \to \mathrm{Diff}^K(N)$ be a homomorphism and $h : M \to \mathrm{Diff}^K(N)$ a transfer map such that:

$$\beta(b, x) = h(bx)\pi(b)h(x)^{-1}, \quad b \in \mathcal{A}^k, x \in M.$$

Let $\mathcal{C} = (x_1, \ldots, x_l, x_{l+1})$, $x_{l+1} = x_1$, be a $\mathcal{F}_{1,\ldots,r}$-cycle. Assume that the foliation $\mathcal{F}_{j(i)}$ is contracting (the proof for expanding is similar) under the action of a subflow $\mathcal{A}a$. Then

$$\begin{aligned}\gamma_{x_i}^{\mathcal{F}_{j(i)}}(x_{i+1}) &= \lim_{t\to\infty} \beta(ta, x_{i+1})^{-1}\beta(ta, x_i)\\ &= \lim_{t\to\infty} h(x_{i+1})\pi(ta)^{-1}h(tax_{i+1})^{-1}h(tax_i)\pi(ta)h(x_i)^{-1}\\ &= h(x_{i+1})h(x_i)^{-1},\end{aligned} \tag{5.8.2}$$

where for the last equality we use the continuity of h and that

$$\lim_{t\to\infty} \mathrm{dist}_M(tax_i, tax_{i+1}) = 0.$$

Thus

$$\begin{aligned}H(\beta, \mathcal{C}) &= \gamma_{x_l}^{\mathcal{F}_{j(l)}}(x_1)\gamma_{x_{l-1}}^{\mathcal{F}_{j(l-1)}}(x_l)\cdots\gamma_{x_2}^{\mathcal{F}_{j(2)}}(x_3)\gamma_{x_1}^{\mathcal{F}_{j(1)}}(x_2)\\ &= h(x_1)h(x_l)^{-1}h(x_l)h(x_{l-1})^{-1}\cdots h(x_3)h(x_2)^{-1}h(x_2)h(x_1)^{-1}\\ &= Id_N.\end{aligned} \tag{5.8.3}$$

□

Another instance when all the heights are trivial appears when we work with only one foliation.

Proposition 5.8.4 *Let $\mathcal{F}$ be a contracting or expanding foliation under the action of a subflow $\mathcal{A}a$. Then the heights of β over all $\mathcal{F}$-cycles are trivial.*

Proof We assume that $\mathcal{F}$ is contracting. Let $\mathcal{C} = (x_1, \ldots, x_l, x_{l+1}), x_{l+1} = x_1$, be a $\mathcal{F}$-cycle. Then

$$\begin{aligned}H(\beta, \mathcal{C}) &= \gamma_{x_l}^{\mathcal{F}}(x_1)\gamma_{x_{l-1}}^{\mathcal{F}}(x_l)\cdots\gamma_{x_2}^{\mathcal{F}}(x_3)\gamma_{x_1}^{\mathcal{F}}(x_2)\\ &= \lim_{t\to\infty} \beta(ta, x_1)^{-1}\beta(ta, x_l)\beta(ta, x_l)^{-1}\beta(ta, x_{l-1})\ldots\beta(ta, x_3)^{-1}\\ &\quad \beta(ta, x_2)\beta(ta, x_2)^{-1}\beta(ta, x_1)\\ &= Id_N.\end{aligned} \tag{5.8.4}$$

□

Under additional assumptions on the family of foliations, the necessary condition presented in Proposition 5.8.3 is also sufficient for the cocycle to be cohomologous to a constant.

Definition 5.8.5 Let $\mathcal{F}_1, \ldots, \mathcal{F}_r$ be a family of foliations of M. The family is called *transitive* if for any $x, y \in M$ there exists $(x, x_2, \ldots, x_l, y)$ an $\mathcal{F}_{1,\ldots,r}$-path joining x and y. The family is called *locally transitive* if there exists an integer $N \geq 1$ such that for any $\epsilon > 0$ there exists $\delta > 0$ such that for any $x \in M$, $y \in B_M(x, \delta)$, there is a $\mathcal{F}_{1,\ldots,r}$-path $(x = x_1, \ldots, x_l = y), l \leq N$, such that $d_{\mathcal{F}_{j(i)}(x_i)}(x_{i+1}, x_i) < \epsilon$ for $i = 1, \ldots, l$ and $j(i) \in \{1, \ldots, r\}$.

Proposition 5.8.6 *Let $\mathcal{F}_1, \dots, \mathcal{F}_r$ be a family of transitive locally transitive foliations, each foliation $\mathcal{F}_i$ either contracting or expanding under the action of a subflow $\mathcal{A}a_i, a_i \in S$. Assume that $H(\beta, \mathcal{C}) = Id_N$ for all cycles $\mathcal{C}$ determined by the family. Then β is cohomologous to a constant cocycle.*

Proof Let $x \in M$ be fixed and $y \in M$ be arbitrary. Since the family of foliations is transitive, there is a $\mathcal{F}_{1,\dots,r}$-path $\mathcal{C}$ connecting x and y. Define the function $h : M \to \mathrm{Diff}^K(N)$ by

$$h(y) = H(\beta, \mathcal{C}). \tag{5.8.5}$$

From $H(\beta, \mathcal{C}) = Id_N$ for all cycles $\mathcal{C}$ it follows that the function h is well defined. Indeed, if $\mathcal{C}'$ is another path connecting x and y, then the concatenation of $\mathcal{C}$, listed from x to y, and $\mathcal{C}'$, listed from y to x, gives a cycle. Thus $H(\beta, \mathcal{C}')^{-1}H(\beta, \mathcal{C}) = Id_N$, and $H(\beta, \mathcal{C}) = H(\beta, \mathcal{C}')$.

Continuity of h is a consequence of the local transitivity of the family of continuous foliations with smooth leaves and the fact that the cocycle β is Hölder.

We verify now that h is a transfer map. Let $a \in \mathcal{A}^k$. Note that if $\mathcal{C} = (x = x_1, \dots, x_l = y)$ is an $\mathcal{F}_{1,\dots,r}$-path connecting x and y, then it follows from the α-invariance of the family of foliations that $a\mathcal{C} = (ax_1, \dots, ax_l)$ is a $\mathcal{F}_{1,\dots,r}$-path connecting ax and ay. Hence

$$\begin{aligned} h(ay) &= H(\beta, a\mathcal{C})h(ax) \\ &= \gamma_{ax_{l-1}}^{\mathcal{F}_{j(l-1)}}(ax_l) \cdots \gamma_{ax_2}^{\mathcal{F}_{j(2)}}(ax_3)\gamma_{ax_1}^{\mathcal{F}_{j(1)}}(ax_2)h(ax) \\ &= \lim_{t\to\infty} \beta(\epsilon_{l-1}ta_{j(l-1)}, ax_l)^{-1}\beta(\epsilon_{l-1}ta_{j(l-1)}, ax_{l-1}) \\ &\qquad \cdots \beta(\epsilon_2 ta_{j(2)}, ax_3)^{-1}\beta(\epsilon_2 ta_{j(2)}, ax_2) \\ &\quad \times \beta(\epsilon_1 ta_{j(1)}, ax_2)^{-1}\beta(\epsilon_1 ta_{j(1)}, ax_1)h(ax_1), \end{aligned} \tag{5.8.6}$$

where $\epsilon_i \in \{\pm 1\}$, depending on the foliation $\mathcal{F}_{j(i)}$ being contracting or expanding.

Observe now that

$$\begin{aligned} &\beta(ta_{j(i)}, ax_{i+1})^{-1}\beta(ta_{j(i)}, ax_i) \\ &\quad = \beta(a, x_{i+1})\beta(ta_{j(i)} + a, x_{i+1})^{-1}\beta(ta_{j(i)} + a, x_i)\beta(a, x_i)^{-1}, \end{aligned} \tag{5.8.7}$$

for $1 \le i \le l - 1$. So (5.8.6) becomes

$$\begin{aligned} h(ay) = \beta(a, x_l) \lim_{t\to\infty} \Big(&\beta(\epsilon_{l-1}ta_{j(l-1)} + a, x_l)^{-1}\beta(\epsilon_{l-1}ta_{j(l-1)} + a, x_{l-1}) \\ &\dots \beta(\epsilon_2 ta_{j(2)} + a, x_3)^{-1}\beta(\epsilon_2 ta_{j(2)} + a, x_2) \end{aligned}$$

$$\times\, \beta(\epsilon_1 t a_{j(1)} + a, x_2)^{-1} \beta(\epsilon_1 t a_{j(1)} + a, x_1)\Big) \\ \times\, \beta(a, x_1)^{-1} h(a x_1). \tag{5.8.8}$$

Note that

$$\begin{aligned} &\lim_{t\to\infty} \beta(\epsilon_{l-1} t a_{j(l-1)} + a, x_l)^{-1} \beta(\epsilon_{l-1} t a_{j(l-1)} + a, x_{l-1}) \\ &\quad = \lim_{t\to\infty} \beta(\epsilon_{l-1} t a_{j(l-1)}, x_l)^{-1} \beta(a, \epsilon_{l-1} t a_{j(l-1)} x_l)^{-1} \\ &\quad\quad \times \beta(\epsilon_{l-1} t a_{j(l-1)}, x_{l-1}) \beta(\epsilon_{l-1} t a_{j(l-1)}, x_{l-1}) \\ &\quad = \lim_{t\to\infty} \beta(\epsilon_{l-1} t a_{j(l-1)}, x_l)^{-1} \beta(\epsilon_{l-1} t a_{j(l-1)}, x_{l-1}), \end{aligned}$$

because β θ-Hölder implies that

$$\lim_{t\to\infty} \beta(a, \epsilon_{l-1} t a_{j(l-1)} x_l)^{-1} \beta(a, \epsilon_{l-1} t a_{j(l-1)} x_{l-1}) = Id_N.$$

Similar identities hold for the other products on the right-hand side of (5.8.8), so (5.8.8) becomes

$$\begin{aligned} h(ay) &= \beta(a, x_l) h(y) \beta(a, x_1)^{-1} h(a x_1) \\ &= \beta(a, y) h(y) \beta(a, x)^{-1} h(ax). \end{aligned} \tag{5.8.9}$$

Define $\pi : \mathcal{A}^k \to \mathrm{Diff}^K(N)$ by

$$\pi(a) = h(ax)^{-1} \beta(a, x). \tag{5.8.10}$$

Note that π is well defined because x is fixed. We show that π is a representation, that is,

$$\pi(a + b) = \pi(a)\pi(b). \tag{5.8.11}$$

Formula (5.8.11) is equivalent to

$$h((a+b)x)^{-1} \beta(a+b, x) = h(a)^{-1} \beta(a, x) h(b)^{-1} \beta(b, x), \tag{5.8.12}$$

which follows immediately from (5.8.9) if we replace y by bx and take into account that $\beta(a+b, x)\beta(b, x)^{-1} = \beta(a, b)$.

We finish the proof by observing that (5.8.9) is equivalent to

$$\beta(a, y) = h(ay)\pi(a)h(y)^{-1}, \tag{5.8.13}$$

that is, β is cohomologous to a constant cocycle. □

Remark 5.8.7 Under better accessibility properties for the foliations, [71] presents Hölder regularity results for h.

5.9 Non-abelian cocycles over TNS actions

In this section we extend to non-abelian cocycles the cohomological results for TNS actions obtained in Section 4.4.3.

Theorem 5.9.1 *Let G a connected Lie group. Let α be a $\mathbb{Z}^k$ linear TNS-action on a torus $\mathbb{T}^d$, and $\beta : \mathbb{Z}^k \times \mathbb{T}^d \to G$ be a small δ-Holder cocycle over α. Then β is cohomologous to a constant cocycle via a Hölder transfer map. Moreover, if α and β are C^∞, then the transfer map is C^∞.*

The smallness of the cocycle means in particular that $\lambda(a)^\delta \mu(a) < 1$ to hold for all $a \in S \cup (-S)$, where S is the set of generators from Definition 4.4.9. Further smallness requirements will be imposed by Lemma 5.9.5.

We observed in Section 5.1 that the smallness condition cannot be eliminated from Theorem 5.9.1 even if the group G is compact. Moreover, the center-bunching condition is not sufficient in order for a cocycle to be cohomologous to a constant one. Nevertheless, center-bunching will allow for a simple description of the cohomology classes.

As usual, we define the extended action $\widetilde{\alpha} : \mathbb{Z}^k \times (\mathbb{T}^d \times G) \to \mathbb{T}^d \times G$ by

$$\widetilde{\alpha}(a)(x, g) = (\alpha(a)x, \beta(a, x)g).$$

The main step in the proof of Theorem 5.9.1 is to construct an $\widetilde{\alpha}$-invariant topological foliation $\mathcal{F}_\beta$ of $\mathbb{T}^d \times G$ with leaves of dimension equal to d. It is here where the TNS property plays a role. Indeed, using Proposition 5.4.4 we start by building foliations of $\mathbb{T}^d \times G$ that projects on the stable and unstable foliations that we have in the base. The TNS property is used to show that these foliations of $\mathbb{T}^d \times G$ are jointly integrable. Using a holonomy argument and the hyperbolicity of the action, we show that the integral foliation has all the leaves closed manifolds, which cover $\mathbb{T}^d$ simply. This fact, and the invariance of the foliation under the action, allow us to find the representation π and the transfer map P.

We recall the results of Proposition 5.4.4. Let a be a partially hyperbolic diffeomorphism of $\mathbb{T}^d$, β a cocycle over a and $\{W(x)\}_{x \in \mathbb{T}^d}$ an a-invariant foliation of $\mathbb{T}^d$ whose leaves are included in the stable foliation of a. Let $\beta(a, \cdot)$ be δ-Hölder and $\lambda(a)$ and $\mu(a)$ as in Definition 5.2.1. We assume that

$$\lambda(a)^\delta \mu(a) < 1,$$

i.e., $\beta(a, \cdot)$ is center bunched. Then, for any $x \in \mathbb{T}^d$, there is a δ-Hölder function $\gamma_x^{a,W} : W(x) \to G$ such that:

(i) $\gamma_x^{a,W}(x) = I$;
(ii) the family of "graphs" $\mathbf{W}(x; g) := \{(t, \gamma_x^{a,W}(t)g) | t \in W(x)\}$, $x \in \mathbb{T}^d$, $g \in G$, gives an $\widetilde{\alpha}(a)$-invariant foliation of $\mathbb{T}^d \times G$;
(iii) if the cocycle β is C^∞ and the foliation $\{W(x)\}$ has smooth leaves varying continuously in the C^∞-topology, then each function $\gamma_x^{a,W}$ is smooth along $W(x)$, with derivatives varying continuously on $\mathbb{T}^d$.

The functions $\gamma_x^{a,W}$ are defined by the formula

$$\gamma_x^{a,W}(t) = \lim_{n\to\infty} \beta(na, t)^{-1}\beta(na, x), t \in W(x), \tag{5.9.1}$$

and depend continuously on the point $x \in \mathbb{T}^d$. Moreover, these are the only functions that are uniformly δ-Hölder on W_{loc} and satisfy conditions (i) and (ii).

In what follows, if the foliation $\{W(x)\}$ is the stable foliation of a, then we denote $\gamma_x^{a,W}$ by γ_x^a. By the last statement of the lemma, the functions $\gamma_x^{a,W}$ are the restrictions of γ_x^a to $W(x)$.

The invariance property of the family $\{\mathbf{W}(x; g)\}_{x,g}$ is equivalent to the relation

$$\beta(a, t)\gamma_x^{a,W}(t) = \gamma_{ax}^{a,W}(at)\beta(a, x),\ t \in W(x). \tag{5.9.2}$$

In the following lemma we prove some properties of γ_x^a.

Lemma 5.9.2 *Let a and b be two commuting diffeomorphisms which generate an abelian group $\mathcal{A}$ in* Diff$^1(\mathbb{T}^d)$*. Let $\beta : \mathcal{A} \times \mathbb{T}^d \to G$ be a δ-Hölder cocycle. Assume that a is Anosov and $\lambda(a)^\delta\mu(a) < 1$.*

(i) If b is Anosov and $\lambda_-(b)^\delta\mu(b) < 1$, then

$$\gamma_x^a|_{W^s(x;a)\cap W^s(x;b)} = \gamma_x^b|_{W^s(x;a)\cap W^s(x;b)};$$

(ii) $\beta(b, t)\gamma_x^a(t) = \gamma_{bx}^a(bt)\beta(b, x)$ for $t \in W^s(x; a)$; and
(iii) $\gamma_{x_1}^a(x_n) = \gamma_{x_{n-1}}^a(x_n)\cdots\gamma_{x_k}^a(x_{k+1})\cdots\gamma_{x_1}^a(x_2)$ for $x_1, x_2, \ldots, x_n \in W^s(x; a)$.

Proof We derive first (ii). Consider the family $\widetilde{\gamma}_x : W^s(x; a) \to G$ given by

$$\widetilde{\gamma}_x(t) := \beta(b, t)^{-1}\gamma_{bx}^a(bt)\beta(b, x)$$

(since b commutes with a, it preserves the stable foliation of a). Clearly $\widetilde{\gamma}_x(x) = I$. We will show that $\widetilde{\gamma}_x$ satisfies (5.9.2) and then the uniqueness part of Lemma 5.4.4 implies that $\widetilde{\gamma}_x = \gamma_x^a$, i.e. (ii).

Indeed, since $ab = ba$, the cocycle equation gives

$$\beta(b, ax)\beta(a, x) = \beta(a, bx)\beta(b, x).$$

Together with 5.9.2, this yields

$$\begin{aligned}
&\beta(a,t)\widetilde{\gamma}_x(t)\\
&= \left[\beta(a,t)\beta(b,t)^{-1}\right]\gamma^a_{bx}(bt)\beta(b,x)\\
&= \beta(b,at)^{-1}\left[\beta(a,bt)\gamma^a_{bx}(bt)\right]\beta(b,x)\\
&= \beta(b,at)^{-1}\gamma^a_{ab(x)}(ab(t))\left[\beta(a,bx)\beta(b,x)\right]\\
&= \left[\beta(b,at)^{-1}\gamma^a_{ab(x)}(ab(t))\beta(b,ax)\right]\beta(a,x)\\
&= \widetilde{\gamma}_{ax}(at)\beta(a,x),
\end{aligned} \tag{5.9.3}$$

as claimed.

To prove (i), notice that γ^a_x satisfies condition (2) (i.e., equation (5.9.2)) in the characterization of γ^b_x: indeed, this is exactly (ii). Therefore, we obtain the equality (i) by applying again the uniqueness part of Lemma 5.4.4 for b, the b-invariant foliation $W := W^s(x;a) \cap W^s(x;b) \subset W^s(x;b)$ and $\gamma^a_x|_W$.

Finally, (iii) follows from formula (5.9.1), the definition of γ^a_x. □

After these preliminaries, we describe the construction of the foliation mentioned at the beginning of this section. More precisely, we will construct a family of plaques. This construction requires only one Anosov diffeomorphism. The TNS condition is used to show that the result is indeed a foliation.

Let a be an Anosov element in S. Let dist be a distance on $\mathbb{T}^d$ induced by a Riemannian metric. Due to the product structure of the stable and unstable foliations of a, the following holds:

(P0) There are $K_0 > 0$ and a size δ_0 of the local foliation such that if $x, y \in \mathbb{T}^d$ and $\mathrm{dist}(x,y) < \delta_0$, then $W^s_{\mathrm{loc}}(x;a) \bigcap W^u_{\mathrm{loc}}(x;a)$ contains a unique point, and its distance to both x and y is at most $K_0\mathrm{dist}(x,y)$.

We want to obtain a continuous $\mathcal{A}$-invariant foliation $\mathcal{F}_\beta$ of $\mathbb{T}^d \times G$. The leaves are determined locally by graphs of functions $\{F_{U,x}\}_{x\in U}$ to be introduced as follows.

Let $U \subset \mathbb{T}^d$ be a open set of diameter less than δ_0; U is foliated by the local (un)stable manifolds of a. By (P0) for any $x \in U$, $W^s_{\mathrm{loc}}(x;a)$ intersects any local unstable manifold foliating U (not necessarily at a point in U). Then the function $F_{U,x} : U \to G$ is defined by: if $z \in U$, let u be the unique point in $W^s_{\mathrm{loc}}(x;a) \cap W^u_{\mathrm{loc}}(z;a)$ and set

$$F_{U,x}(z) := \gamma_u^{-a}(z)\gamma^a_x(u).$$

Note that $F_{U,x}(x) = I$ and $F_{U,x}$ is continuous.

Consider the foliation chart whose plaques (local leaves) are given by the graphs of the functions $F_{U,x}(\cdot)h$ where $h \in G$. The local leaves can be extended to a global foliation if the standard cocycle condition is satisfied by the foliation charts (see [149]). In our case this is equivalent to the following fact: let $U \subset \mathbb{T}^d$ be as above, $x, y \in U$ and $g_1, g_2 \in G$; if the graphs of the functions $F_{U,x} \cdot g_1$ and $F_{U,y} \cdot g_2$ have a common point, then the two functions coincide on U.

In order to prove this it is enough to consider the case when $g_1 = I$ and the common point is the center of one of the plaques. Assume therefore that the common point is $(y, F_{U,x}(y))$; then $g_2 = F_{U,x}(y)$, and one has to show that $F_{U,x}(z) = F_{U,y}(z)F_{U,x}(y)$ for $z \in U$.

Let $z \in U$. Denote $u := W^s_{\mathrm{loc}}(x; a) \cap W^u_{\mathrm{loc}}(z; a)$, $w_1 := W^u_{\mathrm{loc}}(y; a) \cap W^s_{\mathrm{loc}}(x; a)$ and $w_2 := W^s_{\mathrm{loc}}(y; a) \cap W^u_{\mathrm{loc}}(z; a)$ (see Figure 5.1). Using Lemma 5.9.2, (iii),

$$F_{U,x}(z) = \gamma_u^{-a}(z)\gamma_x^a(u) = \gamma_{w_2}^{-a}(z)\gamma_u^{-a}(w_2)\gamma_{w1}^a(u)\gamma_x^a(w_1)$$

and

$$F_{U,x}(y) = \gamma_{w_1}^{-a}(y)\gamma_x^a(w_1), \qquad F_{U,y}(z) = \gamma_{w_2}^{-a}(z)\gamma_x^a(w_2).$$

Hence the identity $F_{U,x}(z) = F_{U,y}(z)F_{U,x}(y)$ is equivalent to

$$\gamma_u^{-a}(w_2)\gamma_{w_1}^a(u) = \gamma_y^a(w_2)\gamma_{w_1}^{-a}(y). \tag{5.9.4}$$

Our goal is to obtain a foliation by integrating the foliations described by Lemma 5.4.4 for a and $-a$ (these can be seen as the stable, respectively

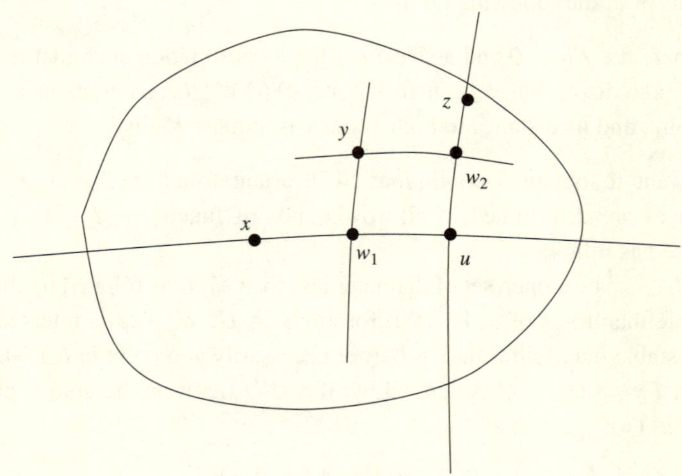

Figure 5.1 Local product structure of stable and unstable foliations.

unstable foliations of $\widetilde{\alpha}(a)$). The functions $F_{U,x}$ describe plaques obtained by stacking the unstable leaves along one stable leaf. The equation (5.9.4) is the standard condition for two foliations to commute, and hence to span together a new foliation.

We postpone the proof of (5.9.4), respectively of the fact that the above construction yields a foliation. See Lemma 5.9.9.

For the rest of this section we assume that equation (5.9.4) holds if $\operatorname{diam}(U)$ is small enough. We denote the obtained foliation by $\mathcal{F}_\beta$.

Once we obtained the foliation, we want to deduce that the cocycle β is constant. The first observation is the following:

Lemma 5.9.3 *Assume that equation* (5.9.4) *holds. Then the foliation $\mathcal{F}_\beta$ is $\mathcal{A}$-invariant and has δ-Hölder local leaves.*

Proof The invariance of $\mathcal{F}_\beta$ is the consequence of the fact that it is obtained by integrating two $\mathcal{A}$-invariant foliations. Indeed, let $b \in \mathcal{A}$. Using Lemma 5.9.2, (ii) for $-a$ and b, respectively a and b, we have

$$\begin{aligned} \widetilde{\alpha}(b)(z, F_{U,x}(z)h) &= (bz, \beta(b, z)\gamma_u^{-a}(z)\gamma_x^a(u)h) \\ &= (bz, \gamma_{bu}^{-a}(bz)\beta(b, u)\gamma_x^a(u)h) \\ &= (bz, \gamma_{bu}^{-a}(bz)\gamma_{bx}^a(bu)\beta(b, x)h) \\ &= (bz, F_{bU,bx}(bz)\beta(b, x)h), \end{aligned} \tag{5.9.5}$$

where $z \in U$, $u = W^s_{\text{loc}}(x; a) \cap W^u_{\text{loc}}(z; a)$ and $h \in G$. Therefore, the local leaves are carried by $\widetilde{\alpha}(b)$ into local leaves, which shows that $\mathcal{F}_\beta$ is $\mathcal{A}$-invariant.

The remaining statement follows from the fact that $F_{U,x}$ is δ-Hölder. To see this, in view of (P0), it is enough to show that $F_{U,x}$ is Hölder when restricted to either $W^s_{\text{loc}}(z; a)$ or $W^u_{\text{loc}}(z; a)$, for any $z \in U$. For the restriction to $W^u_{\text{loc}}(z; a)$ use the definition of $F_{U,x}$ and the fact that γ_u^{-a} is Hölder (see Lemma 5.4.4). For the restriction to $W^s_{\text{loc}}(z; a)$ use the commutation relation (5.9.4) to write $F_{U,x}(z) = \gamma_v^a(z)\gamma_x^{-a}(v)$ where $v := W^u_{\text{loc}}(x; a) \cap W^s_{\text{loc}}(z; a)$ and then apply the same argument. □

The next step is to show that $\mathcal{F}_\beta$ has closed leaves. Moreover, these leaves cover simply $\mathbb{T}^d$ under the projection $\mathbb{T}^d \times G \to \mathbb{T}^d$.

A leaf is a component of $\mathbb{T}^d \times G$ in the leaf topology, i.e., the topology induced by the topology of the local leaves. Pick a point $x_0 \in \mathbb{T}^d$ which is fixed by some hyperbolic element of $\mathcal{A}$. Due to the way the foliation $\mathcal{F}_\beta$ was constructed, it is clear that each leaf is a covering space of $\mathbb{T}^d$. Therefore one can define a group homomorphism $H : \pi_1(\mathbb{T}^d, x_0) \to \operatorname{Maps}(G_{x_0}, G_{x_0})$, where G_{x_0} stands for the fiber over x_0. This map is obtained by associating to a loop

$\gamma \in \Omega(\mathbb{T}^d, x_0)$ and $h \in G_{x_0}$ the endpoint of the lift of γ in $\mathcal{F}_\beta(h)$ starting at h. Since $\mathcal{F}_\beta$ is invariant under right multiplication by G, the range of the above map is actually in $\{\phi : G \to G | \phi(h) = \phi(I)h\} \cong G$. Hence there is a well-defined holonomy map $H : \pi_1(\mathbb{T}^d, x_0) \to G$.

Our next goal is to show that H is the trivial homomorphism, in view of the following lemma.

Lemma 5.9.4 *The cocycle β is cohomologous to a constant cocycle via a δ-Hölder transfer map if and only if equation (5.9.4) holds and the holonomy of the foliation $\mathcal{F}_\beta$ is trivial.*

If the foliation $\mathcal{F}_\beta$ has smooth leaves, then the transfer map is also smooth.

Proof Assume first that β is cohomologous to a constant cocycle via a δ-Hölder transfer map. This gives an invariant Hölder foliation which, by the uniqueness result of Proposition 5.4.4, has to coincide with $\mathcal{F}_\beta$. The statement about the holonomy follows.

For the converse implication, assume that the holonomy H is trivial. Then one can find a global horizontal section $F : \mathbb{T}^d \to G$ of $\mathcal{F}_\beta$, given by a Hölder function. If the leaves of the foliation are smooth, then F will be smooth too. Note that up to right multiplication by appropriate elements of G, $F|_U$ coincides with $F_{U,x}$ for any $x \in U \subset \mathbb{T}^d$. This F will be the transfer map P. The desired conclusion follows from the invariance of $\mathcal{F}_\beta$ under the action $\widetilde{\alpha}$.

Indeed, let $a \in \mathcal{A}$ and $x, y \in \mathbb{T}^d$. Since $(x, F(x))$ and $(y, F(y))$ are in the same leaf of $\mathcal{F}_\beta$, so are their images under $\widetilde{\alpha}(a)$. That is, there is some $t \in G$ such that

$$\widetilde{\alpha}(a)(x, F(x)) = (ax, \beta(a, x)F(x)) = (ax, F(ax)t),$$

and

$$\widetilde{\alpha}(a)(y, F(y)) = (ay, \beta(a, y)F(y)) = (ay, F(ay)t),$$

which shows that

$$F(ax)^{-1}\beta(a, x)F(x) = F(ay)^{-1}\beta(a, y)F(y).$$

Therefore, $\pi : \mathcal{A} \to G$ defined by

$$\pi(a) := F(ax)^{-1}\beta(a, x)F(x)$$

does not depend on x and satisfies $\beta(a, x) = F(ax)\pi(a)F(x)^{-1}$. □

Since $\mathcal{F}_\beta$ is $\widetilde{\alpha}$-invariant, the holonomy map is equivariant in the sense that

$$H(\alpha(a')_*\gamma) = \beta(a', x_0)H(\gamma)\beta(a', x_0)^{-1},$$

for any $\gamma \in \pi_1(\mathbb{T}^d, x_0)$ and $a' \in \mathcal{A}_{x_0} := \{a \in \mathcal{A} | a(x_0) = x_0\}$.

Moreover, it is clear from the construction of the holonomy map and the Hölder estimates on the foliation $\mathcal{F}_\beta$ that by requiring the cocycle β to be close enough to the identity one can obtain that a set of generators of $\pi_1(\mathbb{T}^d)$ be mapped by H into an arbitrarily small neighborhood of the identity in G.

These last two properties of the holonomy map imply that H has to be trivial for β small. We prove this below. Note that $\mathcal{A}_{x_0}$ does not have to be of rank higher than one.

We apply next lemma for $H : \pi_1(\mathbb{T}^d, x_0) \to G$ the holonomy of $\mathcal{F}_\beta$, ρ the action induced by some Anosov element $a \in \mathcal{A}_{x_0}$ on $\pi_1(\mathbb{T}^d, x_0)$ and $\bar{g} = \beta(a, x_0)$. We denote an inner automorphism of a group by $\mathrm{Int}_g : h \mapsto ghg^{-1}$.

Lemma 5.9.5 *Let $a \in \mathrm{Diff}(\mathbb{T}^d)$ be an Anosov diffeomorphism which fixes a point $x_0 \in \mathbb{T}^d$. Fix a set T of generators of $\pi_1(\mathbb{T}^d, x_0)$. Consider the automorphism $\rho \in \mathrm{Aut}(\pi_1(\mathbb{T}^d, x_0))$ induced by a (by the Franks–Manning classification, ρ is hyperbolic).*

Given a finite dimensional Lie group G, there is a neighborhood U of the identity in G with the following property: if $\widetilde{\rho} := \mathrm{Int}_{\bar{g}} \in \mathrm{Aut}(G)$ with $\bar{g} \in U$ and $H : \pi_1(\mathbb{T}^d, x_0) \to G$ is a ρ-$\widetilde{\rho}$ equivariant homomorphism (i.e., $H \circ \rho = \widetilde{\rho} \circ H$) which maps T into U, then H is the trivial homomorphism.

Proof Note that it is enough to prove the conclusion for the canonical set of generators of $\mathbb{Z}^n \cong \pi_1(\mathbb{T}^n)$. We denote it by $T = \{f_i\}_{i=1,\dots,n}$, and let $A = (a_{ij})_{i,j}$ be the hyperbolic matrix $\rho \in \mathrm{Aut}(\pi_1(\mathbb{T}^n)) \cong GL(n, \mathbb{Z})$.

Denote by $\mathcal{G}$ the Lie algebra of G. By Ado's theorem ([141], Lecture 10), we may assume that $\mathcal{G}$ is the Lie algebra of a matrix Lie group to which G is locally isomorphic. Choose a neighborhood $\mathfrak{U}_0 \subset \mathcal{G}$ of O, the origin in $\mathcal{G}$, such that the exponential map $\exp : \mathfrak{U}_0 \subset \mathcal{G} \to G$ is a diffeomorphism onto its image and its inverse, $\log := \exp^{-1} : \exp(\mathfrak{U}_0) \to \mathfrak{U}_0$, admits a power series expansion on $\exp(\mathfrak{U}_0)$. Let $\mathfrak{U}_1 \subset \frac{1}{2}\mathfrak{U}_0$ be a neighborhood of O such that

$$X_i \in \mathfrak{U}_1 \text{ for all } i = 1, \dots, n \implies \sum_{i=1}^{n} a_{ij} X_i \in \mathfrak{U}_0 \text{ for all } j = 1, \dots, n,$$

and set $U := \exp(\mathfrak{U}_1) \cap \{g \in G | \mathrm{spec}(\mathrm{Ad}_g) \cap \mathrm{spec}(A) = \emptyset, \mathrm{Ad}_g(\mathfrak{U}_1) \subset \mathfrak{U}_0\}$, where $\mathrm{Ad}_g \in \mathrm{Aut}(\mathcal{G})$ denotes the differential of Int_g.

Assume now that $H(T) \subset U$ and $\bar{g} \in U$. Let $\mathfrak{f}_i := \log(H(f_i)) \in \mathfrak{U}_1$ and define a linear map $\mathfrak{H} : \mathbb{Z}^n \to \mathcal{G}$ by $\mathfrak{H}(f_i) = \mathfrak{f}_i$. Since log is given by a power series, $\{\mathfrak{f}_i\}_i \subset \mathcal{G}$ is a commutative family, hence $\exp \circ \mathfrak{H} = H$. Using the identity $\mathrm{Int}_g(\exp X) = \exp(\mathrm{Ad}_g(X))$ for $g \in G$, $X \in \mathcal{G}$, the equivariance property of H yields

$$\begin{aligned}\exp(\mathfrak{H}(Af_j)) &= H(Af_j) = \mathrm{Int}_{\bar{g}}(Hf_j) = \mathrm{Int}_{\bar{g}}(\exp(\mathfrak{f}_j)) \\ &= \exp(\mathrm{Ad}_{\bar{g}}(\mathfrak{f}_j)) = \exp(\mathrm{Ad}_{\bar{g}}(\mathfrak{H} f_j)),\end{aligned} \tag{5.9.6}$$

for $j = 1, \ldots, n$. By our choice of U this implies that $\mathfrak{H} \circ A = \mathrm{Ad}_{\bar{g}} \circ \mathfrak{H}$ (note that $\mathfrak{H}(Af_j) = \sum_{i=1}^n a_{ij}\mathfrak{f}_i$). But this is possible only if $\mathfrak{H} = 0$ because $\mathfrak{H}$ intertwines the linear mappings A and $\mathrm{Ad}_{\bar{g}}$ that have disjoint spectra.

Indeed, assume that the linear maps $A \in \mathrm{End}(E)$, $B \in \mathrm{End}(F)$, $C : E \to F$ satisfy $CA = BC$. Consider the induced maps $\hat{A} \in \mathrm{End}(E/\operatorname{Ker} C)$, $\hat{B} \in \mathrm{End}(\operatorname{Im} C)$ and $\hat{C} : E/\operatorname{Ker} C \to \operatorname{Im} C$. If $C \neq 0$ then $\hat{C}$ is invertible, therefore $\hat{C}\hat{A} = \hat{B}\hat{C}$ implies that $\mathrm{spec}(\hat{A}) = \mathrm{spec}(\hat{B})$, and clearly $\mathrm{spec}(\hat{A}) \subset \mathrm{spec}(A)$, $\mathrm{spec}(\hat{B}) \subset \mathrm{spec}(B)$. □

What remains to be proven is the relation (5.9.4), i.e., the existence of the foliation integrating the stable and unstable foliations of $\widetilde{\alpha}(a)$ for some $a \in S$ Anosov.

The distributions E_i are constant. Call the foliations of $\mathbb{T}^d$ corresponding to the E_is *minimal foliations.*

For a torus, any subset of minimal foliations $\mathcal{F}_1, \mathcal{F}_2, \ldots, \mathcal{F}_k$ generates an integrable foliation. If the integrable foliation is $\mathcal{F}$, we write $\mathcal{F} = \{\mathcal{F}_1, \mathcal{F}_2, \ldots, \mathcal{F}_k\}$.

The following lemma is immediate.

Lemma 5.9.6 *Denote by $N(\alpha)$ the number of minimal foliations of the action α. There are constants $K_1 > K_0 > 1$, $\varepsilon_0 > 0$, $\delta_1 > 0$ and a size $\delta_0 > 0$ for the local foliations, $\varepsilon_0 < \delta_0 < \delta_1/N(\alpha)$, such that given two disjoint families $\mathcal{F}_1, \mathcal{F}_2, \ldots, \mathcal{F}_k$ and $\mathcal{G}_1, \mathcal{G}_2, \ldots, \mathcal{G}_l$ of minimal foliations and $\mathcal{F} := \{\mathcal{F}_1, \ldots, \mathcal{F}_k\}$, $\mathcal{G} := \{\mathcal{G}_1, \ldots, \mathcal{G}_l\}$, $\mathcal{H} := \{\mathcal{F}, \mathcal{G}\}$, the following properties hold:*

(P1) For any $x \in \mathbb{T}^d$, $y \in \mathcal{F}^{loc}(x)$, $z \in \mathcal{G}^{\mathrm{loc}}(x)$ such that $\mathrm{dist}(y, z) < \varepsilon_0$, *there is a unique $w := \mathcal{F}^{\mathrm{loc}}(z) \cap \mathcal{G}^{\mathrm{loc}}(y)$, and*

$$\max\{\mathrm{dist}(w, y), \mathrm{dist}(w, z)\} \leq K_1 \mathrm{dist}(y, z).$$

(P2) For any $x, y \in \mathbb{T}^d$ such that $\mathrm{dist}(x, y) < \varepsilon_0$ and $y \in \mathcal{H}^{\mathrm{loc}}(x)$, there is a unique $w := \mathcal{F}^{\mathrm{loc}}(x) \cap \mathcal{G}^{\mathrm{loc}}(y)$, and

$$\max\{\mathrm{dist}(w, x), \mathrm{dist}(w, y)\} \leq K_1 \mathrm{dist}(x, y).$$

(P3) If $x, y \in \mathbb{T}^d$ are such that $y \in \mathcal{H}(x)$, x and y can be joined in $\mathcal{H}(x)$ by a path of length less than δ_1 and $\mathrm{dist}(x, y) < \delta_0$, then $y \in \mathcal{H}^{\mathrm{loc}}(x)$.

In the rest of the proof the size of the local foliations will be the δ_0 given by the above lemma.

Lemma 5.9.7 *There is a constant ε_1, $0 < \varepsilon_1 < \varepsilon_0$ such that the following holds: given any foliation $\mathcal{F} = \{\mathcal{F}_1, \mathcal{F}_2, \ldots, \mathcal{F}_k\}$, where the $\mathcal{F}_i$s are minimal foliations, and $x \in \mathbb{T}^d$, $z \in \mathcal{F}^{\text{loc}}(x)$ with dist$(x, z) < \varepsilon_1$, there exist $y_1 \in \mathcal{F}_1^{\text{loc}}(x)$, $y_2 \in \mathcal{F}_2^{\text{loc}}(y_1), \ldots, y_{k-1} \in \mathcal{F}_{k-1}^{\text{loc}}(y_{k-2})$ such that $z \in \mathcal{F}_k^{\text{loc}}(y_{k-1})$. Moreover,*

$$\max_i \{\text{dist(x, y}_\text{i}\text{), dist(z, y}_\text{i}\text{)}\} \leq K_1 \text{dist(x, z)}. \tag{5.9.7}$$

Proof Let $\varepsilon_1 \leq \varepsilon_0/(K_1^2 + K_1)$.

We construct the points y_i as follows. The families of foliations $\mathcal{F}_1, \mathcal{F}_2, \ldots, \mathcal{F}_i$ and $\mathcal{F}_{i+1}, \mathcal{F}_{i+2}, \ldots, \mathcal{F}_k$ are both integrable; denote

$$\mathcal{G}_i := \{\mathcal{F}_1, \mathcal{F}_2, \ldots, \mathcal{F}_i\} \text{ and } \widetilde{\mathcal{G}}_i := \{\mathcal{F}_{i+1}, \mathcal{F}_{i+2}, \ldots, \mathcal{F}_k\}.$$

Then, by (P2) of Lemma 5.9.6, the local leaves $\mathcal{G}_i^{\text{loc}}(x)$ and $\widetilde{\mathcal{G}}_i^{\text{loc}}(z)$, which are both included in $\mathcal{F}(x)$, intersect in a unique point y_i, which satisfies (5.9.7) as well.

It remains to show that $y_i \in \mathcal{F}_i^{\text{loc}}(y_{i-1})$. By our choice of ε_1 the distance between the points z and y_{i-1} is smaller than ε_0. Apply property (P2) for the foliations $\mathcal{F}_i$ and $\widetilde{\mathcal{G}}_i$, which span $\widetilde{\mathcal{G}}_{i-1}(z)$, and $y_{i-1} \in \widetilde{\mathcal{G}}_{i-1}^{\text{loc}}(z)$. It follows that $\mathcal{F}_i^{\text{loc}}(y_{i-1})$ and $\widetilde{\mathcal{G}}_i^{\text{loc}}(z)$ intersect in a unique point ζ. Moreover,

$$\begin{aligned} \text{dist}(\zeta, x) &\leq \text{dist}(\zeta, y_{i-1}) + \text{dist}(y_{i-1}, x) \\ &\leq K_1 \text{dist}(y_{i-1}, z) + K_1 \text{dist}(x, z) \\ &\leq (K_1^2 + K_1) \text{dist}(x, z) \\ &\leq (K_1^2 + K_1)\varepsilon_1 \leq \varepsilon_0 < \delta_0. \end{aligned} \tag{5.9.8}$$

Since $\zeta \in \mathcal{F}_i^{\text{loc}}(y_{i-1})$ and $y_{i-1} \in \mathcal{G}_{i-1}^{\text{loc}}(x)$, it follows that there is a path in $\mathcal{G}_i(x)$ between x and ζ of length at most $2\delta_0$, and then (P3) implies that $\zeta \in \mathcal{G}_i^{\text{loc}}(x)$. But $\mathcal{G}_i^{\text{loc}}(x)$ intersects $\widetilde{\mathcal{G}}_i^{\text{loc}}(z)$ in a unique point, y_i. Hence ζ has to coincide with y_i, and therefore $y_i \in \mathcal{F}_i^{\text{loc}}(y_{i-1})$. □

Before we prove the main lemma, let us notice that a commutation formula similar to (5.9.4) automatically holds in some other cases.

Lemma 5.9.8 *Let $a, b, c \in S$ be partially hyperbolic diffeomorphisms. Assume $\mathcal{F}_1$ and $\mathcal{F}_2$ are minimal foliations such that $\mathcal{F}_1 \subset W^s(a) \cap W^s(c)$ and $\mathcal{F}_2 \subset W^s(b) \cap W^s(c)$. Then for any $x \in \mathbb{T}^d$, $y \in \mathcal{F}_1^{\text{loc}}(x)$, $z \in \mathcal{F}_2^{\text{loc}}(x)$ and $w = \mathcal{F}_2^{\text{loc}}(y) \cap \mathcal{F}_1^{\text{loc}}(z)$ we have*

$$\gamma_x^c(w) = \gamma_y^b(w)\gamma_x^a(y) = \gamma_z^a(w)\gamma_x^b(z).$$

Proof Apply first Lemma 5.9.2 (iii) for γ^c and the families of points $\{x, y, w\}$ and $\{x, z, w\}$, and then use Lemma 5.9.2 (i). □

Lemma 5.9.9 *There is a constant ε_2, $0 < \varepsilon_2 < \varepsilon_1$ with the following property: let $a \in S$ be any Anosov diffeomorphism and $y, u \in \mathbb{T}^d$ such that dist$(y, u) < \varepsilon_2$. If $w_1 = W^u_{\rm loc}(y; a) \cap W^s_{\rm loc}(u; a)$ and $w_2 = W^s_{\rm loc}(y; a) \cap W^u_{\rm loc}(u; a)$, then*

$$\gamma_u^{-a}(w_2)\gamma_{w_1}^a(u) = \gamma_y^a(w_2)\gamma_{w_1}^{-a}(y).$$

Proof Let $\{\mathcal{F}_1, \mathcal{F}_2, \dots, \mathcal{F}_k\}$ and $\{\mathcal{G}_1, \mathcal{G}_2, \dots, \mathcal{G}_l\}$ be the two disjoint families of minimal foliations such that $W^s(a) = \{\mathcal{F}_1, \mathcal{F}_2, \dots, \mathcal{F}_k\}$ and $W^u(a) = \{\mathcal{G}_1, \mathcal{G}_2, \dots, \mathcal{G}_l\}$.

Use Lemma 5.9.7 to find $x_{11} = w_1, x_{12}, \dots, x_{1,k+1} = u$ such that

$$x_{12} \in \mathcal{F}_1^{\rm loc}(w_1), x_{13} \in \mathcal{F}_2^{\rm loc}(x_{12}), \dots, x_{1,k+1} \in \mathcal{F}_k^{\rm loc}(x_{1k}).$$

and $x_{21}, \dots, x_{l+1,1} = y$ such that

$$x_{21} \in \mathcal{G}_1^{\rm loc}(x_{11}), x_{31} \in \mathcal{G}_2^{\rm loc}(x_{21}), \dots, x_{l+1,1} \in \mathcal{G}_l^{\rm loc}(x_{l1}).$$

We define recurrently the points x_{ij} for all $1 \le i \le l+1$ and $1 \le j \le k+1$ (Figure 5.2 illustrates the case $l = 2$ and $k = 3$): given the points x_{ij}, $x_{i+1,j} \in \mathcal{G}_i^{\rm loc}(x_{ij})$ and $x_{i,j+1} \in \mathcal{F}_j^{\rm loc}(x_{ij})$, we apply (P1) of Lemma 5.9.6 to define $x_{i+1,j+1} := \mathcal{F}_j^{\rm loc}(x_{i+1,j}) \cap \mathcal{G}_i^{\rm loc}(x_{i,j+1})$. Since there are only a finite number of minimal foliations, by taking ε_2 small enough we can assume that all the points $\{x_{ij}\}$ are in a neighborhood of diameter ε_0 of u and y.

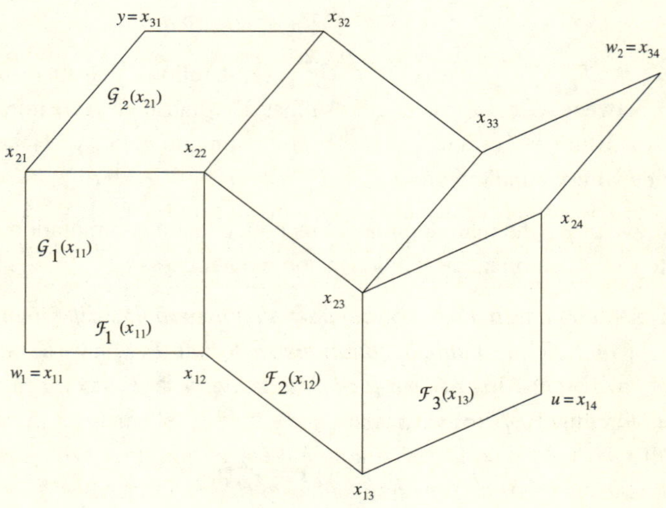

Figure 5.2 Joint integrability of stable and unstable foliations.

We claim that $x_{l+1,k+1} = w_2$. Indeed, the family of local leaves

$$\mathcal{F}_1^{\mathrm{loc}}(x_{l+1,1}), \mathcal{F}_2^{\mathrm{loc}}(x_{l+1,2}), \dots, \mathcal{F}_k^{\mathrm{loc}}(x_{l+1,k})$$

is contained in $W^s(y; a)$, hence there is a path in $W^s(y; a)$ of length less than $N(\alpha)\delta_0$ connecting $y = x_{l+1,1}$ and $x_{l+1,k+1}$. Since $\mathrm{dist}_M(x_{l+1,k+1}, y) < \varepsilon_0 < \delta_0$, property (P3) implies that $x_{l+1,k+1} \in W^s_{\mathrm{loc}}(y; a)$. Similarly, the family of local leaves

$$\mathcal{G}_1^{\mathrm{loc}}(x_{1,k+1}), \mathcal{G}_2^{\mathrm{loc}}(x_{2,k+1}), \dots, \mathcal{G}_l^{\mathrm{loc}}(x_{l,k+1})$$

is contained in $W^u(u; a)$, hence (P3) and $\mathrm{dist}_M(x_{l+1,k+1}, u) < \varepsilon_0 < \delta_0$ implies that $x_{l+1,k+1} \in W^u_{\mathrm{loc}}(u; a)$. But $W^u_{\mathrm{loc}}(u; a)$ and $W^s_{\mathrm{loc}}(y; a)$ have w_2 as the unique point of intersection, which shows that $x_{l+1,k+1}$ coincides with w_2, as claimed.

Since the action is TNS, for each pair of minimal foliations $\mathcal{F}_j$, $\mathcal{G}_i$ there is a partially hyperbolic diffeomorphism whose stable manifold contains both of them. Therefore each quadruple $\{x_{ij}, x_{i+1,j}, x_{i,j+1}, x_{i+1,j+1}\}$ satisfies the hypothesis of Lemma 5.9.8 and we obtain that

$$\gamma^{-a}_{x_{i,j+1}}(x_{i+1,j+1})\gamma^a_{x_{ij}}(x_{i,j+1}) = \gamma^a_{x_{i+1,j}}(x_{i+1,j+1})\gamma^{-a}_{x_{ij}}(x_{i+1,j}). \tag{5.9.9}$$

But equation (5.9.9) implies

$$\begin{aligned} &\gamma^{-a}_{x_{l,k+1}}(x_{l+1,k+1})\gamma^{-a}_{x_{l-1,k+1}}(x_{l,k+1})\cdots\gamma^{-a}_{x_{1,k+1}}(x_{2,k+1}) \\ &\quad\times \gamma^a_{x_{1k}}(x_{1,k+1})\gamma^a_{x_{1,k-1}}(x_{1k})\cdots\gamma^a_{x_{11}}(x_{12}) \\ &= \gamma^a_{x_{l+1,k}}(x_{l+1,k+1})\gamma^a_{x_{l+1,k-1}}(x_{l+1,k})\cdots\gamma^a_{x_{l+1,1}}(x_{l+1,2}) \\ &\quad\times \gamma^{-a}_{x_{l1}}(x_{l+1,1})\gamma^{-a}_{x_{l-1,1}}(x_{l1})\dots\gamma^{-a}_{x_{11}}(x_{21}). \end{aligned} \tag{5.9.10}$$

To see this, define a total order on the set $\{(i, j) | 1 \le i \le l+1,\ 1 \le j \le k+1\}$ of indices by: $(i_1, j_1) \prec (i_2, j_2) \iff$ either $j_1 > j_2$ or $j_1 = j_2$ and $i_1 < i_2$. Now transform the left-hand side of (5.9.10) as follows: for indices ordered increasingly with respect to "$\prec$", at each step substitute the left-hand side of (5.9.9) by its right-hand side.

Finally, Lemma 5.9.2 (iii) shows that formula (5.9.10) is equivalent to (5.9.4). □

5.10 Rigidity of non-abelian cocycles over Cartan actions: *K*-theory approach

In this section we generalize to cocycles with non-abelian range the results in Section 4.4.5. The material in this section appeared in [72].

Definition 5.10.1 Let Γ be a finitely generated discrete group, H a connected Lie group and N a compact manifold. On the set of representations from Γ into H, or on the set of group representations from Γ into $\mathrm{Diff}^K(N)$, one introduces the compact-open topology: two representations $\rho_1, \rho_2 : \Gamma \to H$ (respectively $\rho_1, \rho_2 : \Gamma \to \mathrm{Diff}^K(N)$) are close in this topology if they are close on a finite set of generators of Γ.

Definition 5.10.2 A representation $\rho_0 : \Gamma \to H$ is called locally rigid if for any representation $\rho : \Gamma \to H$ that is close to ρ_0, there exists an element in $h \in H$ that conjugates ρ and ρ_0, that is,

$$\rho(\gamma) = h\rho_0(\gamma)h^{-1}, \quad g \in \Gamma.$$

Definition 5.10.3 Let $k_1, k_2 \geq 0$, be integers. A smooth representation $\rho_0 : \Gamma \to \mathrm{Diff}^\infty(N)$ is called C^{K,k_1,k_2}-locally rigid if for any representation $\rho : \Gamma \to \mathrm{Diff}^K(N)$ that is C^{k_1}-close to ρ_0, there exists a diffeomorphism $h \in C^{k_2}(N)$ that conjugates ρ and ρ_0, that is,

$$\rho(\gamma) = h \circ \rho_0(\gamma) \circ h^{-1}, \quad g \in \Gamma.$$

The following theorem of Fisher–Margulis [35] will be used in the proof of the main result.

Theorem 5.10.4 *Let Γ be a discrete group with property (T). Let M be a compact smooth manifold, and let ρ_0 be a smooth action of Γ on M by Riemannian isometries. Then the action ρ_0 is $C^{K,K,K-\kappa}$ locally rigid for every $\kappa > 0$ for $K > 1$.*

Remark 5.10.5 The previous theorem has a smooth version as well. For our application we need only the C^1 version.

The following theorem of Newman [121] will be used in the proof of the main result. See [10], §9 for a proof.

Theorem 5.10.6 *Let N be a connected topological manifold endowed with a metric. Then there is $\epsilon > 0$ such that any non-trivial action of a finite group on X has an orbit of diameter larger that ϵ.*

Recall that the notions of λ-bunched cocycle with respect to a set of generators were introduced in Definitions 5.2.1 and 5.4.8.

The following theorems are the main results of this section.

Theorem 5.10.7 *Let H be a semisimple real Lie group that satisfies $H^1(H, \mathrm{Lie}(H)) = 0$, or let $H = GL(n, \mathbb{R})$. Let $n \geq 3$, $G = SL(n, \mathbb{K})$, $\Gamma \subset G$ a co-compact torsion free lattice, and $M = G/\Gamma$. Let $\alpha : D_n^+ \times M \to M$ be* *the Cartan action. Let $\beta : D_n^+ \times M \to H$ be a C^K-cocycle that is λ-center bunched with respect to a set of hyperbolic generators with the contraction constants along stable and unstable foliations bounded by λ. Assume that β is close enough to the identity Id_G on the set of generators S.*

Then β is cohomologous to a constant cocycle via a $C^{[K/2-\dim(M)/2]}$ transfer function $h : M \to H$. Moreover, if β is Hölder or smooth, then the transfer function is Hölder, respectively smooth.

Theorem 5.10.8 *Let N be a compact manifold. Let $n \geq 3$, $G = SL(n, \mathbb{K})$, $\Gamma \subset G$ a co-compact torsion free lattice, and $M = G/\Gamma$. Let $\alpha : D_n^+ \times M \to M$ be the Cartan action. Let $\beta : D_n^+ \times M \to \mathrm{Diff}^K(N)$ be a C^K-cocycle that is λ-center bunched with respect to a set of hyperbolic generators with the contraction constants along stable and unstable foliations bounded by λ. Assume that β is close enough to the identity $Id(N)$ on the set of generators S.*

Then β is cohomologous to a constant cocycle via a $C^{[K/2-\dim(M)/2]}$ transfer function $h : M \to \mathrm{Diff}^K(N)$. Moreover, if β is smooth, then the transfer function is smooth.

Remark 5.10.9 (a) Note that it was proved by Weil [174, 175] that

$$H^1(H, \mathrm{Lie}(H)) = 0,$$

for all semisimple Lie groups without compact or three-dimensional factors.

(b) Let $S \subset \mathbb{D}$ be a subspace that contains a two-dimensional subspace in general position, that is, a subspace that intersects each hyperplane given by the equation $t_i = t_j, i \neq j$, along a different line. Using [21], one can show that Theorems 5.10.7 and 5.10.8 hold for abelian actions on M given by $\exp S \subset D_n^+$.

(c) We do not have a result for Hölder cocycles with values in diffeomorphism groups because a counterpart of the Fisher–Margulis result for representations in Hölder homeomorphism groups has not yet been found.

(d) One should compare these results, in particular the regularity for the transfer map, with the Livshitz type results from Section 5.3.2. The loss or regularity there appears in the N direction.

We prove Theorem 5.10.8 in full detail and then explain the changes needed for the proof of Theorem 5.10.7.

Proof of Theorem 5.10.8 Let $F_{i,j}, i \neq j$, be the α-invariant foliations introduced in Section 4.4.5. These foliations are smooth and their brackets generate the whole tangent space. As shown in [12], this facts imply that the system of foliations is locally transitive. Each F_{ij}-path built using these foliations can be described by a product of elements of type $e_{ij}(t)$. Indeed, each piece of an F_{ij}-leaf can be parameterized by $t \to e_{ij}(t)x, t \in I$, for some $x \in G$ and a compact interval I. The path is a cycle if and only if the product of these elements belongs to Γ.

It follows from Proposition 5.8.6 that if the heights $H(\beta, \mathcal{C})$ are equal to Id_N for all cycles $\mathcal{C}$ determined by a family of locally transitive foliations, then the cocycle β is cohomologous to a constant cocycle. Furthermore, it follows from Proposition 5.8.4 that if the cycle $\mathcal{C}$ is included in a stable or unstable leaf then the height $H(\beta, \mathcal{C})$ is equal to $Id(N)$.

The height over a cycle is, so far, dependent of the word in $e_{ij}(t)$s describing the cycle. Changing the word, without changing the value of the product, can produce a different height. We show first that if for a cycle the product of the $e_{ij}(t)$s is equal to identity then the height over such a cycles is trivial. Using the presentation for $SL(n, \mathbb{R})$ from Section 4.4.5, each word in $e_{ij}(t)$s representing the product can be written as a concatenation of conjugates of the basic relations (4.4.22), (4.4.23), and (4.4.24). Each of these relations defines an F_{ij}-cycle.

The relations of type (4.4.22) or (4.4.23) give cycles that are contained in stable leaves for elements of the action α. Indeed, in the case of (4.4.23), the motion along the cycle is described by multiplication by $e_{ij}(t)$, for various values of t, so the cycle is included in the stable leaf of an element $\mathbf{t} \in D_n^+$ with $t_i < t_j$. In the case of (4.4.22), we split the proof into three cases:

(i) If $j \neq k, i \neq l$ then the cycle is contained in the stable leaf of an element $\mathbf{t} \in D_n^+$ with $t_i < t_j, t_k < t_l$.
(ii) If $j = k, i \neq l$ then the cycle is contained in the stable leaf of an element $\mathbf{t} \in D_n^+$ with $t_i < t_j < t_k$.
(iii) If $j \neq k, i = l$ then the cycle is contained in the stable leaf of an element $\mathbf{t} \in D_n^+$ with $t_k < t_l < t_j$.

Consider now relations (4.4.24) for two cases, $\mathbb{K} = \mathbb{R}$ and $\mathbb{K} = \mathbb{C}$, separately.

Assume $\mathbb{K} = \mathbb{R}$.

We fix a small neighborhood $\mathcal{U}$ of the trivial cocycle in which the cocycle β needs to be in order for the argument to work.

Let $\mathcal{V}$ be the neighborhood of Id_N in Homeo(N) that does not contain any map of period 2 (according to Theorem 5.10.6). Let

$$A = \{(1, 1), (-1, 1), (1, -1), (-1, -1)\}.$$

Let $\mathcal{C}(s,t)$ be the cycle given by the product

$$\{t,s\} := h_{12}(t)h_{12}(s)h_{12}(ts)^{-1},$$

for some $s, t \in \mathbb{R}$. Since the functions γ_x depend continuously on the cocycle β, there is a neighborhood $\mathcal{U}$ of the trivial cocycle such that the heights $H(t,s)$ over $\mathcal{C}(t,s)$, $(t,s) \in A$, belongs to $\mathcal{V}$ if β belongs to $\mathcal{U}$.

Recall from Section 4.4.5 that $(t,s) \to \{t,s\}$ is a Steinberg symbol with values in $K_2(\mathbb{R})$. We show that when we vary (t,s) over $\mathbb{R}^* \times \mathbb{R}^*$ the height $H(t,s)$ over $\mathcal{C}(t,s)$ gives a continuous Steinberg symbol with values in an abelian subgroup of $\mathrm{Diff}^K(N)$. For $t_1, t_2, s \in \mathbb{R}^*$ consider the product $\Pi := \{t_1t_2, s\}\{t_2, s\}^{-1}\{t_1, s\}^{-1}$. The Steinberg symbol $\{t,s\}$ is bi-multiplicative in $St_n(\mathbb{R})$, so Π is equal to identity of $St_n(\mathbb{R})$. Using the presentation for $St_n(\mathbb{R})$ from Section 4.4.5, any word representing Π can be written as a concatenation of conjugates of the basic relations (4.4.22) and (4.4.23) (which respectively coincide with (4.4.20) and (4.4.21)). So using the discussion above, the height over the cycle determined by Π is trivial. The height over a concatenation of two cycles is the product of the heights over the cycles. This implies $H(t_1t_2, s) = H(t_1, s)H(t_2, s)$. In a similar way one can show $H(s,t)$ is multiplicative in the second variable. To show that the height satisfies skew-symmetry, consider the relation $\{t, 1-t\}$. This relation is equal to identity in the Steinberg group, because the Steinberg symbol $\{t,s\}$ is skew-symmetric, so any word representing $\{t, 1-t\}$ is a product of conjugates of the standard relations (4.4.22) and (4.4.23). Using the discussion above it follows that the height $H(t, 1-t)$ over the cycle determined by $\{t, 1-t\}$ is trivial. We show that $H(s,t)$ takes values in an abelian group. This follows from the fact that any two symbols $\{t_1, s_1\}$ and $\{t_2, s_2\}$ belong to the abelian group $K_2(\mathbb{R})$, so the following equality $\{t_1, s_1\}\{t_2, s_2\} = \{t_2, s_2\}\{t_1, s_1\}$ holds in $St_n(\mathbb{R})$. As before, the identity in $St_n(\mathbb{R})$ implies $H(t_1, s_1)H(t_2, s_2) = H(t_2, s_2)H(t_1, s_1)$. The continuity of the symbol $H(t,s)$ follows from its definition and from the fact that the abelian group in which the symbol takes values has a Hausdorff topology induced from $\mathrm{Diff}^K(N)$.

So $(t,s) \to H(t,s)$ is a continuous Steinberg symbol. Due to Theorem 4.4.20(b), the only possible values for the height are $Id(N)$ or an element of order 2 in $\mathrm{Diff}^K(N)$.

We show now that if the cocycle β belongs to the neighborhood $\mathcal{U}$ described above, then the height is trivial. The height is continuous in $(s,t) \in \mathbb{R}^* \times \mathbb{R}^*$. We look at each connected component of $\mathbb{R}^* \times \mathbb{R}^*$. Let $(t,s) \in (0,\infty) \times (0,\infty)$. The other cases are similar. Since $(0,\infty) \times (0,\infty)$ is connected, the image of the height is connected. The image belongs to the union of the sets $\{h \in \mathrm{Homeo}(N) | h = Id_N\}$ and $\{h \in \mathrm{Homeo}(N) | h^2 = Id_N, h \neq Id_N\}$, which are both closed. It follows from Theorem 5.10.6 that the sets are disjoint. So the image is included in one of the sets. If the image is included in the first set, we are done. Otherwise, let $(1, 1) \in (0, \infty) \times (0, \infty)$. Since $\beta \in \mathcal{U}$, the height over the cycle $\{1, 1\}$ belongs to $\mathcal{V}$. But $\mathcal{V}$ does not contain any map of period 2, in contradiction to our assumption.

If $\mathbb{K} = \mathbb{C}$ the proof is similar, but simpler, because Theorem 4.4.20(a), implies that the continuous Steinberg symbol is trivial in this case.

After eliminating the contribution to the height that appears due to the relations, the product contains only the elements that conjugate the relations. Cancelations of type $e_{ij}(t)e_{ij}(-t) = Id_N$ do not change the height because the cycle determined by the product $e_{ij}(t)e_{ij}(-t)$ is contained in a stable leaf of an element of the action α, so the height over it has to be trivial.

We consider now the height over an arbitrary cycle, not necessarily with the product of the e_{ij}s trivial. Any cycle induces an element in the first fundamental group $\pi_1(G/\Gamma)$. One has an exact sequence

$$1 \to \pi_1(G) \to \pi_1(G/\Gamma) \to \pi_1(\Gamma) \to 1.$$

It is well known that for $n \geq 3$ one has $\pi_1(SL(n, \mathbb{R})) = \mathbb{Z}_2$ and $\pi_1(SL(n, \mathbb{C})$ is trivial. If $\mathbb{K} = \mathbb{R}$ then the cycles that induce the nontrivial element in $\pi_1(SL(n, \mathbb{R}))$ are homotopic to the cycle determined by the extra relation. See [115].

Two cycles that induce the same element in the fundamental group have the same height. Indeed, if their products are Π_1 and Π_2, then the concatenated product $\Pi_1\Pi_2^{-1}$ gives a word that is equal to identity, so the height over the cycle determined by $\Pi_1\Pi_2^{-1}$ is trivial, so the heights over Π_1 and Π_2 are equal. Since the height over a concatenation of two cycles is the product of the heights over the cycles, the height determines a homomorphism ψ from $\pi_1(G)$ into $\text{Diff}^K(N)$. The above remarks about the fundamental group of G imply that ψ factors to a homomorphisms from Γ into $\text{Diff}^K(N)$.

If the cocycle β is C^K-small on a set of generators S, then ψ is C^K close to identity on a set of generators of Γ. Indeed, this follows from the fact that the functions γ_x used to construct the height are C^K continuous as functions of β and from the fact that the height depends only on the homotopy class of the cycle. Consider now the trivial representation π_0 of Γ into $\text{Diff}^K(N)$ as an isometric action on the smooth manifold N. Then ψ is C^K close to π_0. Note that any co-compact lattice in $SL(n, \mathbb{K}), n \geq 3$, has property (T) [45]. Thus the Fisher–Margulis local rigidity result for isometric actions (Theorem 5.10.4) can be applied, and ψ is $C^{K-\kappa}$ conjugate to π_0. But this implies that ψ coincides with π_0. So all the heights over the cycles are trivial, and the cocycle is cohomologous to a constant cocycle. Note that for this argument we only need the C^1-version of the Fisher–Margulis result, that is we can assume $K = 1$. So this argument works for smooth cocycles even though we do not have a result on smooth dependence of the stable foliation on the cocycle β.

So far, the transfer map $h : M \to \mathrm{Diff}^K(N)$ is only continuous. To show higher regularity for h we employ standard results in rigidity. Look at h as a map $M \times N \to N$. It is standard to show that for any partially hyperbolic element in D_n^+, h is C^K along its stable and unstable directions. See for example [126]. This gives C^K regularity along a finite set of directions, that have the vectors tangent to their distributions, and their length 2 commutators, generating the whole tangent space TM. The commutators needed to consider are of type $[e_{ij}(t), e_{ji}(s)]$. Now Theorem 3.7.2, with $r = 2$, implies that h is $C^{[K/2-\dim(M)/2]}$ in the M direction. The statement about smooth cocycles also follows from Theorem 3.7.2. □

Proof of Theorem 5.10.7 The proof is similar to the proof of Theorem 5.10.8. One starts by constructing the functions γ_x using Proposition 5.4.4. To show that the height over the extra relation is not an element of order 2 in H we use the fact that the cocycle is small on a set of generators. As before, this implies that the height belongs to a small neighborhood of identity in H. But H is a Lie group and consequently does not have small subgroups. So the height is trivial.

When we study the height over general cycles, instead of a homomorphism from $\pi_1(M)$ into $\mathrm{Diff}^K(N)$ we have a homomorphism from Γ into the fiber H. Note that continuity of the functions γ_x, which follows from Proposition 5.4.4, is enough to guarantee the smallness of this representation. This is why we obtain here a Hölder result as well. The Fisher–Margulis result is replaced either by the rigidity result of Andre Weil [174],[175]: a homomorphism π from a finitely generated group Γ to a semisimple Lie group H is locally rigid whenever the cohomology group $H^1(H, Lie(H)) = 0$; or by the result of Margulis [110] that $H^1(\Gamma, V) = 0$ for every homomorphism of Γ to $\mathrm{GL}(V)$, where V is finite dimensional and Γ is a lattice in a higher rank connected semisimple algebraic $\mathbb{R}$-group without compact factors. In our case Γ is finitely generated because it has property (T), and it is also a lattice in $\mathrm{SL}(n, \mathbb{R})$. Since the homomorphism is close to identity on a set of generators it has to be trivial.

The smooth and C^K regularity results for h follows as before. For the Hölder regularity result one can apply the Hölder regularity result from [71]. □

6
Higher order cohomology

6.1 Introduction to higher cohomology of group actions

The notion of cohomology of a group action is a generalization of the notion of group cohomology and it reduces to the latter in the case of a group acting on itself by translations. We will define smooth cohomology of a smooth action of a group G with coefficients in $\mathbb{R}^l$. For our purposes we only need to consider the cases $G = \mathbb{Z}^k$ or $\mathbb{R}^k$.

Definition 6.1.1 Let α be a $\mathbb{Z}^k$-action on a compact differentiable manifold M generated by commuting C^∞ maps $F_1, \ldots, F_k$. Let $1 \leq n \leq k$. A *n-cochain* on M with values in $\mathbb{R}^\ell$ ($\ell \geq 1$) is a function

$$\varphi : M \times (\mathbb{Z}_+^k)^n \to \mathbb{R}^\ell$$

that is multi-linear and skew-symmetric in the last n variables, and C^∞ in the first variable. Since every such function is determined by its coefficients, it can also be viewed as a smooth vector-function

$$\varphi : M \to (\mathbb{R}^\ell)^{\binom{k}{n}},$$

whose components are indexed by $i_1 < \cdots < i_n, \quad i_1, \ldots, i_n \in \{1, \ldots, k\}$. The coboundary operator $\mathcal{D}$ is given by the formula

$$(\mathcal{D}\varphi)_{i_1,\ldots,i_{n+1}}(x) = \sum_{j=1}^{n+1} (-1)^{j+1} \Delta_{i_j} \varphi_{i_1,\ldots,\hat{i_j},\ldots,i_{n+1}}(x),$$

where the operators Δ_i are defined on functions $\psi : M \to \mathbb{R}^\ell$ by

$$\Delta_i \psi = \psi \circ F_i - \psi$$

The cohomology of this cochain complex is called the *smooth n-cohomology* of the action α.

Notice that for a co-chain φ independent of x all operators Δ_i vanish and hence such a co-chain is always a cocycle. We will call those cocycles *constant cocycles*. Coboundary operators commute with the action and the integration with respect to any invariant measure. Hence the integral of a coboundary with respect to any α-invariant measure vanishes and a non-zero constant cocycle is not a coboundary. Thus smooth n-th cohomology group with $\mathbb{R}^l$ coefficients of any $\mathbb{Z}^k$-action always contains a $(\mathbb{R}^\ell)^{\binom{k}{n}}$ component coming from the constant cocycles that actually corresponds to the cohomology of the group $\mathbb{Z}^k$ itself. Notice that integration of a cocycle with respect to an invariant measure produces a constant cocycle.

The smooth cohomology for an action of $\mathbb{R}^k$ is defined similarly. In that case, which is somewhat easier to visualize geometrically, n-cochains are smooth fields of differential n-forms along the orbits of the action, cocycles correspond to the fields of closed forms, and coboundaries are given by the restrictions to the orbit foliation of differentials of smooth globally defined $(n-1)$-forms.

One defines other classes of n-cohomology such ar C^r, Hölder, or analytic similarly.

Let A be $\mathbb{Z}_+^k$, $\mathbb{Z}^k$, or $\mathbb{R}^k$.

Definition 6.1.2 C^∞ (C^r, Hölder, analytic) n-cohomology for an A-action α *trivializes* if any n-cocycle in the corresponding category is cohomologous to a constant cocycle.

Let α be an A-action on M. Let us consider a closed orbit $\mathcal{C}$ of α. Restrictions of cochains, cocycles, and coboundaries to $\mathcal{C}$ form cochains, cocycles, and coboundaries for the transitive action on the orbit. Thus, the cohomology class of the restriction of a smooth cocycle φ to $\mathcal{C}$ is an obvious cohomology invariant of φ. This invariant is conveniently described as follows. The orbit $\mathcal{C}$ carries the unique normalized α-invariant measure $\sigma_{\mathcal{C}}$. Integrating φ with respect to this measure produces a n-cocycle independent on the first variable which determines an element $[\varphi]_{\mathcal{C}} \in H^n(\mathcal{A}; \mathbb{R}^\ell)$.

If n-cohomology of an action trivializes then classes $[\varphi]_{\mathcal{C}}$ are equal for all closed orbits $\mathcal{C}$.

The contrast between rank-one and higher rank results for the first cohomology discussed in Chapter 4 can be explained if one remembers that for a $\mathbb{Z}$ or $\mathbb{R}$ action the first cohomology is also the highest. For the highest cohomology every cochain is a cocycle, and hence there is one-to-one correspondence between cocycles and $\mathbb{R}^\ell$-valued functions on M. In particular, the classes $[\varphi]_{\mathcal{C}}$ are "independent" since for any finite set of closed orbits one can find a C^∞ function with prescribed averages over those orbits. Thus a "true" generalization of the Livshitz theorem should deal with the highest cohomology of an action. Namely:

(i) For classes of hyperbolic and partially hyperbolic algebraic actions of $\mathbb{Z}^k$ and $\mathbb{R}^k$ described in Chapter 2 smooth n-cohomology for $1 \le n \le k-1$ should trivialize; for hyperbolic (Anosov) actions among those the same should hold for C^1 and Hölder cohomology.
(ii) For hyperbolic actions among those, the vanishing of the classes $[\varphi]_C$ for a k-cocycle for all closed orbits should imply that φ is a coboundary; we will call this the *Livshitz property*.

Let us point out an essential difference between the first and higher cohomology. In the former case a regular solution Φ of the coboundary equation $\varphi = \mathcal{D}\Phi$ is unique up to a constant and its existence is equivalent to the coincidence of two distribution solutions Φ^+ and Φ^- which always exist and which are obtained by integrating or summing the values of the cocycle in the positive and negative direction along a one-parameter subgroup. For $n \ge 2$ the solution is not unique and when it exists, there is no canonical procedure for constructing it. In the standard de Rham cohomology theory the issue of non-uniqueness is handled by introducing a Hodge structure. This method does not work in the case of cohomology of ergodic group actions. Roughly speaking, the reason is the following. Normalization given by Hodge theory requires choosing a particular solution on each closed orbit, but those solutions cannot be glued together into a global solution. This is already apparent in the Livshitz case ($n = k = 1$) where the solution is unique up to a constant so the normalization on every closed orbit is uniquely determined once it has been fixed on a single orbit. From a different viewpoint, one may say that Hodge theory requires solving second-order differential equations whereas in the case of cohomology of an action one must stick to the first-order equations since no ellipticity is present which would guarantee solvability of higher-order equations. We will resolve this difficulty for actions by toral automorphism. Assuming vanishing of the classes $[\varphi]_C$ we construct a particular solution of the coboundary equation in the case $n = k \ge 2$ which depends on choosing a point on each orbit of the dual action on the group of characters. The same method will be used in the proof of trivialization of the intermediate cohomology.

Let us note that a positive answer to (ii) in the Weyl chamber flow case would provide a crucial step in the construction of the spanning sets for cusp forms on some locally symmetric spaces of higher rank. Those cusp forms are generalizations of relative Poincaré series associated with closed geodesics [82], which in this case are associated with maximal compact flats.

6.2 Cohomology for partially hyperbolic actions by toral automorphisms

In this section we present a complete description of smooth untwisted cohomology with coefficient in $\mathbb{R}^l$ for $\mathbb{Z}^k$-actions by ergodic toral automorphisms; in agreement with (i) and (ii) above we show trivialization of the smooth cohomology groups except the highest and the Livshitz property for the highest cohomology. Our presentation closely follows [68] and [69].

6.2.1 Formulation of results

Theorem 6.2.1 *Let α be an action of $\mathbb{Z}^k$ by partially hyperbolic automorphisms of $\mathbb{T}^N$, and φ be a C^∞ k-cocycle over α with values in $\mathbb{R}^\ell$ $(\ell \geq 1)$ that vanishes on all periodic orbits of $\mathbb{Z}^k$, i.e., $[\varphi]_C = 0$ for each $C \in \mathcal{P}(\alpha)$. Then for $x \in \mathbb{T}^N$, $t \in (\mathbb{Z}^k)^k$*

$$\varphi(x, t) = \mathcal{D}\Phi(x, t), \tag{6.2.1}$$

where Φ is a C^∞ $(k-1)$-cochain.

Theorem 6.2.2 *Let α be an action of $\mathbb{Z}^k$ by partially hyperbolic automorphisms of $\mathbb{T}^N$, and φ be a C^∞ n-cocycle over α with values in $\mathbb{R}^\ell$ $(\ell \geq 1)$ and $1 \leq n \leq k-1$. Then φ is C^∞-cohomologous to a constant cocycle ψ, i.e., for $x \in \mathbb{T}^N$, $t \in (\mathbb{Z}^k)^n$*

$$\varphi(x, t) = \psi(t) + \mathcal{D}\Phi(x, t), \tag{6.2.2}$$

where Φ is a C^∞ $(n-1)$-cochain.

For $n > 1$ these results were proved in [68] for actions by hyperbolic toral automorphisms, and in [69] for actions by partially hyperbolic toral automorphisms.

6.2.2 Periodic orbits and orbits for the dual action by toral endomorphisms

The following theorem is a generalization of a result of Veech [169] to the case of several commuting toral endomorphisms. It will be used in the proof of Theorem 6.2.1.

Theorem 6.2.3 *Let α be a faithful action of $\mathbb{Z}_+^k$ by partially hyperbolic endomorphisms of $\mathbb{T}^N$. Then the linear combinations of invariant δ-measures concentrated on periodic orbits of the action α are dense in the space of all* *invariant pseudomeasures in the weak-* topology of pseudomeasures, i.e., the dual space to the space of functions on $\mathbb{T}^N$ with absolutely convergent Fourier series.*

Let $\mathbb{A} = \mathbb{A}(\mathbb{T}^N)$ be the space of absolutely convergent Fourier series on the torus $\mathbb{T}^N$

$$f \sim \sum_{n \in \mathbb{Z}^N} \hat{f}(n) e(n \cdot x),$$

where $e(t) = \exp(2\pi i t)$, with ℓ^1-norm $\|f\|_1 = \sum_{n \in \mathbb{Z}^N} |\hat{f}(n)|$. The dual space to $\mathbb{A}$ is a space of distributions on $\mathbb{T}^N$ denoted by $\mathbb{P} = \mathbb{P}(\mathbb{T}^N)$, equipped with the ℓ^∞-norm $\|\mu\|_\infty = \sup_{n \in \mathbb{Z}^N} |\hat{\mu}(n)|$, where $\hat{\mu}(n) = \mu(e(n \cdot x))$. We shall call these distributions *pseudomeasures*.

Any surjective endomorphism $A : \mathbb{T}^N \to \mathbb{T}^N$ is given by a non-singular integral matrix which we will also denote by A; it induces a map $f \to f \circ A$ of $\mathbb{A}$ given by $f(x) \to f(Ax)$. The dual endomorphism $A^* : \mathbb{Z}^N \to \mathbb{Z}^N$ is given by the transpose matrix ${}^tA$. It induces a dual map on the characters:

$$e(n \cdot x) \to e(A^* n \cdot x).$$

In terms of Fourier coefficients A sends

$$f \sim \sum_{n \in \mathbb{Z}^N} \hat{f}(n) e(n \cdot x) \quad \text{to} \quad f \circ A \sim \sum_{m \in \mathbb{Z}^N} (f \hat{\circ} A)(m) e(m \cdot x),$$

where

$$(f \hat{\circ} A)(m) = \begin{cases} \hat{f}(n) & \text{if } m = {}^tAn \text{ for some } n \in \mathbb{Z}^N \\ 0 & \text{otherwise} \end{cases} \tag{6.2.3}$$

This map is not expanding with respect to ℓ^1-norm, and if A is an automorphism, then it is an isometry. Similarly, it induces a mapping $\mu \to A\mu$ of $\mathbb{P}$ which is non-expanding with respect to the ℓ^∞-norm.

Recall that an endomorphism A satisfying these properties as well as the matrix A are called *partially hyperbolic*.

Let α be a faithful action by partially hyperbolic commuting endomorphisms of $\mathbb{T}^N$ given by $N \times N$ integer matrices $A_1, \ldots, A_k$, $k \geq 1$, with determinants $\Delta_1, \ldots, \Delta_k \neq 0$, and β be the dual action on $\mathbb{Z}^N$ by transpose matrices $B_i = {}^tA_i$.

We write

$$\beta^t m^* = B_1^{t_1} B_2^{t_2} \cdots B_k^{t_k} m^*,$$

for $t = (t_1, \ldots, t_k)$ and $m^* \in \mathbb{Z}^N$.

The following definition is useful while considering the non-invertible case.

Definition 6.2.4 A dual semiorbit of a given vector $m^* \in \mathbb{Z}^N$ is a subset of $\mathbb{Z}^N$:

$$\mathcal{O}^+(m^*) = \{m = \beta^t m^*, t \in \mathbb{Z}^k_+\}.$$

A dual orbit of a given vector $m^* \in \mathbb{Z}^N$ is a subset of $\mathbb{Z}^N$:

$$\mathcal{O}(m^*) = \{m \in \mathbb{Z}^N \mid m = \beta^t m^*, t \in \mathbb{Z}^k\}.$$

Dual orbits form an equivalence relation on $\mathbb{Z}^N$.

A pseudomeasure μ is invariant under α if for any $A \in \alpha$, $A\mu = \mu$. For any dual orbit $\mathcal{O}(m^*)$ with the initial point $m^* \in \mathbb{Z}^N$ we construct an α-invariant pseudomeasure $\mu_{\mathcal{O}(m^*)}$ by setting its Fourier coefficients:

$$\hat{\mu}_{\mathcal{O}(m^*)}(m) = \begin{cases} 1, & \text{if } m \in \mathcal{O}(m^*), \\ 0, & \text{otherwise.} \end{cases} \tag{6.2.4}$$

Obviously $\hat{\mu}_{\mathcal{O}(m^*)}$ does not depend on the choice of the initial point m^* and hence can be denoted by $\hat{\mu}_{\mathcal{O}}$. Pseudomeasures $\mu_{\mathcal{O}}$ for different dual orbits $\mathcal{O}$ form a basis in the space of all α-invariant pseudomeasure denoted by $\mathbb{IP}(\alpha)$.

To each periodic (finite) orbit $\mathcal{C}$ of α one associates an α-invariant measure $\sigma_{\mathcal{C}}$ concentrated on that orbit:

$$\sigma_{\mathcal{C}} = \frac{1}{|\mathcal{C}|} \sum_{x \in \mathcal{C}} \delta_x.$$

Let us point out that in the non-invertible case not every point whose α-orbit is finite belongs to a periodic orbit. Let $\mathcal{P}(\alpha)$ be the set of all periodic orbits of $\mathbb{T}^N$. We denote by $< \mathcal{P}(\alpha) >$ the $\mathbb{C}$-linear span of $\{\sigma_{\mathcal{C}} \mid \mathcal{C} \in \mathcal{P}(\alpha)\}$.

Let $q > 0$ be an integer relatively prime to all determinants $\Delta_1, \ldots, \Delta_k$. Notice that B_i are invertible if considered as matrices with entries in $\mathbb{Z}/q\mathbb{Z}$, so $G_q =< B_1, \ldots, B_k >$ as a subgroup of $GL(N, \mathbb{Z}/q\mathbb{Z})$ is finite. Fix $m^* \in \mathbb{Z}^N$ and consider it as an element of $(\mathbb{Z}/q\mathbb{Z})^N$. Then the factor group $G_q/C(m^*)$ where $C(m^*)$ is the stabilizer of m^* in G_q, is finite, and the representatives can be always chosen of the form

$$B_1{}^{\tau_1} \cdots B_k{}^{\tau_k} \quad \text{with} \quad \tau_1, \ldots \tau_k \geq 0. \tag{6.2.5}$$

We pick a fundamental domain for $G_q/C(m^*)$ satisfying (6.2.5) and denote it by $\mathcal{F}$.

Let $\Gamma(q)$ be the set of points in the torus $\mathbb{T}^N$ whose coordinates are rational numbers whose denominators are divisors of q, i.e., the image in $\mathbb{T}^N$ of $q^{-1}\mathbb{Z}^N$

under the canonical projection. Then $\Gamma(q) \subset \mathcal{P}(\alpha)$. We can define now a complex valued measure, $\mu = \mu(m^*, q)$ by

$$\mu = \mu(m^*, q) = q^{-N} \sum_{x \in \Gamma(q)} \Big(\sum_{B_1^{j_1} \dots B_k^{j_k} \in \mathcal{F}} e(-B_1^{j_1} \cdots B_k^{j_k} m^* \cdot x) \Big) \delta_x. \quad (6.2.6)$$

Notice that its Fourier coefficients are given by the formula

$$\hat{\mu}(m) = \begin{cases} 1, & \text{if } m \equiv B_1^{j_1} \cdots B_k^{j_k} m^* \pmod{q}, \\ & \quad \text{for some } B_1^{j_1} \cdots B_k^{j_k} \in \mathcal{F}, \\ 0, & \text{otherwise.} \end{cases} \quad (6.2.7)$$

Since the preudomeasures $\hat{\mu}_{\mathcal{O}(m^*)}$ form a basis in the space of all α-invariant preudomeasures Theorem 6.2.3 immediately follows from the following result.

Theorem 6.2.5 *Let α be an action by commuting partially hyperbolic endomorphisms of $\mathbb{T}^N$ given by $N \times N$ matrices $A_1, \dots, A_k$ with determinants $\Delta_1, \dots, \Delta_k$, and let $m^* \in \mathbb{Z}^N$. If $p^{(n)}$ is the product of the first n primes numbers relatively prime to $\Delta_1, \dots, \Delta_k$, then in the notations of* (6.2.6) *and* (6.2.4),

$$\lim_{n\to\infty} \mu(m^*, p^{(n)}) = \mu_{\mathcal{O}(m^*)}$$

in the weak- topology of* $\mathbb{P}(\mathbb{T}^N)$.

This result is a consequence of the following theorem.

Theorem 6.2.6 *Let $B_1, \dots, B_k \in GL(N, \mathbb{Z})$ be commuting partially hyperbolic matrices with determinants $\Delta_1, \dots, \Delta_k$, $p^{(n)}$ the product of the first n primes numbers relatively prime to $\Delta_1, \dots, \Delta_k$. If $m^*, m \in \mathbb{Z}^N$ and there are k sequences $\{j_i^{(n)}, 1 \le i \le k\}$ of integers such that*

$$B_1^{j_1^{(n)}} \cdots B_k^{j_k^{(n)}} m^* \equiv m \pmod{p^{(n)}}, \quad (6.2.8)$$

then there exists a vector $(j_1^{(0)}, \dots, j_k^{(0)}) \in \mathbb{Z}^k$ such that

$$B_1^{j_1^{(0)}} \cdots B_k^{j_k^{(0)}} m^* = m,$$

and for each $0 < r \in \mathbb{Z}$ and $1 \le i \le k$

$$\lim_{n\to\infty} j_i^{(n)} \equiv j_i^{(0)} \pmod{r}.$$

Proof Following Veech [169] we let $\Lambda(m^*)$ be the smallest subgroup of $\mathbb{Z}^N$ containing the semiorbit $\mathcal{O}^+(m^*)$, $\Omega(m^*)$ be the $\mathbb{Q}$-span of $\Lambda(m^*)$ and

$\Lambda^*(m^*) = \Omega(m^*) \cap \mathbb{Z}^N$. Then $\Lambda^*(m^*)$ is a β-invariant subspace and without loss of generality we may assume that $\Lambda^*(m^*) = \mathbb{Z}^N$ and m^* is a cyclic vector of the action β on $\mathbb{Z}^N$.

The transpose matrices ${}^tB_1, \ldots, {}^tB_k$ commute, hence they have a common eigenvector $\xi = (\xi_1, \ldots, \xi_N)$ such that ${}^tB_i\xi = \lambda_i\xi$ for $1 \leq i \leq k$. Let $K = \mathbb{Q}(\lambda_1)$. Then by the Cramer's rule, all $\xi_i \in K$, and by solving the equation

$$({}^tB_i - \lambda_i)\xi = 0,$$

for $i = 2, \ldots, k$, we obtain that all $\lambda_i \in K$. Moreover, there exists a co-finite set of non-archimedean prime divisors of K, $\mathcal{S}_0$, such that $\xi_j \in \mathcal{O}^*(\mathcal{S}_0)$ for $1 \leq j \leq N$ and $\lambda_i \in \mathcal{O}^*(\mathcal{S}_0)$ for $1 \leq i \leq k$, where $\mathcal{O}^*(\mathcal{S}_0)$ is the group of units of the ring of integers $\mathcal{O}(\mathcal{S}_0)$ in K with respect to the set $\mathcal{S}_0$. (For all necessary references see, e.g., [177, Chapter 5].) Choose a non-singular matrix C having ξ as the last row and other integer entries. Then $C \in GL(N, \mathcal{O}(\mathcal{S}_0))$. Let $\zeta_0 = Cm^*$, $\zeta = Cm$.

Let p be a rational prime and assume that the congruence

$$B_1^{j_1} \cdots B_k^{j_k} m^* \equiv m \pmod{p}$$

admits a solution $j = (j_1, \ldots, j_k) \in \mathbb{Z}^N$. Then for the same j we have

$$CB_1^{j_1} \cdots B_k^{j_k} C^{-1}\zeta_0 \equiv \zeta \pmod{p\mathcal{O}(\mathcal{S}_0)^N}.$$

The last row of the matrix $CB_1^{j_1} \cdots B_k^{j_k} C^{-1}$ will have the form

$$(0, \ldots, 0, \lambda_1^{j_1} \cdots \lambda_k^{j_k}),$$

and we obtain the following congruence:

$$\lambda_1^{j_1} \cdots \lambda_k^{j_k} a \equiv b \pmod{p\mathcal{O}(\mathcal{S}_0)},$$

where a is the Nth component of ζ_0 and b is the Nth component of ζ. Since m^* is a cyclic vector, $a \neq 0$. Condition (6.2.8) implies that there exists a co-finite set of non-archimedean prime divisors of K, $\mathcal{S} \subset \mathcal{S}_0$ such that $a, b \in \mathcal{O}^*(\mathcal{S})$ and

$$\lambda_1^{j_1} \cdots \lambda_k^{j_k} a \equiv b \pmod{\pi(P)}, \tag{6.2.9}$$

admits a solution $j = (j_1, \ldots, j_k)$ for each finite set of $P \in \mathcal{S}$. Since matrices $B_1, \ldots, B_k \in GL(N, \mathbb{Z})$ and have no eigenvalues that are roots of unity, neither of $\lambda_1, \ldots, \lambda_k$ is a root of unity. The rest of the proof verbatim follows the scheme of Veech [169]. It makes use of Chevalley's theorem [15, 169] in order to interpret (6.2.9) as a limit in the pro-finite topology of the group $\mathcal{O}^*(\mathcal{S})$ and the fact that the set

$$\{\lambda_1^{j_1} \cdots \lambda_k^{j_k} a \mid (j_1 \cdots j_k) \in \mathbb{Z}^k\},$$

which is a sublattice in $\mathcal{O}^*(\mathcal{S}) \sim \mathbb{Z}_m \times \mathbb{Z}^s$, is closed in the pro-finite topology on $\mathcal{O}^*(\mathcal{S})$. □

Theorem 6.2.5 can be applied to k-cocycles over a $\mathbb{Z}_k^+$-action α by toral endomorphisms in the following way. A k-cocycle $\varphi(x, t)$ is uniquely determined by a function

$$\varphi(x) : \mathbb{T}^N \to \mathbb{R}^\ell \quad (\ell \geq 1)$$

(see the Introduction for general definitions). Recall that a k-cocycle vanishes on a periodic orbit $\mathcal{C}$ if

$$[\varphi]_{\mathcal{C}} = \int_{\mathbb{T}^N} \varphi d\sigma_{\mathcal{C}} = 0. \tag{6.2.10}$$

Corollary 6.2.7 *Let α be an action by commuting ergodic endomorphisms of $\mathbb{T}^N$ given by $N \times N$ matrices $A_1, \ldots, A_k$ with determinants $\Delta_1, \ldots, \Delta_k$. Let φ be a C^∞ k-cocycle over α with values in $\mathbb{R}^\ell$ ($\ell \geq 1$) that vanishes on all periodic orbits of α. Then, for any dual orbit $\mathcal{O}(m^*)$,*

$$\sum_{m \in \mathcal{O}(m^*)} \hat{\varphi}(m) = 0.$$

Proof Since φ vanishes on all periodic orbits of α, for any $n \geq 1$,

$$\int_{\mathbb{T}^N} \varphi d\mu(m^*, p^{(n)}) = 0,$$

where $\mu(m^*, \cdot)$ was defined in (6.2.6). By Theorem 6.2.5

$$\mu_{\mathcal{O}(m^*)}(\varphi) = 0.$$

But

$$\mu_{\mathcal{O}(m^*)}(\varphi) = \sum_{m \in \mathbb{Z}^N} \hat{\varphi}(m) \hat{\mu}_{\mathcal{O}(m^*)}(m) = \sum_{m \in \mathcal{O}(m^*)} \hat{\varphi}(m).$$

Thus if the condition (6.2.10) holds for any $\mathcal{C} \in \mathcal{P}(\alpha)$, for any dual orbit $\mathcal{O}(m^*)$

$$\sum_{m \in \mathcal{O}(m^*)} \hat{\varphi}(m) = 0.$$

□

In particular, since $\mathcal{O}(0) = 0$ we obtain the following.

Corollary 6.2.8 *Let α be an action by commuting ergodic endomorphisms of $\mathbb{T}^N$ given by $N \times N$ matrices $A_1, \dots, A_k$ with determinants $\Delta_1, \dots, \Delta_k$. Let φ be a C^∞ k-cocycle over α with values in $\mathbb{R}^\ell$ ($\ell \geq 1$) that vanishes on all periodic orbits of α. Then $\hat{\varphi}(0) = \int_{\mathbb{T}^N} \varphi d\lambda = 0$, where λ is the Lebesgue measure on $\mathbb{T}^N$.*

6.2.3 The ℓ^1 n-cohomology of $\mathbb{Z}^k$

Now we proceed to the consideration of the dual cohomology problem which appears in the space of Fourier coefficients. The case of the highest cohomology plays a special role since, together with the first cohomology, it is a base of the induction in the general case. So we begin by discussing the case $n = k$. Let φ be a C^∞ k-cocycle over a $\mathbb{Z}^k_+$-action α by ergodic toral endomorphisms. Let us fix an initial point $m^* \in \mathbb{Z}^N$. The dual orbit $\mathcal{O}(m^*)$ can be identified with a subset of $\mathbb{Z}^k$ via the correspondence

$$m = B_1^{m_1} \cdots B_k^{m_k} m^* \to (m_1, \dots, m_k).$$

If $m^* = 0$ we obtain a trivial orbit. Throughout this section we assume that $m^* \neq 0$. Since all elements are ergodic, $\mathcal{O}(m^*)$ has rank k. Since the original cocycle φ is C^∞, its Fourier coefficients decay *super-polynomially*, i.e.,

$$\forall j \in \mathbb{Z}_+ \quad \exists C(j) \quad \text{such that} \quad \forall m \in \mathbb{Z}^N \quad | \hat{\varphi}(m) | \leq C(j) \|m\|^{-j},$$

where $\| \cdot \|$ is any norm in $\mathbb{R}^N$. Hence on each dual orbit $\mathcal{O}(m^*)$ we have

$$\sum_{m \in \mathcal{O}(m^*)} | \hat{\varphi}(m) | < \infty.$$

If we restrict $\hat{\varphi}$ to $\mathcal{O}(m^*)$, i.e., assign $\hat{\varphi}(m) = 0$ for $m \notin \mathcal{O}(m^*)$ we obtain an (untwisted) ℓ^1 k-cocycle on $\mathbb{Z}^k$ dual to φ,

$$\hat{\varphi} : \mathbb{Z}^k \to \mathbb{R}^\ell, \quad \sum_{m \in \mathbb{Z}^k} | \hat{\varphi}(m) | < \infty.$$

We shall ignore its dependence on m^* for the moment.

A $(k-1)$-cochain Φ over α can be identified with a vector function

$$\Phi : \mathbb{T}^N \to (\mathbb{R}^\ell)^k, \quad \Phi = (\Phi_1, \dots, \Phi_k).$$

The coboundary operator is given by the formula

$$\mathcal{D}\Phi = \sum_{i=1}^{k} (-1)^{i+1} \Delta_i \Phi_i,$$

where

$$\Delta_i \varphi = \varphi \circ A_i - \varphi. \tag{6.2.11}$$

According to the formula (6.2.3) A_i acts on Fourier coefficients corresponding to the points of the dual orbit exactly the way the ith left coordinate shift

$$\sigma_i(m_1, \ldots, m_k) = (m_1, \ldots, m_{i-1}, m_i - 1, m_{i+1}, \ldots, m_k).$$

acts on the dual cocycle $\hat{\varphi}$:

$$(\varphi \hat{\circ} A_i) = \hat{\varphi} \circ \sigma_i.$$

Consequently, if we identify a dual $(k-1)$-cochain with a vector-function

$$\hat{\Phi} : \mathbb{Z}^k \to (\mathbb{R}^\ell)^k, \quad \hat{\Phi} = (\hat{\Phi}_1, \ldots, \hat{\Phi}_k),$$

the coboundary operator for dual cochains will be given by the formula

$$\mathcal{D}\hat{\Phi} = \sum_{i=1}^{k} (-1)^{i+1} \Delta_i \hat{\Phi}_i,$$

where

$$\Delta_i \hat{\varphi} = \hat{\varphi} \circ \sigma_i - \hat{\varphi}. \tag{6.2.12}$$

We shall discuss here elementary properties of untwisted ℓ^1-k-cocycles on $\mathbb{Z}^k$. Now let

$$(m_1, \ldots, m_k) = m \in \mathbb{Z}^k.$$

If $\hat{\varphi}$ is a ℓ^1-k-cocycle on $\mathbb{Z}^k$, it vanishes at ∞, i.e., $|\ \hat{\varphi}(m)\ | \to 0$ as $\|m\| \to \infty$.

We use the following notations:

For $\hat{\varphi} \in \ell^1(\mathbb{Z}^k)$ let

$$\bar{\hat{\varphi}} = \sum_{m \in \mathbb{Z}^k} \hat{\varphi}(m).$$

For $\hat{\varphi} \in \ell^1(\mathbb{Z}^k), \quad i = 1, \ldots, k$

$$(\Sigma_i \hat{\varphi})(m_1, \ldots, m_k) = \sum_{j=-\infty}^{\infty} \hat{\varphi}(m_1, \ldots, m_{i-1}, j, m_{i+1}, \ldots, m_k),$$

$$(\Sigma_i^- \hat{\varphi})(m_1, \ldots, m_k) = \sum_{j=-\infty}^{m_i - 1} \hat{\varphi}(m_1, \ldots, m_{i-1}, j, m_{i+1}, \ldots, m_k),$$

$$(\Sigma_i^+ \hat{\varphi})(m_1, \ldots, m_k) = -\sum_{j=m_i}^{\infty} \hat{\varphi}(m_1, \ldots, m_{i-1}, j, m_{i+1}, \ldots, m_k). \tag{6.2.13}$$

Obviously $\Sigma_i^- - \Sigma_i^+ = \Sigma_i$ (as operators on functions). Thus $\Sigma_i^- \hat{\varphi} = \Sigma_i^+ \hat{\varphi}$ if and only if $\Sigma_i \hat{\varphi} \equiv 0$. Note that the operators Σ_i, Σ_i^-, Σ_i^+ do not preserve the ℓ^1-condition.

Lemma 6.2.9 *$\Sigma_i^- \hat{\varphi}$ and $\Sigma_i^+ \hat{\varphi}$ vanish at ∞ if and only if they coincide.*

Proof $(\Sigma_i^- \hat{\varphi})(m_1, \dots, m_i, \dots, m_k)$ converges to $(\Sigma_i \hat{\varphi})(m_1, \dots, m_k)$ as $m_i \to \infty$, and similarly for Σ_i^+. □

Furthermore, we have

$$\Sigma_i^+ \Delta_i = \Sigma_i^- \Delta_i = \mathrm{Id} \quad \text{and} \quad \Delta_i \Sigma_i^+ = \Delta_i \Sigma_i^- = \mathrm{Id} \tag{6.2.14}$$

as operators on ℓ^1-functions.

Let $(\delta_i \hat{\varphi})(m_1, \dots, m_k) = \delta(m_i) \hat{\varphi}(m_1, \dots, m_k)$, where

$$\delta(n) = \begin{cases} 1, & \text{if } n = 0, \\ 0, & \text{otherwise}, \end{cases}$$

is the usual δ-function. Operators $\Sigma_1, \dots, \Sigma_k$ commute, hence we can define composition operators

$$\Sigma_{i_1, \dots, i_j} = \Sigma_{i_1} \cdots \Sigma_{i_j}$$

for any set of pairwise distinct indices. Similarly, $\delta_1, \dots, \delta_k$ commute and we define

$$\delta_{i_1, \dots, i_j} = \delta_{i_1} \cdots \delta_{i_j}.$$

Operator $\Sigma_1 \cdots \Sigma_k$ associates to every ℓ^1-function $\hat{\varphi}$ the constant function equal to $\overline{\hat{\varphi}}$.

Proposition 6.2.10 *An ℓ^1 k-cocycle $\hat{\varphi}$ satisfies*

$$\hat{\varphi} = \mathcal{D} \hat{\Phi}, \tag{6.2.15}$$

where $\hat{\Phi}$ vanishes at ∞ ($\hat{\Phi}$ may not be ℓ^1 itself) if and only if $\overline{\hat{\varphi}} = 0$. If $\overline{\hat{\varphi}} = 0$ a solution of (6.2.15) $\hat{\Phi} = (\hat{\Phi}_1, \dots, \hat{\Phi}_k)$ *is given in the form* (6.2.16) *below.*

Proof Let $\hat{\varphi} \in \ell^1(\mathbb{Z}^k)$. Then the function $\delta_1 \Sigma_1 \hat{\varphi} \in \ell^1(\mathbb{Z}^k)$. An easy calculation shows that $\Sigma_1(\hat{\varphi} - \delta_1 \Sigma_1 \hat{\varphi}) \equiv 0$. Let

$$\hat{\Phi}_1(\hat{\varphi}) = \Sigma_1^-(\hat{\varphi} - \delta_1 \Sigma_1 \hat{\varphi}).$$

By Lemma 2.1 $\hat{\Phi}_1(\hat{\varphi})$ vanishes at ∞, and by (6.2.14)

$$\hat{\varphi} - \delta_1 \Sigma_1 \hat{\varphi} = \Delta_1 \hat{\Phi}_1(\hat{\varphi}).$$

Since the function $\delta_i \Sigma_i \hat{\varphi}$ vanishes outside the hyperplane $m_i = 0$ one can proceed by induction. Thus

$$\begin{aligned}\hat{\varphi} - \delta_{1,\dots,k}\Sigma_{1,\dots,k}\hat{\varphi} &= \sum_{j=1}^{k}(\delta_{1,\dots,j-1}\Sigma_{1,\dots,j-1}\hat{\varphi} - \delta_{1,\dots,j}\Sigma_{1,\dots,j}\hat{\varphi})\\ &= \sum_{j=1}^{k}\delta_{1,\dots,j-1}(\Sigma_{1,\dots,j-1}\hat{\varphi} - \delta_j\Sigma_j(\Sigma_{1,\dots,j-1}\hat{\varphi})\\ &= \sum_{j=1}^{k}(-1)^{j+1}\Delta_j\hat{\Phi}_j(\hat{\varphi}),\end{aligned}$$

where

$$\hat{\Phi}_j(\hat{\varphi}) = (-1)^{j+1}\Sigma_j^{-}\delta_1\cdots\delta_{j-1}(\Sigma_{1,\dots,j-1}\hat{\varphi} - \delta_j\Sigma_j(\Sigma_{1,\dots,j-1}\hat{\varphi})). \quad (6.2.16)$$

$\hat{\Phi}_j$ vanishes at ∞ since

$$\Sigma_j(\delta_1,\dots,\delta_{j-1}(\Sigma_{1,\dots,j-1}\hat{\varphi} - \delta_j\Sigma_j(\Sigma_{1,\dots,j-1}\hat{\varphi}))) \equiv 0.$$

Thus, $\hat{\Phi} = (\hat{\Phi}_1,\dots,\hat{\Phi}_k)$ is a solution of (6.2.15) if and only if $\overline{\hat{\varphi}} = \Sigma_{1,\dots,k} = 0$. In the latter case formula (6.2.16) gives a solution. □

The map $\hat{\varphi} \to (\hat{\Phi}_1(\hat{\varphi}),\dots,\hat{\Phi}_k(\hat{\varphi}))$ is linear. It is not bounded in ℓ^1-norm; in fact, $\hat{\Phi}$ may not be ℓ^1. However, if $\hat{\varphi}$ decreases at ∞ fast enough so do $\hat{\Phi}_i$s. The last observation will be used in the next section.

Now let φ be a C^∞ n-cochain on $\mathbb{T}^N$ with $1 \le n \le k-1$. As before, the restriction of its Fourier coefficients $\hat{\varphi}$ to a dual orbit $\mathcal{O}(m^*)$ gives us an ℓ^1 n-cochain on $\mathbb{Z}^k$. It can be viewed as a vector-function

$$\hat{\varphi} : \mathbb{Z}^k \to (\mathbb{R}^\ell)^{\binom{k}{n}},$$

whose components $\hat{\varphi}_{i_1\dots i_n}$ are indexed by $i_1 < \cdots < i_n$, $i_1,\dots,i_n \in \{1,\dots,k\}$. The cocycle equation $\mathcal{D}\varphi = 0$ gives us the following equations for components of φ:

$$\sum_{j=1}^{n+1}(-1)^{j+1}\Delta_{i_j}\varphi_{i_1\dots\hat{i_j}\dots i_{n+1}} = 0, \quad (6.2.17)$$

where Δ_i is given by the formula (6.2.11). Then on each non-trivial dual orbit the components of the dual cocycle $\hat{\varphi}$ satisfy to the same equation

$$\sum_{j=1}^{n+1}(-1)^{j+1}\Delta_{i_j}\hat{\varphi}_{i_1\dots\hat{i_j}\dots i_{n+1}} = 0, \quad (6.2.18)$$

with Δ_i given by the formula (6.2.12).

The following proposition is the generalization to the intermediate cohomology case of the key argument for the first cohomology of the higher rank abelian actions which first appeared in [78]. It demonstrates why for smooth cocycles the dual obstructions vanish. We are going to use elementary facts about n-cocycles analogous to those proved for k-cocycles.

Proposition 6.2.11 *If $\hat{\varphi}$ is an ℓ^1 n-cocycle on $\mathbb{Z}^k$ with $1 \le n \le k-1$, then for each component we have*

$$\Sigma_{i_1,\dots,i_n}\hat{\varphi}_{i_1\dots i_n} = 0.$$

Proof It is sufficient to consider a component $\hat{\varphi}_{i_1\dots i_n}$ with $i_1 > 1$. From (6.2.18) we have

$$\Delta_1\hat{\varphi}_{i_1\dots i_n} = \sum_{j=1}^{n}(-1)^{j+1}\Delta_{i_j}\hat{\varphi}_{1i_1\dots\hat{i_j}\dots i_n}.$$

Applying $\Sigma_{i_1,\dots,i_n} = \Sigma_{i_1}\cdots\Sigma_{i_n}$ we obtain

$$\Delta_1\Sigma_{i_1}\dots\Sigma_{i_n}\hat{\varphi}_{i_1\dots i_n} = \Sigma_{i_1,\dots,i_n}\Delta_1\hat{\varphi}_{i_1\dots i_n} =$$
$$\Sigma_{i_1,\dots,i_n}\sum_{j=1}^{n}(-1)^{j+1}\Delta_{i_j}\hat{\varphi}_{1i_1\dots\hat{i_j}\dots i_n} = 0.$$

since $\Sigma_{i_1},\dots,\Sigma_{i_n}$ commute and $\Sigma_i\Delta_i = 0$. Therefore for each point $(c_1, c_2, \dots, c_k) \in \mathbb{Z}^k$ we have

$$\Sigma_{i_1}\cdots\Sigma_{i_n}\hat{\varphi}_{i_1\dots i_n}(m_1, c_2, \dots, c_k) = \Sigma_{i_1}\cdots\Sigma_{i_n}\hat{\varphi}_{i_1\dots i_n}(c_1, c_2, \dots, c_k) = C$$

for each $m_1 \in \mathbb{Z}$. But applying Σ_1 and using that $\hat{\varphi}$ is an ℓ^1 cocycle, we see that

$$\Sigma_1\Sigma_{i_1}\cdots\Sigma_{i_n}\hat{\varphi}_{i_1\dots i_n}(c_1, \dots, c_k) < \infty$$

which implies that $C = \Sigma_{i_1}\cdots\Sigma_{i_n}\hat{\varphi}_{i_1\dots i_n}(c_1, c_2, \dots, c_k) = 0.$ □

6.2.4 Growth estimates for the orbits of the dual action

Here we present estimates on the growth of a given dual orbit in terms of the norm of coordinates on the orbit.

Let $B_1, \dots, B_k \in SL(N, \mathbb{Z})$ be generators for the dual action β. Since β is an action by automorphisms, $\mathcal{O} \approx \mathbb{Z}^k$.

Since $B_1, \dots, B_k$ are commuting real matrices, the space $\mathbb{R}^N$ can be decomposed into a direct sum of subspaces invariant under all B_j:

$$\mathbb{R}^N = \mathbb{I}_1 \oplus \cdots \oplus \mathbb{I}_r, \tag{6.2.19}$$

such that the minimal polynomial of B_j on $\mathbb{I}_i$ is a power of an irreducible polynomial $q_{ij}(x)$ over $\mathbb{R}$. According to this decomposition matrices $B_1, \ldots, B_k$ can be simultaneously brought to the following form with square blocks along the diagonal of sizes $N_1, \ldots, N_r$, $N_1 + \cdots + N_r = N$:

$$\Lambda_1 = \begin{pmatrix} \Lambda_{11} & \cdots & 0 \\ \vdots & \ddots & \vdots \\ 0 & \cdots & \Lambda_{r1} \end{pmatrix}, \ldots, \Lambda_k = \begin{pmatrix} \Lambda_{1k} & \cdots & 0 \\ \vdots & \ddots & \vdots \\ 0 & \cdots & \Lambda_{rk} \end{pmatrix}. \tag{6.2.20}$$

If for a given $1 \le i \le r$ the minimal polynomials of all B_j on $\mathbb{I}_i$ are powers of linear polynomials: $q_{ij}(x) = (x - \lambda_{ij})^{N_i}$, all blocks Λ_{ij} $(1 \le j \le k)$ can be simultaneously brought to an upper-triangular form with λ_{ij} on the diagonal. If for at least one j, $1 \le j \le k$, the minimal polynomial of B_j is a power of an irreducible quadratic polynomial with complex conjugate roots $(\lambda_{ij}, \overline{\lambda_{ij}})$, then the blocks Λ_{ij} $(1 \le j \le k)$ can be simultaneously brought to the following form:

$$\begin{pmatrix} |\lambda_{ij}| \begin{pmatrix} \cos\theta_{ij} & \sin\theta_{ij} \\ -\sin\theta_{ij} & \cos\theta_{ij} \end{pmatrix} & \cdots & \star \\ \vdots & \ddots & \vdots \\ 0 & \cdots & |\lambda_{ij}| \begin{pmatrix} \cos\theta_{ij} & \sin\theta_{ij} \\ -\sin\theta_{ij} & \cos\theta_{ij} \end{pmatrix} \end{pmatrix}.$$

Notice that, for some j, θ_{ij} may be equal to 0 or π, so the 2×2 blocks on the diagonal will correspond to a pair of equal real eigenvalues. Each block Λ_{ij} can be canonically represented as a product

$$\Lambda_{ij} = S_{ij} U_{ij} \tag{6.2.21}$$

where S_{ij} is semisimple and U_{ij} is unipotent (identity plus nilpotent) which commute. The unipotent part U_{ij} is upper-triangular, and the semisimple part S_{ij} is either a diagonal matrix with $\lambda_{ij} \in \mathbb{R}$ on the diagonal, or is of the form

$$\begin{pmatrix} |\lambda_{ij}| \begin{pmatrix} \cos\theta_{ij} & \sin\theta_{ij} \\ -\sin\theta_{ij} & \cos\theta_{ij} \end{pmatrix} & \cdots & 0 \\ \vdots & \ddots & \vdots \\ 0 & \cdots & |\lambda_{ij}| \begin{pmatrix} \cos\theta_{ij} & \sin\theta_{ij} \\ -\sin\theta_{ij} & \cos\theta_{ij} \end{pmatrix} \end{pmatrix}.$$

The invariant subspaces $\mathbb{I}_i$ are said to be *of the first* or *of the second kind* depending on the form of the semisimple parts S_{ij} $(1 \le j \le k)$ described above.

Unipotent matrices U_{ij} have the following important property which will play a crucial role in the argument:

$$C^{-1}\|t_j\|^{-N_i} \leq \|U_{ij}^{t_j}\| \leq C\|t_j\|^{N_i}. \tag{6.2.22}$$

for some constant $C > 0$ independent on the choice of the orbit.

For each $t = (t_1, \ldots, t_k) \in \mathbb{Z}^k$, $\beta^t = B_1^{t_1} \cdots B_k^{t_k}$ is a partially hyperbolic automorphism, hence $\mathcal{O}(m)$, the orbit of the point $m \in \mathbb{Z}^N \setminus \{0\}$, is of rank k, i.e., $\mathcal{O}(m) \approx \mathbb{Z}^k$. For each $t \in \mathbb{Z}^k$ we have a decomposition of $\mathbb{R}^N$ into a direct sum of expanding, neutral, and contracting subspaces, $\mathbb{R}^N = V_t^+ \oplus V_t^\circ \oplus V_t^-$ such that $\beta^t(V_t^i) = V_t^i$, $i \in \{+, \circ, -\}$. These subspaces are direct sums of $\mathbb{I}_i$s with positive, zero, and negative Lyapunov exponents

$$\chi_i(t) = \sum_{j=1}^{k} t_j \ln|\lambda_{ij}|, \quad i = 1, \ldots, r,$$

respectively. Both V_t^+ and V_t^- are non-trivial for all $t \in \mathbb{Z}^k \setminus \{0\}$, and $V_t^\circ = \{0\}$ for all $t \in \mathbb{Z}^k \setminus \{0\}$ is equivalent to hyperbolicity of the action. We use the following norms: for $t \in \mathbb{Z}^k$, $\|t\| = \sum_{j=1}^k |t_j|$, and for $x \in \mathbb{Z}^N$ (or $\mathbb{R}^N$), we decompose it according to (6.2.19), $x = (x_1, \ldots, x_r)$ and let $\|x\| = \sum_{i=1}^r \|x_i\|$, where $\|x_i\|$ is a norm on $\mathbb{I}_i$. It is convenient to use the following norm: $\|x\| = \|(x_1, \ldots, x_r)\| = \sum_{i=1}^r \|x_i\|$, where

$$\|x_i\| = \|(x_{i1}, \ldots, x_{iN_i})\| = |x_{i1}| + \cdots + |x_{iN_i}|$$

if $\mathbb{I}_i$ is of the first kind, and

$$\|x_i\| = \|(x_{i1}, \ldots, x_{iN_i})\| = \sqrt{x_{i1}^2 + x_{i2}^2} + \cdots + \sqrt{x_{i(N_i-1)}^2 + x_{iN_i}^2}$$

if $\mathbb{I}_i$ is of the second kind.

Theorem 6.2.12 *Let α be an action by commuting partially hyperbolic automorphisms of $\mathbb{T}^N$, and β be the dual action. Then there exist constants $a, b, C_1, C_2 > 0$ depending on the action only such that for any initial point $m \in \mathbb{Z}^N$*

$$C_1\|m\|^{-N} \exp(b\|t\|) \leq \|\beta^t m\| \leq C_2\|m\| \exp(a\|t\|).$$

We show first how to obtain the estimate from above.

Lemma 6.2.13

$$N_1\chi_1(t) + N_2\chi_2(t) + \cdots + N_r\chi_r(t) = 0. \tag{6.2.23}$$

Proof We have

$$N_i \chi_i(t) = N_i \sum_{j=1}^{k} t_j \ln |\lambda_{ij}| = \sum_{j=1}^{k} t_j \ln |\det \Lambda_{ij}|.$$

Hence

$$\sum_{i=1}^{r} N_i \chi_i(t) = \sum_{i=1}^{r} \sum_{j=1}^{k} t_j \ln |\det \Lambda_{ij}|$$
$$= \sum_{j=1}^{k} t_j \ln \prod_{i=1}^{r} |\det \Lambda_{ij}| = \sum_{j=1}^{k} t_j \ln |\det B_j| = 0,$$

since $|\det B_j| = 1$. □

Lemma 6.2.14

(i) The number of linearly independent linear functionals among $\chi_1(t), \ldots, \chi_r(t)$ *is equal to* k.

(ii) The function $\max \chi_i(t)$ *is a norm in* $\mathbb{R}^k$, *i.e., there exist constants* $a, b > 0$ *such that*

$$b\|t\| \leq \max \chi_i(t) \leq a\|t\|.$$

Proof Since linear functionals $\chi_1(t), \ldots, \chi_r(t)$ are in k variables, the number of linearly independent among them is not greater than k. Suppose it is less than k. Then there is a point $t \neq 0$ such that $\chi_i(t) = 0$ for all $1 \leq i \leq r$. If all t_i are rational, there exists an integer n such that all nt_i are integers. Then

$$B_1^{nt_1} \cdots B_k^{nt_k}$$

has all eigenvalues of absolute value 1, which contradicts the ergodicity of the action unless $t = 0$. If some of t_i are irrational, for any $\epsilon > 0$ there exists $n = (n_1, \ldots, n_k) \in \mathbb{Z}^k$ such that $A = B_1^{n_1} \cdots B_k^{n_k}, A^2, \ldots, A^N$ have all their eigenvalues ϵ-close to 1, and hence their traces ϵ-close to N. Since all traces must be integers, they all are equal to N. Let eigenvalues of A be equal to $\lambda_1, \ldots, \lambda_N$. Then eigenvalues of A^i are equal to $\lambda_1^i, \ldots, \lambda_N^i$, and we have the following system of equations:

$$\begin{cases} \lambda_1 + \cdots + \lambda_N = N \\ \cdots\cdots\cdots\cdots\cdots\cdots \\ \lambda_1^N + \cdots + \lambda_N^N = N. \end{cases}$$

It follows that all symmetric functions in $\lambda_1, \ldots, \lambda_N$, which appear as coefficients of the polynomial $(x - \lambda_1) \ldots (x - \lambda_N)$ are the same as for $\lambda_1 =$

$\lambda_2 = \cdots = \lambda_N = 1$. This proves part (i). It follows from Lemma 6.2.13 that for any point $t \neq 0$ there exists at least one χ_i such that $\chi_i(t) > 0$, hence $\max \chi_i(t) > 0$. Since the functions $\chi_i(t)$ are linear, $\max \chi_i(t)$ is a norm in $\mathbb{R}^k$, hence equivalent to the $\| \cdot \|$, so part (ii) follows. □ □

Remark 6.2.15 It follows immediately from Lemmas 6.2.14 and 6.2.13 that the number r in the decomposition (6.2.19) is greater than the rank k of the abelian action, and $k \leq N - 1$.

Let us consider a function $\omega : \mathbb{R}^k \to \mathbb{R}$ given by the formula

$$\omega(t) = \sum_{i=1}^{r} \|x_i^*\| \exp \chi_i(t).$$

Then

$$\omega(t) = \sum_{i=1}^{r} \prod_{j=1}^{k} |\lambda_{ij}|^{t_j} \|x_i^*\| = \sum_{i=1}^{r} \| \prod_{j=1}^{k} \Lambda_{ij}^{t_j} x_i^* \| = \|\beta^t(x_1^*, \ldots, x_r^*)\|,$$

and an estimate from above follows immediately from Lemma 6.2.14(ii) for any choice of the initial point $x^* \in \mathcal{O}_{\mathbb{R}}$:

$$\omega(t) \leq \|x^*\| \max \exp \chi_i(t) \leq \|x^*\| \exp(a\|t\|). \tag{6.2.24}$$

On the other hand, we have

$$\omega(t) \geq \min \|x_i^*\| \max \exp \chi_i(t) \geq \min \|x_i^*\| \exp(b\|t\|) \geq 0,$$

and since $\min \|x_i^*\|$ may as well be equal to 0, we only get a trivial estimate this way.

Proof of an estimate from below We first establish the estimates in the the semisimple case, i.e., when the matrices $B_1, \ldots, B_k$ are simultaneously diagonalizable over $\mathcal{C}$.

Now we proceed to a proof of the crucial estimate from below. Let $V \subset \mathbb{R}^N$ be a β-invariant subspace, and $\Lambda = V \cap \mathbb{Z}^N$. Then Λ is either trivial or infinite. In the latter case $\Lambda \approx \mathbb{Z}^d$ for some $1 \leq d \leq N$, and $\beta|_\Lambda$ is dual to the restriction of α to an invariant d-dimensional subtorus. Hence it is also partially hyperbolic. This is because for each $t \in \mathbb{Z}^k$ each eigenvalue of $\beta^t|_\Lambda$ is also an eigenvalue of $\beta^t|_V$ and β^t, and if $\beta^t|_\Lambda$ has an eigenvalue which is a root of unity, then so does β^t. Moreover, $\mathbb{R}^d$ spanned by Λ is decomposed into a direct sum of β-invariant subspaces

$$\mathbb{R}^d = \oplus_{i \in I} \mathbb{I}_i',$$

where $I \subset \{1, 2, \dots, r\}$ and $\mathbb{I}_i' \subset \mathbb{I}_i$ so that the minimal polynomial of B_j on $\mathbb{I}_i'$ divides the minimal polynomial of B_j on $\mathbb{I}_i$. Thus we have $|I|$ Lyapunov exponents for $\beta|_\Lambda$:

$$\chi_i(t) = \sum_{j=1}^{k} t_j \ln |\lambda_{ij}|, \quad i \in I.$$

Each non-trivial β-invariant lattice Λ gives rise to a subset $I \subset \{1, \dots, r\}$, and hence there are only finitely many types of such lattices. (Notice that there may be infinitely many lattices of the same type.)

We denote the collection of all subsets $I \subset \{1, 2, \dots, r\}$ obtained by non-trivial β-invariant lattices by $\mathcal{I}_0$; $\mathcal{I}_0 \neq \emptyset$ since it includes $\{1, \dots, r\}$.

For each Λ we make the following construction. Since $\beta|_\Lambda$ is a partially hyperbolic action, not all χ_i, $i \in I$, are identically 0, and hence, as follows from Lemma 6.2.13, for any $t \in \mathbb{R}^k$ there exists $i \in I$ such that $\chi_i(t) > 0$. The function $M(t) = \max_i \chi_i(t)$ is continuous and achieves its minimum on the unit sphere $S^{k-1} \subset \mathbb{R}^k$ which must be positive by the above argument. Let $b_I = \min_{S^{k-1}} M(t)$. Then for any $t/\|t\| \in S^{k-1}$, $\max_{i \in I} \chi_i(t/\|t\|) \geq b_I$, hence there exists a $i \in I$ for which $\chi_i(t) \geq b_I \|t\|$. Let $b = \min_{I \in \mathcal{I}_0} b_I$; $b > 0$.

Now let $m \in \mathbb{Z}^N$ be any non-zero initial point. It belongs to a β-invariant lattice Λ of minimal dimension d, $\Lambda \approx \mathbb{Z}^d$, therefore β is irreducible over $\mathbb{Q}$ on $\mathbb{R}^d$ spanned by Λ. Hence $\beta|_{\mathbb{R}^d}$ is separable (has no repeated eigenvalues) since otherwise the minimal polynomial of $\beta|_\Lambda$ would not be relatively prime with its derivative, i.e., the minimal polynomial factors over $\mathbb{Q}$, and since it is monic, by Gauss' lemma, it factors over $\mathbb{Z}$, which contradicts the fact that irreducibility of the action implies that the action contains a matrix with irreducible characteristic polynomial [7].

Now, for $I \subset \mathcal{I}_0$ corresponding to the lattice Λ we choose i as above, so that $\chi_i(t) \geq b_I \|t\| \geq b\|t\| > 0$, and take the corresponding eigenspace $\mathbb{I}_i'$. Then $\mathbb{R}^d = \mathbb{I}_i' \oplus \bigoplus_{j \in I-\{i\}} \mathbb{I}_j'$, where $\beta^t|_{\mathbb{I}_i'}$ and $\beta^t|\bigoplus_{j \in I-\{i\}} \mathbb{I}_j'$ have no common eigenvalues, and also $\bigoplus_{j \in I-\{i\}} \mathbb{I}_j' \cap \mathbb{Z}^d = \{0\}$.

Let m_i be a projection of m to $\mathbb{I}_i'$. Then, by Katznelson's lemma [83, Lemma 3], there exists a constant $\mathcal{G}_i$ such that

$$\|m_i\| \geq d(m, V) \geq \mathcal{G}_I \|m\|^{-N},$$

where d is the Euclidean distance, and the constant $\mathcal{G}_i$ depends only on the splitting (6.2.19) for the action β. Thus, we have

$$\|\beta^t m\| = \sum_{i=1}^{r} \exp \chi_i(t)\|m_i\| \geq \exp \chi_i(t)\|m_i\| \geq \exp(b\|t\|)\|m_i\| \geq \mathcal{G}_i\|m\|^{-N}. \tag{6.2.25}$$

So, our estimate holds with $C_1 = \mathcal{G}_i$ for any initial point m.

If the action is not semisimple, only the polynomial growth in $\|t\|$ may occur in addition due to the presence of unipotent factor. Thus. the same estimates will hold with slightly smaller b and slightly larger a. This completes the proof of the theorem. □

Remark 6.2.16 Construction of dual solutions in relies on the invertibility of the action. If it is literally carried out to the non-invertible case, i.e., an action by toral endomorphisms, the resulting solution would, in general, not be defined on the torus, but on a solenoid where the natural extension of our action operates.

6.2.5 Proofs of Theorems 6.2.1 and 6.2.2

We have seen that the average over a dual orbit is an obstruction for solving a coboundary equation on an individual dual orbit (Proposition 6.2.10). We shall show that the vanishing of this obstruction on *all* dual orbits allows us to obtain a global C^∞ solution of the original coboundary equation. The following proposition plays a crucial role in the proof of Theorems 6.2.1 and 6.2.2.

Proposition 6.2.17 *Let α be an action of $\mathbb{Z}^k$ by hyperbolic automorphisms of $\mathbb{T}^N$, and φ be a C^∞ k-cocycle over α with values in $\mathbb{R}^\ell$ $(\ell \geq 1)$ such that for any non-trivial dual orbit $\mathcal{O}$, $\sum_{m\in\mathcal{O}} \hat{\varphi}(m) = 0$. Then φ is C^∞-cohomologous to a constant cocycle ψ, i.e., for $x \in \mathbb{T}^N$, $t \in (\mathbb{Z}^k)^k$*

$$\varphi(x, t) = \psi(t) + \mathcal{D}\Phi(x, t), \tag{6.2.26}$$

where Φ is a C^∞ $(k-1)$-cochain.

Proof First we apply Proposition 6.2.10 to construct a dual cochain $\hat{\Phi}$ on each non-trivial dual orbit. Since the cocycle φ is C^∞ we have the following estimate on the decay of the dual cocycle $\hat{\varphi}$: for any $B \in \mathbb{Z}_+$ there exists $C = C(B)$ such that

$$|\hat{\varphi}(m)| \leq C\|m\|^{-B}. \tag{6.2.27}$$

We want to obtain a similar estimate on the decay of each component of the dual cochain $\hat{\Phi}_j$ $(1 \leq j \leq k)$. Each $0 \neq m \in \mathbb{Z}^N$ belongs to some dual orbit

$\mathcal{O}(m^*)$ where m^* is chosen to be "the lowest": $\|m^*\| = \min_{s\in\mathbb{Z}^k} \|\beta^s(m^*)\|$; then $m = \beta^t m^*$ for some $t \in \mathbb{Z}^k$.

Let $t = (t_1, \dots, t_k)$. Formula (6.2.16) shows that $\hat{\Phi}_j(\beta^t m^*) = 0$ if at least one of the coordinates $t_1, \dots, t_{j-1}$ is not equal to 0, hence it is sufficient only to consider the case when $t_1 = \cdots = t_{j-1} = 0$. Let $s = (0, \dots, 0, t_j, \dots, t_k)$ be fixed and consider the following half-lattice:

$$\mathbb{H}^j = \{r \in \mathbb{Z}^k \mid r = (r_1, \dots, r_{j-1}, r_j, 0, \dots, 0),\\ r_j \geq t_j \text{ if } t_j \geq 0,\ r_j < t_j \text{ if } t_j < 0\}.$$

Then again by formula (6.2.16)

$$|\hat{\Phi}_j(\beta^s m^*)| \leq \sum_{r\in\mathbb{H}^j} |\hat{\varphi}(\beta^{r+s} m^*)|. \tag{6.2.28}$$

If for $t = r + s$ we put $\|t\| = \sum_{i=1}^k |t_i|$, then $\|r + s\| = \|r\| + \|s\|$. We split the right-hand side of (6.2.28) into two sums, $S_1(\beta^s m^*)$ and $S_2(\beta^s m^*)$ where $S_1(\beta^s m^*)$ is a finite sum over $\|t\| < t_0$, where we are going to use a simple estimate $\|\beta^t m^*\| \geq \|m^*\|$, and $S_2(\beta^s m^*)$ is the infinite one over $\|t\| \geq t_0$, where the exponential estimates of Theorem 6.2.12 prevail and become uniform. Namely, for $\|t\| \geq t_0 > \frac{N+1}{b} \ln \|m^*\|$ we have $\exp(bt_0)\|m^*\|^{-N} \geq \|m^*\|$, therefore

$$\begin{aligned}\|\beta^t m^*\| &\geq C_1\|m^*\| \exp(b(\|t\| - t_0))\\ &= C_1\|m^*\| \exp(b(\|s\|)) \exp(b(\|r\| - t_0)).\end{aligned} \tag{6.2.29}$$

We first estimate $S_2(\beta^s m^*)$. We use (6.2.29) and the estimate from above of Theorem 6.2.12

$$\|\beta^s m^*\| \leq C_2\|m^*\| \exp(a\|s\|).$$

For some constant $C_3 > 0$

$$\|\beta^s m^*\|^{\frac{b}{a}} \leq C_3\|m^*\|^{\frac{b}{a}} \exp(b\|s\|) \leq C_3\|m^*\| \exp(b\|s\|),$$

since $\|m^*\| \geq 1$, so that

$$\|\beta^t m^*\| \geq C_4\|\beta^s m^*\|^{\frac{b}{a}} \exp(b\|r\|)e^{-bt_0},$$

for yet another constant $C_4 > 0$. By (6.2.27) we have

$$|\hat{\varphi}(\beta^t m^*)| \leq C\|\beta^t m^*\|^{-B} \leq CC_4^{-B}\|\beta^s m^*\|^{-B\frac{b}{a}} \exp(-Bb\|r\|).$$

Then, for some constants $C_5, C_6 > 0$, and $m = \beta^s m^*$, we obtain a super-polynomial estimate for $S_2(m)$:

$$S_2(m) \leq C_5\|m\|^{-B\frac{b}{a}} \sum_{r\in\mathbb{H}^j} \exp(-Bb\|r\|) \leq C_6\|m\|^{-B\frac{b}{a}}. \tag{6.2.30}$$

In order to estimate $S_1(\beta^s m^*)$ we write for $C_7 = C_2^{-1}$

$$\|\beta^t m^*\| \geq \|m^*\| \geq C_7 \|\beta^s m^*\| \exp(-a\|s\|),$$

therefore

$$|\hat{\varphi}(\beta^t m^*)| \leq C\|\beta^t m^*\|^{-B} \leq CC_7^{-B}\|\beta^s m^*\|^{-B} \exp(-Ba\|s\|).$$

Since the number of terms in this finite sum is $\leq t_0^k$, we obtain for some constant $C_8, C_9 > 0$

$$S_1(m) \leq C_9\|m\|^{-B}(\ln \|m^*\|)^k \leq C_9\|m\|^{-B}(\ln \|m\|)^k, \qquad (6.2.31)$$

but since $(\ln \|m\|)^k \geq \|m\|^\epsilon$ for every $\epsilon > 0$, taking $\epsilon = B(1 - \frac{b}{a})$ we obtain

$$S_1(m) \leq C_9\|m\|^{-B+\epsilon} = C_9\|m\|^{-B\frac{b}{a}}.$$

Combining this with (6.2.30) we obtain a super-polynomial estimate for $\hat{\Phi}$ for some $C_{10} > 0$.

$$|\hat{\Phi}_j(m)| \leq C_{10}\|m\|^{-B\frac{b}{a}}.$$

Thus we obtained global estimates on the decay of $\hat{\Phi}_j$. Letting $\hat{\Phi}_j(0) = 0$ and using formulas (6.2.11) and (6.2.12) one therefore obtains a C^∞ $(k-1)$-cochain $\Phi = (\Phi_1, \ldots, \Phi_k)$ such that

$$\mathcal{D}\Phi = \varphi - \hat{\varphi}(0),$$

i.e., a solution of the equation (6.2.26). □

Proof of Theorem 6.2.1 First we apply Corollary 6.2.7 to conclude that if a C^∞ k-cocycle over α, φ, vanishes on all periodic orbits of α, then for any dual orbit $\mathcal{O}$, including 0, $\sum_{m\in\mathcal{O}} \hat{\varphi}(m) = 0$. Now, by Proposition 6.2.17 $\mathcal{D}\Phi = \varphi - \hat{\varphi}(0)$, and since $\hat{\varphi}(0) = 0$ (see Corollary 6.2.8), one obtains a solution of the equation (6.2.1). □

Proof of Theorem 6.2.2 First, let φ be a C^∞ 1-cocycle on $\mathbb{T}^N$: $\varphi = (\varphi_1, \ldots, \varphi_k)$ and $\hat{\varphi} = (\hat{\varphi}_1, \ldots, \hat{\varphi}_k)$ be a dual cocycle. By Proposition 6.2.11 $\Sigma_i\hat{\varphi}_i = 0$, and we can write a solution of the dual coboundary equation

$$\hat{\varphi} = \mathcal{D}\hat{\Phi},$$

$\hat{\Phi}(\hat{\varphi}) = \Sigma_i^- \hat{\varphi}_i$. The cocycle equations

$$\Delta_i\hat{\varphi}_j = \Delta_j\hat{\varphi}_i$$

for $i \neq j$ imply that $\Sigma_i^- \hat{\varphi}_i = \Sigma_j^- \hat{\varphi}_j$, hence $\hat{\Phi}$ is well-defined. It is unique up to an additive constant. As before, we construct a solution $\hat{\Phi}$ on each non-trivial dual orbit. Now we recall that the solution of the coboundary equation in the case $n = k$ was constructed inductively (6.2.16). Hence the estimates (6.2.27) and (6.2.28) with $j = 1$ actually give the super-polynomial decay for 1-cocycles. Thus we obtain a C^∞ solution of the equation

$$\mathcal{D}\Phi = \varphi - \hat{\varphi}(0).$$

We are going to proceed by induction on k. Our hypothesis holds for the highest cocycles for which their dual cocycles have zero average over each dual orbit (Proposition 6.2.17) and for 1-cocycles. These cases will be considered as the base step in our induction argument. Suppose the equation (6.2.2) has a C^∞ solution for n-cocycles on $\mathbb{Z}^p$, where $2 \leq p \leq k-1$ and $1 \leq n \leq p-1$. Let φ be a C^∞ n-cocycle on $\mathbb{T}^N$, $1 \leq n \leq k-1$, i.e., a vector-function with $\binom{k}{n}$ components satisfying cocycle equations (6.2.17). The $\binom{k-1}{n}$ components not containing index 1 may be regarded as an n-cocycle of the $\mathbb{Z}^{k-1}$-action by $A_2, \ldots, A_k$ since for components with $i_1 > 1$ $(\mathcal{D}\varphi)_{i_1,\ldots,i_{n+1}} = 0$ are just part of the cocycle equations for φ. If $n < k-1$, by the induction hypothesis we can find C^∞ solutions for the first $\binom{k-1}{n}$ equations. If $n = k-1$, then by Proposition 6.2.11 the dual cocycle has zero average over each non-trivial dual orbit, hence a C^∞ solution of (6.2.26) can be found by Proposition 6.2.17. The remaining $\binom{k-1}{n-1}$ components contain index 1. Let

$$\phi_{i_2,\ldots,i_n} = \varphi_{1,i_2,\ldots,i_n} - \Delta_1 \mathrm{h}_{i_2,\ldots,i_n},$$

where $\mathrm{h}_{i_2,\ldots,i_n}$ are already obtained from the first $\binom{k-1}{n}$ equations. We need to show that ϕ is a C^∞ $(n-1)$-cocycle of the $\mathbb{Z}^{k-1}$-action by $A_2, \ldots, A_k$. Indeed,

$$\begin{aligned}(\mathcal{D}\phi)_{i_2,\ldots,i_{n+1}} &= \sum_{j=2}^{n+1} (-1)^j \Delta_{i_j} (\varphi_{1,i_2,\ldots,\hat{i_j},\ldots,i_{n+1}} - \Delta_1 \mathrm{h}_{i_2,\ldots,\hat{i_j},\ldots,i_{n+1}}) \\ &= \Delta_1 \varphi_{i_2,\ldots,i_{n+1}} - \Delta_1 \sum_{j=2}^{n+1} (-1)^j \Delta_{i_j} \mathrm{h}_{i_2,\ldots,\hat{i_j},\ldots,i_{n+1}} \\ &= \Delta_1 \varphi_{i_2,\ldots,i_{n+1}} - \Delta_1 \varphi_{i_2,\ldots,i_{n+1}} = 0.\end{aligned}$$

We used the cocycle equation for φ:

$$\sum_{j=2}^{n+1} (-1)^j \Delta_{i_j} \varphi_{1,i_2,\ldots,\hat{i_j},\ldots,i_{n+1}} = \Delta_1 \varphi_{i_2,\ldots,i_{n+1}},$$

and that

$$\varphi_{i_2,\dots,i_{n+1}} = \sum_{j=2}^{n+1}(-1)^j \Delta_{i_j} \mathrm{h}_{i_2,\dots,\hat{i_j},\dots,i_{n+1}} + \hat{\varphi}_{i_2,\dots,i_{n+1}}(0).$$

Notice that $\phi_{i_2,\dots,i_n}$ is C^∞ since $\varphi_{1,i_2,\dots,i_n}$ is, $\mathrm{h}_{i_2,\dots,i_n}$ is by the induction hypothesis, and the operator Δ_1 preserves the smoothness. Then by the induction hypothesis we can solve the coboundary equation for $(n-1)$-cocycles of the $\mathbb{Z}^{k-1}$-action:

$$\begin{aligned}\phi_{i_2,\dots,i_n} &= (\mathcal{D}g)_{i_2,\dots,i_n} + \hat{\varphi}_{i_2,\dots,i_n}(0)\\ &= \sum_{j=2}^{n}(-1)^{j+1} \Delta_{i_j} g_{i_2,\dots,\hat{i_j},\dots,i_n} + \hat{\varphi}_{i_2,\dots,i_n}(0).\end{aligned}$$

But then, if we define components of h that have index 1 by the formula

$$\mathrm{h}_{1,i_2,\dots,i_{n-1}}(x) = g_{i_2,\dots,i_{n-1}}(x)$$

and the components of the constant cocycle by

$$\psi_{1,i_2,\dots,i_{n-1}} = \hat{\varphi}_{1,i_2,\dots,i_{n-1}}(0)$$

we get

$$\varphi_{1,i_2,\dots,i_n} = \phi_{i_2,\dots,i_n} + \Delta_1 \mathrm{h}_{i_2,\dots,i_n} = (\mathcal{D}\mathrm{h})_{1,i_2,\dots,i_n} + \psi_{1,i_2,\dots,i_{n-1}}. \qquad \square$$

6.3 Cohomology for Weyl chamber flows

In this section we present a partial extension of the Katok–Spatzier cohomological result to higher cohomology. We will only present a sketch of the proof and refer to Section 4.4.2 for some missing details.

Let W be a compact manifold foliated by a smooth foliation $\mathcal{F}$. Denote by $H^*(W,\mathcal{F};\mathbb{R})$ be the tangential (or leafwise) de Rham cohomology of the foliated manifold $(W,\mathcal{F})$, that is, the cohomology of the tangential de Rham complex $(\Omega^*(W,\mathcal{F};\mathbb{R}), d_{\mathcal{F}})$ of the foliated manifold $(W,\mathcal{F})$. An element of $(\Omega^*(W,\mathcal{F};\mathbb{R}), d_{\mathcal{F}})$ is an $\mathbb{R}$-valued tangential p-form, which is by definition a C^∞ section of the vector bundle $\bigwedge^p T^*\mathcal{F}$ over W with $T^*\mathcal{F}$ the cotangent bundle of $\mathcal{F}$. The coboundary operator is the tangential exterior derivative $d_{\mathcal{F}}$ that is defined in the same manner as the ordinary exterior derivative except that the differential is performed only in the direction tangent to $\mathcal{F}$. More precisely, for $\omega \in \Omega^p(W,\mathcal{F};\mathbb{R})$, its tangential exterior derivative $d_{\mathcal{F}}\omega \in \Omega^{p+1}(W,\mathcal{F};\mathbb{R})$ is defined by

$$(d_{\mathcal{F}}\omega)(X_0,\dots,X_p) = \sum_q (-1)^q X_q \cdot \omega(X_0,\dots,\hat{X}_q,\dots,X_p) + \sum_{q<r} \omega([X_q,X_r],X_0,\dots,\hat{X}_q,\dots,\hat{X}_r,\dots,X_p), \tag{6.3.1}$$

where $X_0,\dots,X_p$ are vectors fields on W that are tangent to the foliation $\mathcal{F}$.

Let G be a connected semisimple Lie group with finite center, Lie algebra $\mathfrak{g}$ and real rank $k \geq 2$. Assume that none of the simple factors of $\mathfrak{g}$ is compact. Let G/Γ be a torsion-free irreducible co-compact lattice and form the vector space $V = G/\Gamma$ on which G acts from the left. This action induces an identification of the elements of the Lie algebra $\mathfrak{g}$ with smooth vector fields on V. We denote the element of $\mathfrak{g}$ and the corresponding vector field by the same symbol.

Take a Cartan decomposition $\mathfrak{g} = \mathfrak{k} + \mathfrak{p}$ with $\mathfrak{k}$ compact subalgebra. Let $\mathfrak{a}$ be a maximal abelian subalgebra of $\mathfrak{p}$, Λ denote the restricted root system of G, and $\mathfrak{m}$ be the centralizer of $\mathfrak{a}$ in $\mathfrak{k}$. Then (see Section 2.3.3) the Lie algebra $\mathfrak{g}$ of G has the root space decomposition

$$\mathfrak{g} = \mathfrak{m} + \mathfrak{a} + \sum_{\lambda\in\Lambda} \mathfrak{g}_\lambda,$$

where $\mathfrak{g}_\lambda$ is the root space of λ and $\mathfrak{m}$ and $\mathfrak{a}$ are the Lie algebras of the Lie groups M and A.

We denote the elements of the restricted root space $\mathfrak{g}_\lambda$ by X_λ and the elements of $\mathfrak{a}$ by $H_1, H_2, \dots$.

Theorem 6.3.1 *Let $\mathcal{F}$ be the orbit foliation of the action of A on V. Suppose that the assumptions in Theorem 4.4.2 are satisfied. Let $1 \leq n \leq k-1$. Let f be an $\mathbb{R}$-valued C^∞ tangential n-form of $(V,\mathcal{F})$ such that:*

(i) $d_{\mathcal{F}} f = 0$; and
(ii) $\int_V f(H_1,\dots,H_n) = 0$ for all $H_1,\dots,H_n \in \mathfrak{a}$.

Then there exists an $\mathbb{R}$-valued C^∞ tangential $n-1$-form u of $(V,\mathcal{F})$ satisfying the differential equation:

$$d_{\mathcal{F}} u = f. \tag{6.3.2}$$

We sketch the proof of the theorem. Fix a regular element $H_j \in \mathfrak{a}$. Denote by $\phi^t_{H_j}$ the flow on V generated by the vector field H_j. The differential equation (6.3.2) has the formal solutions u^+ and u^- along the orbits of the flow $\phi^t_{H_j}$:

$$u^+(H_1, \ldots, \hat{H}_j, \ldots, H_n) = -\int_0^{\infty} f(H_1, \ldots, H_n) \circ \phi^t_{H_j}\, dt,$$

$$u^-(H_1, \ldots, \hat{H}_j, \ldots, H_n) = \int_0^{-\infty} f(H_1, \ldots, H_n) \circ \phi^t_{H_j}\, dt.$$

Due to the exponential decay of the matrix coefficients, these formal solutions are well defined as distributions and both satisfy equation (6.3.2) in the distribution sense.

Partial differentiability of these distributions follows from the hyperbolicity of the Weyl chamber flow: u^+ (respectively u^-) is differentiable repeatedly in the directions X_λ ($\lambda \in \Lambda$) with $\lambda(H_0) > 0$ (respectively $\lambda(H_0) < 0$). The main point of the proof is the equality $u^+ = u^-$, which in addition will imply the differentiability, and eventually smoothness, of $u = u^+ = u^-$ in any direction X_λ with $\lambda \in \Lambda$. This is a consequence of subelliptic regularity theorem, Theorem 3.7.2.

The equality $u^+ = u^-$ follows using the matrix coefficients estimate from Theorem 4.4.3 and the higher rank condition.

References

[1] R. L. Adler, B. Weiss. Entropy, a complete metric invariant for automorphisms of the torus. *Proc. Nat. Acad. Sci.* **57** (1967) 1575–1576.

[2] D. V. Anosov. Geodesic flows on closed Riemannian manifolds with negative curvature. *Proc. Stek. Inst.* **90** (1967) 1–235.

[3] N. Aoki. A simple proof of Bernoullicity of automorphisms of compact abelian groups. *Isr. J. Math.* **38** (1981) 189–198.

[4] L. Auslander, J. Scheuneman. On certain automorphisms of nilpotent Lie groups. *Proc. Symp. Pure Math.* **14** (1970) 9–15.

[5] H. Bercovici, V. Niţică. A Banach algebra version of the Livšic theorem. *Discrete Contin. Dynam. Systems* **4** (1998) 523–534.

[6] L. Barreira, Ya. Pesin. *Nonuniform Hyperbolicity: Dynamics of Systems with Nonzero Lyapunov Exponents*. Cambridge: Cambridge University Press, 2007.

[7] D. Berend. Multi-invariant sets on tori. *Trans. Amer. Math. Soc.* **280** (1983) 509–532.

[8] Z. I. Borevich, I. R. Shafarevich. *Number Theory*. New York: Academic Press, 1966.

[9] R. Bowen. *Equilibrium States and the Ergodic Theory of Anosov Diffeomorphisms*. Lecture Notes in Math. **470**. New York: Springer, 1975.

[10] G. E. Bredon. *Introduction to Compact Transformations Groups*. New York: Academic Press, 1972.

[11] M. I. Brin. Topological transitivity of a class of dynamical systems and frame flow on manifolds of negative curvature. *Func. Anal. and Appl.* **9** (1975) 9–19.

[12] M. I. Brin, Y. A. Pesin. Partially hyperbolic dynamical systems. (Russian) *Izv. Akad. Nauk SSSR Ser. Mat.* **38** (1974) 170–212.

[13] K. S. Brown. *Cohomology of Groups*. Graduate Texts in Math. **87**. Berlin: Springer-Verlag, 1982.

[14] S. Campanato. Proprieta di una famiglia di spazi functionali. *Ann. Scuola Norm. Sup. Pisa* **18** (1964) 137–160.

[15] C. Chevalley. Deux théorèmes d'Arithmétique. *J. Math. Soc. of Japan* **3** (1951) 36–44.

[16] H. Cohen. *A Course in Computational Algebraic Number Theory*. Berlin-Heidelberg-New York: Springer, 1996.

[17] P. Collet, H. Epstein, G. Gallavotti. Perturbations of geodesic flows on surfaces of constant negative curvature and their mixing properties. *Com. Math. Phys.* **95** (1984) 61–112.

[18] M. Cowling. Sur les Coeficients des Representations Unitaires des Groupes de Lie Simple. Lecture Notes in Math. **739**. Berlin: Springer-Verlag, 1979, pp. 132–178.

[19] D. Damjanovic. Central extensions of simple Lie groups and rigidity of some abelian partially hyperbolic algebraic actions. *J. Mod. Dyn.* **1** (2007) 665–688.

[20] D. Damjanovic, A. Katok. Local rigidity of actions of higher rank abelian groups and KAM method. *Electronic Res. Announce. Amer. Math. Soc.* **10** (2004) 142–154.

[21] D. Damjanovic, A. Katok. Periodic cycle functionals and cocycle rigidity for certain partially hyperbolic actions. *Discrete Contin. Dynam. Systems* **13** (2005) 985–1005.

[22] D. Damjanovic, A. Katok. Local rigidity of partially hyperbolic actions I. KAM method and $\mathbb{Z}^k$ actions on the torus. *Annals Math.* (2010), to appear.

[23] D. Damjanovic, A. Katok. Local rigidity of partially hyperbolic actions. II. The geometric method and restrictions of Weyl chamber flows on $SL(n, \mathbb{R})/\Gamma$, available online at http://www.math.psu.edu/katok_a.

[24] D. Damjanovic, A. Katok. Local rigidity of homogeneous parabolic actions: I. A model case, preprint, 2010.

[25] S. G. Dani, M. G. Mainkar. Anosov automorphisms on compact nilmanifolds associated with graphs. *Trans Amer. Math. Soc.* **357** (2005) 2235–2251.

[26] P. Didier. Stability of accessibility. *Ergodic Theory Dynam. Systems* **23** (2003) 1717–1731.

[27] D. Dolgopyat, A. Wilkinson. Stable accessibility is C^1 dense. *Geometric methods in dynamics. II.* Astérisque **287** (2003) 33–60.

[28] M. Einsiedler, A. Katok. Invariant measures on G/Γ for split simple Lie Groups G. *Comm. Pure. Appl. Math.* (Moser memorial issue) **56** (2003) 1184–1221.

[29] M. Einsiedler, A. Katok. Rigidity of measures – the high entropy case and non-commuting foliations. *Israel Math. J.* **148** (2005) 169–238.

[30] M. Einsiedler, A. Katok, E. Lindenstrauss. Invariant measures and the set of exceptions to Littlewood's conjecture. *Annals of Math.* **164** (2006) 513–560.

[31] M. Einsiedler, E. Lindenstrauss. Rigidity properties of $\mathbb{Z}^d$-actions on tori and solenoids. *Electronic Res. Announce. Math. Soc.* **9** (2004) 99–110.

[32] R. Feres, A. Katok. Ergodic theory and dynamics of G-spaces, in *Handbook of Dynamical Systems*, vol. 1A. Amsterdam: Elsevier, 2002, pp. 665–763.

[33] S. Ferleger, A. Katok. Non-commutative first cohomology rigidity of the Weyl chamber flows, preprint, 1997.

[34] D. Fisher. Local rigidity of group actions: past, present, future, in *Dynamics, Ergodic Theory and Geometry*. Cambridge: Cambridge University Press, 2007.

[35] D. Fisher, G. Margulis. Almost isometric actions, property T, and local rigidity. *Inventiones Math.* **162** (2005) 19-80.

[36] D. Fisher, G. Margulis. Local rigidity of affine actions of higher rank Lie groups and their lattices. *Annals Math.* **170** (2009) 67–122.

[37] L. Flaminio, G. Forni. Invariant distributions and time averages for horocycle flows. *Duke Math. J.* **119** (2003) 465–526.

[38] G. Forni. Solutions of the cohomological equation for area-preserving flows on compact surfaces of higher genus. *Ann. Math.* **146** (1997) 295–344.

[39] J. Franks. Anosov diffeomorphisms on tori. *Trans. Amer. Math. Soc.* **145** (1969) 117–124.

[40] H. Furstenberg. Disjointness in ergodic theory, minimal sets, and a problem in Diophantine approximation. *Math. Systems Theory* **1** (1967) 1–49.

[41] E. Goetze, R. J. Spatzier. Smooth classification of Cartan actions of higher rank semisimple Lie groups and their lattices. *Ann. Math.* **150** (1999) 743–773.

[42] E. Goetze, R. J. Spatzier. On Livsic's theorem, superrigidity, and Anosov actions of semisimple Lie groups. *Duke Math. J.* **88** (1997) 1–27.

[43] V. Guillemin, D. Kazhdan. Some inverse spectral results for negatively curved 2-manifolds. *Topology* **19** (1980) 301–313.

[44] R. Hamilton. The inverse limit theorem of Nash and Moser. *Bull. Amer. Math. Soc.* **7** (1982) 65–222.

[45] P. de la Harpe, A. Valette. La propriété (T) de Kazhdan pour les groupes localement compacts. *Astérisque* **175** (1989).

[46] B. Hasselblatt. Hyperbolic dynamical systems, in *Handbook of Dynamical Systems*, vol. 1A. Amsterdam: Elsevier, 2002, pp. 239–319.

[47] B. Hasselblatt, A. Katok. Principal structures, in *Handbook of Dynamical Systems*, vol. 1A. Amsterdam: Elsevier, 2002, pp. 1–203.

[48] S. Helgason. *Differential Geometry, Lie Groups and Symmetric Spaces*. New York: Academic Press, 1978.

[49] F. R. Hertz, M. A. R. Hertz, R. Ures. Accessibility and stable ergodicity for partially hyperbolic diffeomorphisms with 1d-center bundle. *Inventiones Math.* **172** (2008) 353–381.

[50] M. Hirsch. *Differential Topology*. Graduate Texts in Math. **33**. New York: Springer-Verlag 1976.

[51] M. Hirsch, C. Pugh, M. Shub. *Invariant Manifolds*. Lecture Notes in Math. **583**. Berlin: Springer Verlag, 1977.

[52] L. Hörmander. Hypoelliptic second order differential equations. *Acta Mathematica* **119** (1967) 147–171.

[53] R. Howe. A notion of rank for unitary representations of the classical groups, in A. Figa Talamanca (ed.), *Harmonic Analysis and Group Representations*. Firenze, Italy: CIME, 1980.

[54] S. Hurder. Affine Anosov actions. *Michigan Math. J.* **40** (1993) 561–575.

[55] S. Hurder. Rigidity of Anosov actions of higher rank lattices. *Annals Math.* **135** (1992) 361–410.

[56] S. Hurder, A Katok. Differentiability, rigidity and Godbillon–Vey classes for Anosov flows. *Publ. Math. IHES* **72** (1990) 5–61.

[57] H. C. Im Hof. An Anosov action on the bundle of Weyl chambers. *Ergodic Theory Dynam. Systems* **5** (1985) 587–593.

[58] J.-L. Journé. A regularity lemma for functions of several variables. *Revista Matematica Iberoamericana* **4** (1988) 187–193.

[59] B. Kalinin. Livšic theorem for matrix cocycles. *Annals of Math*, to appear.

[60] B. Kalinin, A. Katok. Invariant measures for actions of higher rank abelian groups, in *Smooth Ergodic Theory and its Applications. Proc. Symp. Pure Math* **69**. Providence, RI: Amer. Math. Soc., 2001, pp. 593–637.

[61] B. Kalinin, A. Katok. Measure rigidity beyond uniform hyperbolicity: invariant measures for Cartan actions on tori. *J. Mod. Dyn.* **1** (2007) 123–146.

[62] B. Kalinin, A. Katok, F. Rodriguez Hertz. Nonuniform measure rigidity. *Annals of Math.*, to appear.

[63] B. Kalinin, V. Sadovskaya. On local and global rigidity of quasi-conformal Anosov diffeomorphisms. *J. Inst. Math. Jussieu* **2–4** (2003) 567–582.

[64] B. Kalinin, R. Spatzier. On the classification of Cartan actions. *Geom. Func. Anal.* **17** (2007) 468–490.

[65] M. Kanai. Rigidity of Weil chamber flow, and vanishing theorems of Matsushima and Weil. *Ergod. Theory Dynam. Systems* **29** (2009) 1273–1288.

[66] A. Katok. *Combinatorial Constructions in Ergodic Theory and Dynamics*. University Lecture Series **30**. Providence, RI: American Mathematical Society, 2003.

[67] A. Katok, B. Hasselblatt. *Introduction to the Modern Theory of Dynamical Systems*. Cambridge: Cambridge University Press, 1995.

[68] A. Katok, S. Katok. Higher cohomology for abelian groups of toral automorphisms. *Ergodic Theory Dynam. Systems* **15** (1995) 569–592.

[69] A. Katok, S. Katok. Higher cohomology for abelian groups of toral automorphisms. II. The partially hyperbolic case, and corrigendum. *Ergodic Theory Dynam. Systems* **25** (2005) 1909–1917.

[70] A. Katok, S. Katok, K. Schmidt. Rigidity of measurable structure for $\mathbb{Z}^d$-actions by automorphisms of a torus. *Comm. Math. Helvetici* **77** (2002) 718–745.

[71] A. Katok, A. Kononenko. Cocycles' stability for partially hyperbolic systems. *Math. Res. Lett.* **3** (1996) 191–210.

[72] A. Katok, V. Niţică. Rigidity of higher rank abelian cocycles with values in diffeomorphism groups. *Geometriae Dedicata*. **124** (2007) 109–131.

[73] A. Katok, V. Niţică, A. Török. Non-abelian cohomology of abelian Anosov actions. *Ergodic Theory Dynam. Systems* **2** (2000) 259–288.

[74] A. Katok, F. Rodriguez Hertz. Uniqueness of large invariant measures for $\mathbb{Z}^k$ actions with Cartan homotopy data. *J. Mod. Dyn.* **1** (2007) 287–300.

[75] A. Katok, F. Rodriguez Hertz. Measure and cocycle rigidity for certain non-uniformly hyperbolic actions of higher rank abelian groups. *J. Mod. Dyn.* **4** (2010), to appear.

[76] A. Katok, F. Rodriguez Hertz. Rigidity of real-analytic actions of $SL(n, \mathbb{Z})$ on T^n: A case of realization of Zimmer program. *Discrete Contin. Dynam. Systems* **27** (2010) 609–615.

[77] A. Katok, K. Schmidt. The cohomology of expansive $\mathbb{Z}^d$-actions by automorphisms of compact, abelian groups. *Pacific J. Math* **170** (1995) 105–142.

[78] A. Katok, R. Spatzier. First cohomology of Anosov actions of higher rank abelian groups and applications to rigidity. *Inst. Hautes Études Sci. Publ. Math.* **79** (1994) 131–156.

[79] A. Katok, R. Spatzier. Subelliptic estimates of polynomial differential operators and applications to rigidity of abelian actions. *Math. Res. Letters* **1** (1994) 193–202.

[80] A. Katok, R. Spatzier. Invariant measures for higher rank hyperbolic Abelian actions. *Ergodic Theory Dynam. Systems* **16** (1996) 751–778.

[81] A. Katok, R. Spatzier. Differential rigidity of Anosov actions of higher rank abelian groups and algebraic lattice actions. *Tr. Mat. Inst. Steklova* **216** (1997) 292–319.

[82] S. Katok. Finite spanning sets for cusp forms and a related geometric result. *J. Reine Angew. Math.* **395** (1989) 186–195.

[83] Y. Katznelson. Ergodic automorphisms on $\mathbb{T}^n$ are Bernoulli shifts. *Israel J. Math.* **10** (1971) 186–195.

[84] D. Kleinbock, N. Shah, A. Starkov. Dynamics of subgroup actions on homogeneous spaces of Lie groups and applications to number theory in: *Handbook of Dynamical Systems*, vol. 1A. Amsterdam: Elsevier, 2002, pp. 813–930.

[85] A. W. Knapp. *Representation Theory of Semisimple Groups: an Overview Based on Examples*. Princeton, NJ: Princeton University Press, 2001.

[86] A. Kononenko. Twisted cocycles and rigidity problems. *Electron. Res. Announc. Amer. Math. Soc.* **1** (1995) 26–34.

[87] N. Kopell. Commuting diffeomorphisms. *Proc. Symp. Pure Math.* **14** (1970) 165–184.

[88] S. Krantz. Lipschitz spaces, smoothness of functions and approximation theory. *Expo. Math.* **3** (1983) 193–260.

[89] R. Krikorian. Reducibility, differentiable rigidity and Lyapunov exponents for quasi-periodic cocycles on $\mathbb{T} \times SL(2, \mathbb{R})$, preprint.

[90] S. Lang. *Algebra*. Reading, MA: Addison-Wesley, 1984.

[91] S. Lang. *Introduction to Differentiable Manifolds*. New York: Interscience, 1962.

[92] J. Lauret. Examples of Anosov diffeomorphisms. *J. Algebra* **262** (2003) 201–209.

[93] F. Ledrappier. Un champ markovien peut être d'entropie nulle et mélangeant. *C. R. Acad. Sci. Paris Sér. A-B* **287** (1978) A561–A563.

[94] K. B. Lee, F. Raymond. Geometric realization of group extensions by the Seifert construction. *Contemporary Math. AMS* **33** (1984) 353–411.

[95] E. Lindenstrauss. Rigidity of multiparameter actions. *Israel Math. J.* **149** (2005) 199–226.

[96] A. Livšic. Homology properties of U-systems. *Math. Zametki* **10** (1971) 758–763.

[97] A. Livšic. Cohomology of dynamical systems. *Math. USSR Izvestija* **6** (1972) 1278–1301.

[98] R. de la Llave. Invariants for smooth conjugacy of hyperbolic dynamical systems. *I. Comm. Math. Phys.* **109** (1987) 369–378.

[99] R. de la Llave. Smooth conjugacy and S-R-B measures for uniformly and non-uniformly hyperbolic dynamical systems. *Commun. Math. Phys.* **150** (1992) 289–320.

[100] R. de la Llave. Analytic regularity of solutions of Livšic's cohomology equation and some applications to analytic conjugacy of hyperbolic dynamical systems. *Ergodic Theory Dynam. Systems* **17** (1997) 649–662.

[101] R. de la Llave. Remarks on Sobolev regularity in Anosov systems. *Ergodic Theory Dynam. Systems* **21** (2001) 1139–1180.

[102] R. de la Llave. Tutorial on KAM theory, in *Smooth Ergodic Theory and its* Applications *Proc. Symp. Pure Math* **69**. RI: American Mathematical Society, Providence, 2001, pp. –.

[103] R. de la Llave. Bootstrap of regularity for integrable solutions of cohomology equations, in *Modern Dynamical Systems and Applications*, M. Brin, B. Hasselblatt, Ya. B. Pesin (eds). Cambridge: Cambridge University Press, 2004, pp. 405–418.

[104] R. de la Llave, J. Marco, R. Moriyon. Canonical perturbation theory of Anosov systems and regularity results for the Livšic cohomology equation. *Ann. of Math.* **123** (1986) 537–611.

[105] R. de la Llave, R. Moriyon. Invariants for smooth conjugacy of hyperbolic dynamical systems IV. *Comm. Math. Phys.* **116** (1988) 185–192.

[106] C. Lobry. Controllability of nonlinear systems on compact manifolds. *SIAM J. Control* **12** (1974) 1–4.

[107] A. Malćev. On a class of homogenous spaces. *Transl. Amer. Math. Soc.* **1** (1962) 276–307.

[108] W. Malfait. An obstruction to the existence of Anosov diffeomorphisms on infra-nilmanifolds. *Contemporary Math.* **262** (2000) 233–251.

[109] A. Manning. There are no new Anosov diffeomorphisms on tori. *Amer. J. Math.* **96** (1974) 422–429.

[110] G. A. Margulis. *Discrete Subgroups of Semisimple Lie Groups.* Berlin: Springer Verlag, 1991.

[111] G. A. Margulis, N. Qian. Local rigidity of weakly hyperbolic actions of higher rank real Lie groups and their lattices, *Ergodic Theory Dynam. Systems* **21** (2001), 121–164.

[112] H. Matsumoto. Sur les sous-groupes arithmétiques des groupes semi-simples déployés. *Ann. Sci. Éc. Norm. Sup.* **4**, serie 2 (1969) 1–62.

[113] D. Mieczkowski. The first cohomology of parabolic actions for some higher-rank abelian groups and representation theory. *J. Modern Dynam.* **1** (2007) 61–92.

[114] G. Metivier. Function spectrale et valeur propres d'une classe d'operateurs non elliptiques. *Communi. PDE* **1** (1976), 467–519.

[115] J. Milnor. *Introduction to Algebraic K-theory*. Princeton, NJ: Princeton University Press, 1971.

[116] D. Montgomery, L. Zippin. A theorem on Lie groups. *Bull. Amer. Math. Soc.* **48** (1942) 448–452.

[117] C. C. Moore. Exponential decay of correlation coefficients for geodesic flows, in C. C. Moore (ed.), *Group Representations, Ergodic Theory, Operator Algebras, and Mathematical Physics. Proceedings of a Conference in Honor of George Mackey*. MSRI publications, Springer Verlag, New York: 1987.

[118] C. C. Moore. Decomposition of unitary representations defined by discrete subgroups of nilpotent groups. *Annals of Math.* **82** (1965), 146–182.

[119] N. Mok, Y. T. Siu, S. K. Yeung. Geometric superrigidity. *Invent. Math.* **113** (1993), 57–83.

[120] D. Montgomery, L. Zippin, *Topological Transformation Groups*. New York: Interscience Publishers, 1955.

[121] M. H. A. Newman. A theorem on periodic transformations of spaces. *Quart. J. Math. Oxford Ser.* **2** (1931), 1–9.

[122] M. Nicol, M. Pollicott. Measurable cocycle rigidity for some non-compact groups. *Bull. London Math. Soc.* **311** (1999) 529–600.

[123] M. Nicol, M. Pollicott. Livšic's theorem for semisimple Lie groups. *Ergodic Theory Dynam. Systems* **21** (2001) 1501–1509.

[124] M. Nicol, A. Török. Whitney regularity for the solutions of the coboundary equations on Cantor sets. *Math. Phys. Electronic J.* **13** (2007) paper 6.

[125] V. Niţică, A. Török. Cohomology of dynamical systems and rigidity of partially hyperbolic actions of higher rank lattices. *Duke Math. J.* **79** (1995) 751–810.

[126] V. Niţică, A. Török. Regularity results for the solutions of the Livshits cohomology equation with values in diffeomorphism groups. *Ergodic Theory Dynam. Systems* **16** (1996) 325–333.

[127] V. Niţică, A. Török. Regularity of the transfer map for cohomologous cocycles. *Ergodic Theory Dynam. Systems* **18** (1998) 1187–1209.

[128] V. Niţică, A. Török. On the cohomology of Anosov actions, in *Rigidity in Dynamics and Geometry*. Berlin: Springer, 2000, pp. 345–361.

[129] V. Niţică, A. Török. An open dense set of stably ergodic diffeomorphisms in a neighborhood of a non-ergodic one. *Topology* **40** (2001) 259–278.

[130] V. Niţică, A. Török. Cocycles over abelian TNS actions. *Geometriae Dedicata* **102** (2003) 65–90.

[131] V. Niţică. Journé's theorem for $C^{n,\omega}$ regularity. *Discrete Contin. Dynam. Systems* **22** (2008) 413–425.

[132] A. L. Onishchik, E. B. Vinberg. *Lie Groups and Lie Algebras*. Berlin: Springer-Verlag, 1994.

[133] D. Ornstein. Bernoulli shifts with the same entropy are isomorphic. *Adv. Math.* **4** (1970) 337–352.

[134] http://pari.math.u-bordeaux.fr/

[135] W. Parry. The Livšic periodic point theorem for non-abelian cocycles. *Ergodic Theory Dynam. Systems* **19** (1999) 687–701.

[136] W. Parry, M. Pollicott. The Livšic cocycle equation for compact Lie group extensions of hyperbolic systems. *J. London Math. Soc.* **56** (1997) 405–416.

[137] W. Parry, M. Pollicott. Skew-products and Livsic theory, in *Representation Theory, Dynamical Systems, and Asymptotic Combinatorics*, V. A. Kaimanovich, A. Lodkin (eds). *Adv. Math. Sci. Series 2*, **217**. Providence, RI: American Mathematical Society, 2006.

[138] M. Pollicott, C. P. Walkden. Livšic theorems for connected Lie groups. *Trans. Amer. Math. Soc.* **353** (2001) 2879–2895.

[139] M. Pollicott, M. Yuri. Regularity of solutions to the measurable Livshits equation. *Trans. Amer. Math. Soc.* **351** (1999) 559–568.

[140] H. L. Porteous. Anosov difeomorphisms of flat manifolds. *Topology* **11** (1972) 307–315.

[141] M. Postnikov. *Lie Groups and Lie Algebras*. Moskow: Mir Publishers, 1986.

[142] G. Prasad, M. S. Raghunathan. Cartan subgroups and lattices in semi-simple groups. *Ann. Math.* **96** (1972) 296–317.

[143] C. Pugh, M. Shub. Ergodicity of Anosov actions. *Invent. Math.* **15** (1972) 1–23.

[144] C. Pugh, M. Shub. Stable ergodicity and julienne quasi-conformality. *J. Eur. Math. Soc. (JEMS)* **2** (2000) 1–52.

[145] N. Qian. Rigidity Phenomena of group actions on a class of nilmanifolds and Anosov $\mathbb{R}^n$ actions. Unpublished Ph.D. thesis, California Insitute of Technology, 1992.

[146] M. S. Raghunathan. *Discrete Subgroups of Lie Groups*. Berlin: Springer-Verlag, 1972.

[147] M. Ratner. The rate of mixing for geodesic and horocycle flows. *Ergodic Theory Dynam. Systems* 7 (1987) 267–288.

[148] M. Ratner. On Raghunathan's measure conjecture. *Ann. of Math.* **134** (1991) 545–607.

[149] B. L. Reinhart *Differential Geometry of Foliations*. Berlin: Springer-Verlag, 1983.

[150] C. Rockland. Hypoellipticity on the Heisenberg group: representation theoretic criteria. *Trans. Amer. Math. Soc.* **240** (1978) 1–52.

[151] F. Rodriguez Hertz. Global rigidity of certain abelian actions by toral automorphisms. *J. Modern Dynam.*, to appear.

[152] L. P. Rothschild. A criterion for hypoellipticity of operators constructed from vector fields. *Commun. PDE* **4** (1979) 645–699.

[153] L. P. Rothschild, E. Stein. Hypoelliptic differential operators and nilpotent groups. *Acta Math.* **1976** 247–320.

[154] D. Rudolph. $\times 2$ and $\times 3$ invariant measures and entropy. *Ergodic Theory Dynam. Systems* **10** (1990) 395–406.

[155] K. Schmidt. The cohomology of higher-dimensional shifts of finite type. *Pacific J. Math.* **170** (1995) 237–269.

[156] K. Schmidt. Cohomological rigidity of algebraic $\mathbb{Z}^d$-actions. *Ergodic Theory Dynam. Systems* **15** (1995) 759–805.

[157] K. Schmidt. *Dynamical Systems of Algebraic Origin*. Basel-Berlin-Boston: Birkhäuser Verlag, 1995.

[158] K. Schmidt. Remarks on Livšic' theory for non-abelian cocycles. *Ergodic Theory Dynam. Systems* **19** (1999) 703–721.

[159] K. Schmidt, T. Ward. Mixing automorphisms of compact groups and a theorem of Schlickewei. *Inventiones Math.* **111** (1993) 69–76.

[160] M. Shub. Endomorphisms of compact differentiable manifolds. *Amer. J. Math.* **96** (1974) 422–429.

[161] Y. G. Sinai. Gibbs measures in ergodic theory. *Russ. Math. Surv.* **27** (1972) 21–70.

[162] S. Smale. Differentiable dynamical systems. *Bull. Amer. Math. Soc.* **73** (1967) 747–817.

[163] R. Spatzier. *Harmonic Analysis in Rigidity Theory. Ergodic theory and its connections with harmonic analysis*. London Math. Soc. Lecture Notes Ser. **205**. Cambridge: Cambridge University Press, 1995.

[164] A. Starkov. First cohomology group, mixing and minimal sets of commutative group of algebraic action on torus. *J. Math. Sci (New York)* **95** (1999) 2576–2582.

[165] N. Steenrod. *The Topology of Fiber Bundles.* Princeton, NJ: Princeton University Press, 1951.

[166] R. Steinberg. Générateurs, relations et revêtements de groupes algébraiques. *Colloq. Theorie des groupes algebraiques, Bruxelles* (1962) 113–127.

[167] E. M. Stein, G. Weiss. *Introduction to Fourier Analysis on Fourier Spaces.* Princeton, NJ: Princeton University Press, 1971.

[168] A. Unterberger, J. Unterberger. Hölder estimates and hypoellipticity. *Ann. Inst. Fourier* **26** (1976) 35–54.

[169] W. A. Veech. Periodic points and invariant pseudomeasures for toral endomorphisms. *Ergodic Theory Dynam. Systems* **6** (1986) 449–473.

[170] C. P. Walkden. Solutions to the twisted cocycle equation over hyperbolic systems. *Discrete Contin. Dynam. Systems* **6** 2000, 935–946.

[171] Z. J. Wang. Local rigidity of partially hyperbolic actions. *J. Mod. Dyn.* **4** (2010) 271–327.

[172] Z. J. Wang. New cases of differentiable rigidity for partially hyperbolic actions: symplectic groups and resonance directions. *J. Mod. Dyn.*, to appear.

[173] G. Warner. *Harmonic Analysis on Semisimple Lie Groups I.* Berlin: Springer Verlag, 1972.

[174] A. Weil. On discrete subgroups of Lie groups I. *Annals of Math.* **72** (1960) 369–384.

[175] A. Weil. On discrete subgroups of Lie groups II. *Annals of Math.* **75** (1962) 578–602.

[176] A. Weil. *Adels and Algebraic Groups*. Progress in Mathematics **23**. Boston, MA: Birkhäuser,

[177] E. Weiss. *Algebraic Number Theory*. New York: Chelsea Publishing Company, 1963.

[178] M. D. Witte. *Ratner's Theorems on Unipotent flows*. Chicago Lectures in Mathematics, 2005.

[179] R. Zimmer. *Ergodic Theory and Semisimple Groups*. Boston, MA: Birkhäuser, 1984.

Index

DUNELM UNIV

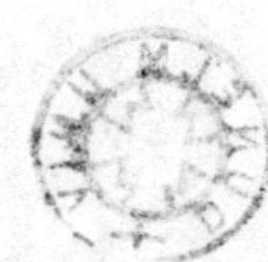